AF829287

To

Dirk

With best Complements.

Thanks for you Support.

Singh

Current Topics in Microbiology and Immunology

Volume 368

For further volumes:
http://www.springer.com/series/82

G. Singh Chhatwal
Editor

Host–Pathogen Interactions in Streptococcal Diseases

Responsible Series Editor: Klaus Aktories

Springer

Editor
Prof. Dr. G. Singh Chhatwal
Department of Medical Microbiology
Helmholtz Centre for Infection Research
Braunschweig
Germany

ISSN 0070-217X
ISBN 978-3-642-36339-9 ISBN 978-3-642-36340-5 (eBook)
DOI 10.1007/978-3-642-36340-5
Springer Heidelberg New York Dordrecht London

Library of Congress Control Number: 2013932461

Printed on acid-free paper

Springer is part of Springer Science+Business Media (www.springer.com)

Preface

Streptococci are Gram-positive bacteria capable of causing a wide spectrum of diseases in humans and animals. Group A streptococci (*Streptococcus pyogenes*) are exclusively human pathogenic bacteria. Group C and group G streptococci, which were traditionally considered as animal pathogenic bacteria, are emerging as causative organisms of human diseases. The diseases caused by streptococci range from self-limiting manifestations such as pharyngitis or impetigo to life-threatening diseases such as necrotizing fasciitis and streptococcal toxic shock syndrome. The disease burden of streptococcal infections is extremely high worldwide. More than 600 million persons, mostly children, suffer from streptococcal pharyngitis every year. There are 600 thousand cases of invasive disease with high mortality. Another problem is the sequelae of streptococcal infections in the form of acute rheumatic fever and rheumatic heart disease. About 15 million children are suffering from rheumatic heart disease, and approximately one million new cases are registered every year. Streptococcal diseases are considered as one of the most important groups of neglected communicable diseases.

Antibiotics alone have not been able to reduce the disease burden and in spite of many efforts no effective vaccine is available. One reason for unsuccessful attempts to develop a vaccine is the complexity of pathogenic mechanisms of streptococci. To establish and maintain an infection, streptococci evade host-immune defenses through their heterogeneity, bind and exploit host proteins for their own advantage, trigger their own internalization by host cell in order to persist and evade action of antibiotics, express surface proteins with similarity to host proteins to cause autoimmune diseases. The list of perplexing properties is far from complete so that streptococci remain a major health hazard and a real challenge for scientists, clinicians, and public health workers.

A prerequisite to develop and design novel combat strategies is a complete understanding of the pathogenic mechanisms. In recent years, the host–pathogen interactions have been shown to play a key role in streptococcal diseases. These interactions therefore represent promising intervention targets. This volume is completely devoted to understand streptococcal diseases. The volume has

10 chapters starting with streptococcal diseases and burden and going on to epidemiology, adaptation and transmission, molecular mechanisms of different diseases as well as sequelae, and ending with vaccine development and clinical management. All the authors are well-known in this field and have contributed enormously to the knowledge beyond the state-of-the-art. This volume will be a useful reference work for clinicians, microbiologists, public health workers, students of medicine and microbiology as well as a large number of scientists working in this field. The volume would provide new avenues for the scientists to meet the challenge of streptococcal diseases and would contribute to developing novel control strategies. The volume will be dedicated to millions of patients who have experienced the streptococcal infections and their sequelae.

I am grateful to Prof. Dr. Klaus Aktories from University of Freiburg for encouraging me to edit this volume. A short while ago, I visited his institute to give a talk after which he thought that it would be an interesting volume for CTMI. I am also thankful to all the contributors to find time from their tight schedules and deliver excellent chapters. All chapters provide state-of-the-art information and there is hardly any overlap among the different chapters. I am grateful to Springer staff, especially Ms. Schlitzberger for their help and to Prof. Manfred Rohde, Dr. Patric Nitsche-Schmitz, and Helga Brink from the Department of Medical Microbiology of our center.

G. S. Chhatwal

Contents

Group A Streptococcal Diseases and Their Global Burden

Anna P. Ralph and Jonathan R. Carapetis

Abstract Group A streptococcus (GAS) or *Streptococcus pyogenes* has been recognised as an important human pathogen since early days of modern microbiology, and it remains among the top ten causes of mortality from an infectious disease. Clinical manifestations attributable to this organism are perhaps the most diverse of any single human pathogen. These encompass invasive GAS infections, with high mortality rates despite effective antimicrobials, toxin-mediated diseases including scarlet fever and streptococcal toxic shock syndrome, the autoimmune sequelae of rheumatic fever and glomerulonephritis with potential for long-term disability, and nuisance manifestations of superficial skin and pharyngeal infection, which continue to consume a sizable proportion of healthcare resources. Although an historical perspective indicates major overall reductions in GAS infection rates in the modern era, chiefly as a result of widespread improvements in socioeconomic circumstances, this pathogen remains as a leading infectious cause of global morbidity and mortality. More than 18 million people globally are estimated to suffer from serious GAS disease. This burden disproportionally affects least affluent populations, and is a major cause of illness and death among children and young adults, including pregnant women, in low-resource settings. We review GAS transmission characteristics and prevention strategies, historical

A. P. Ralph (✉)
Global and Tropical Health, Menzies School of Health Research,
Darwin, NT, Australia
e-mail: anna.ralph@menzies.edu.au

A. P. Ralph
Division of Medicine, Royal Darwin Hospital,
Darwin, NT, Australia

J. R. Carapetis
Telethon Institute for Child Health Research, Centre for Child Health Research,
University of Western Australia, Subiaco, Perth, Western Australia

Current Topics in Microbiology and Immunology (2013) 368: 1–27
DOI: 10.1007/82_2012_280

Published Online: 15 December 2012

and geographical trends and report on the estimated global burden disease attributable to GAS. The lack of systematic reporting makes accurate estimation of rates difficult. This highlights the need to support improved surveillance and epidemiological research in low-resource settings, in order to enable better assessment of national and global disease burdens, target control strategies appropriately and assess the success of control interventions.

Contents

1 Introduction

In the twenty-first century, Group A streptococcus (GAS) retains major importance as a global cause of morbidity and mortality. This Gram-positive bacterium was initially recognised as the causative agent of erysipelas and puerperal sepsis in the late nineteenth century (Bisno and Stevens 2000). Lacking any natural host apart from humans, it is well adapted to transmission within populations characterised by crowding, limited access to hygiene and inadequate medical care. Thus, in the current age, it is chiefly a disease of poverty. Although the name 'pyogenes' (pus-forming) describes the suppurative manifestations of infection, an astonishing range of clinical syndromes is attributable to GAS (Fig. 1).

In addition to illness resulting from direct infection, immunologically and toxin-mediated pathology, the global burden of disease attributable to GAS includes the downstream complications of these conditions, such as stroke and infective endocarditis complicating rheumatic heart disease (RHD) and renal

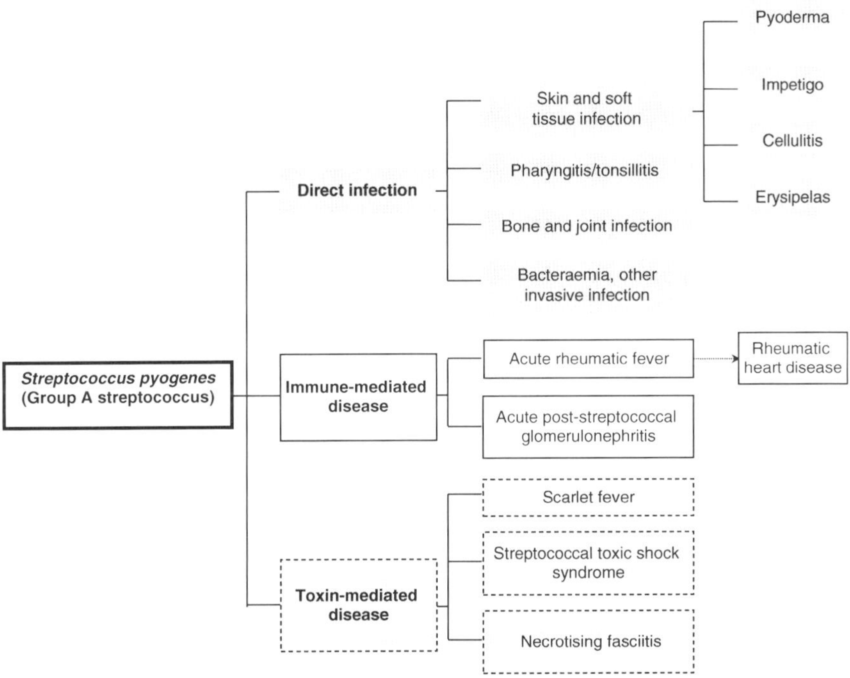

Fig. 1 The Spectrum of GAS Infections. Major manifestations are shown; the list is not intended to be exhaustive, and listed syndromes can co-exist

failure complicating post-streptococcal glomerulonephritis. The young age at which people are often affected by GAS infections leads to a high-disease burden in terms of disability adjusted life years. In 2005, the estimated number of people suffering from a serious GAS disease was approximately 18.1 million, with a further 1.78 million new cases occurring each year, accounting for 517,000 deaths annually (Carapetis et al. 2005a, b; World Health Organization 2005). By way of comparison with another pathogen of major human importance, the combined manifestations of *S. pyogenes* thus outnumber cases of tuberculosis (of which 8.8 million cases were reported globally in 2010) (World Health Organisation 2011). Unlike tuberculosis, GAS-related infections are not universally notifiable and there are few active surveillance systems in developing countries, where the burden is greatest, hence the figures quoted here underestimate the real burden of GAS (World Health Organization 2005).

It is estimated that the various manifestations of GAS infection cause a similar number of deaths as do rotavirus, measles, *Haemophilus influenza* type B and Hepatitis B each (Carapetis et al. 2005a, b). The majority of deaths are attributable to RHD and its complications, followed by invasive group A streptococcal diseases (iGAS), then acute post-streptococcal glomerulonephritis (APSGN).

Also contributing to morbidity and healthcare expenditure is the highly prevalent condition of pharyngitis (a minimum of 616 million incident cases estimated in 2005) and pyoderma (111 million prevalent cases) (World Health Organization 2005) As these are among the commonest causes of doctor consultations in high-income settings, illness burden and healthcare expenditure related to *S. pyogenes* infections in these settings evidently persists.

The incidence rates of GAS-related diseases have fallen in high- and middle-income countries as nations have undergone epidemiological transition from communicable to non-communicable diseases as their predominant causes of morbidity and mortality. However, the relative contribution of *S. pyogenes* to sepsis-related morbidity and mortality remains disproportionately high in such countries which have undergone epidemiological transition, emphasising the ongoing relevance of this organism globally. Factors other than changes in societal socioeconomic status which account for historical trends in diseases arising from GAS include: evolving knowledge of the mode of transmission; implementation of preventative measures; improved access to medical care including antimicrobial therapy and shifts in strain virulence and transmissibility over time. Although rates of serious GAS diseases during the mid twentieth century were mostly low in industrialised countries, the re-emergence of streptococcal toxic shock syndrome (STSS) in the mid-1980s re-focussed attention on the importance of human GAS infections, and fostered research into the molecular epidemiology (Katz and Morens 1992).

S. pyogenes retains robust susceptibility to the most basic of antibiotics, penicillin. There is good understanding of GAS epidemiology and preventive strategies. Despite these facts, controlling GAS transmission in high-burden areas remains challenging. In this chapter, we explore the diversity of manifestations of streptococcal infection, transmission characteristics and prevention strategies relevant to the understanding of disease epidemiology, attributable morbidity and mortality, and overall global burden including temporal and geographical trends.

2 Transmission and Prevention of GAS Infections

An exploration of transmission properties and prevention strategies for GAS infection and carriage provides insight into reasons for the ongoing high GAS disease burden, and for geographical variation in rates.

2.1 Transmission

S. pyogenes is transmitted human to human via direct contact, droplet spread from people with pharyngeal colonisation or carriage and contaminated fomites (Health Protection Agency Group A Streptococcus Working Group 2004; Milne et al. 2011; New South Wales Health 2012; Coburn and Young 1949) in addition to being reportedly food borne (Bisno and Stevens 2000; Williams et al. 1932). The importance of crowding to facilitate transmission has been well documented among personnel residing in military camps with close living arrangements, and this sector is often serviced by high-quality epidemiological units which are able to readily detect and report outbreaks (Coburn and Young 1949; Centers for Disease Control 2002; Wasserzug et al. 2009). Household crowding as a risk factor for acute rheumatic fever (ARF) has been recently confirmed in a large ecological study of over 1,000 rheumatic fever cases in New Zealand (Lennon et al. 2009). Transmission likelihood was mapped clearly in the "barracks studies" of World War II outbreaks among American Navy personnel. Risk of GAS pharyngeal infection in an individual was shown to be inversely proportional to the distance between their bed and that of a colonised or infected case (Wannamaker 1954; Kaplan 2009). The extent of the problem of GAS infections for the US military is indicated by the existence of a specific United States (US) Army Air Force Rheumatic Fever Control Programme during the twentieth century, (Coburn and Young 1949) and routine use of monthly benzathine penicillin prophylaxis among US military recruits (Heggie et al. 1992).

Despite transmission being person to person, *S. pyogenes* has not generally become established as an endemic hospital-acquired organism and rather remains predominantly community acquired.

2.2 Pharyngeal Carriage of S. pyogenes

A major challenge in treating GAS pharyngitis is the existence of the carrier state. A study now four decades old was undertaken by Kaplan and colleagues, enroling 624 children <15 years old with symptomatic pharyngitis, to investigate carriage (Kaplan et al. 1971). The investigators found that *S. pyogenes* was culturable from a throat swab in 35 % of children, yet of these, only 57 % showed a rise in antistreptolysin O or anti-DNAse B titre at follow up at 3 and 6 weeks. The main reason for failure to mount an antibody response was interpreted as *S. pyogenes* carriage rather than infection (Kaplan et al. 1971). A number of more recent studies continue to document high rates of pharyngeal *S. pyogenes* carriage; for instance, carriage was found in 17.8 % of asymptomatic toddlers attending a day care centre in Israel during winter (Yagupsky et al. 1995).

Carriage is thought to be a benign state for the host, contributing neither to infection nor autoimmune sequelae, but to be an important potential source of infection for contacts (Centers for Disease Control, Prevention 1999). This is the basis for the United Kingdom (UK) recommendation to swab staff during nosocomial GAS outbreaks (Milne et al. 2011). Misattribution of colonisation as infection complicates estimates of the likelihood of ARF or APSGN following GAS pharyngitis. Asymptomatic carriers might be important drivers of GAS transmission. Hence, improved diagnostics to distinguish these states would be readily welcomed.

2.3 Prevention

The example of ARF best illustrates the different levels of preventative strategies which can be employed in the control of GAS infection and its consequences. Prevention can be conceptualised at levels described as primordial (improved social health determinants), primary (vaccination or treatment of GAS pharyngitis), secondary (antimicrobial prophylaxis after ARF e.g., monthly benzathine penicillin) and tertiary (medical and surgical management of RHD) (RHD Australia (ARF/RHD writing group) 2012; Seckeler and Hoke 2011).

The success of primordial prevention is evident in overall global trends in GAS rates over the long term as housing and hygiene improved in high- and middle-income countries, prior to antibiotic availability (see Sect. 3.1) (RHD Australia (ARF/RHD writing group) 2012; Quinn 1989). Unfortunately, there are few successful contemporary examples of primordial-level strategies to reduce GAS infection rates in disadvantaged populations, for example through social, economic or environmental initiatives. One possible exception is provided by a before-and-after study undertaken in remote Australian Aboriginal communities, in which the introduction of a community swimming pool was associated with significant reductions in pyoderma rates (pyoderma being chiefly attributable to *S. pyogenes* in this setting) (Lehmann et al. 2003). While increasing affluence is associated with lower ARF rates overall (with notable exceptions) (Pastore et al. 2011; Veasy et al. 1987), it remains largely unknown which specific components of affluence are the most important: housing quantity and quality, healthcare access and quality, education or economic advantage per se (RHD Australia (ARF/RHD writing group) 2012). Some answers are provided by a recent study in New Zealand, which found that the number of people per house was more important than household income in predicting ARF risk (Lennon et al. 2009). Overall, few evidence based and cost-effective prevention strategies at the primordial level have been identified.

Concepts of primary prevention are to pre-empt colonisation or infection (using vaccination or prophylactic antibiotics), or avert the development of adverse immune responses leading to ARF (using antibiotic treatment for established GAS infection). While vaccine developments offer the most important promise for prevention, this is not yet an established option (Steer et al. 2009a, b). Penicillin appears relatively ineffective in achieving sustained eradication of carriage in some studies (Heggie et al. 1992). Other antimicrobials, or combinations thereof, may be more successful in eradicating carriage: in a small study of children with persistent throat carriage of GAS after receiving intramuscular benzathine penicillin, the combination of rifampicin with penicillin eradicated carriage in 93 % versus 23 % in the placebo group and 30 % in the group which received further penicillin alone (Tanz et al. 1985). Another larger study found that azithromycin for 5 days eradicated oropharyngeal GAS in 95 % of school students (Morita et al. 2000). Overall, however, the cost-benefit ratio does not favour generalised screening and treatment for GAS carriage, except in selected settings of high GAS transmission (RHD Australia (ARF/RHD writing group) 2012).

Treatment of GAS pharyngitis (in contrast to carriage) to avert ARF does work, (Del Mar et al. 2006) but is also challenging because of the high proportion of infections which are unrecognised, and the difficulty in distinguishing GAS pharyngitis from other more common, predominantly viral and aetiologies. Among ARF patients, a recent history of symptomatic pharyngitis is absent in two-thirds of cases in some studies, (Veasy et al. 1987) and may be perceived as too trivial to warrant medical attention in others. Even when detection and treatment of GAS pharyngitis is diligent, ARF cases are not completely prevented: a randomised trial of primary prevention in a high-incidence ARF setting (around 60 cases per 100,000 school children) unfortunately failed to show major benefits from assiduous treatment of GAS pharyngitis versus standard care (Lennon et al. 2009). Schools were randomised to receive optimised diagnosis and free treatment of GAS pharyngitis via a school-based 'sore throat' clinic versus standard care; although the ensuing incidence of ARF was around 20 % lower in the intervention schools, this did not reach statistical significance (Lennon et al. 2009).

Observations from Australian Aboriginal communities also implicate GAS-associated skin infection as the precursor for ARF, (McDonald et al. 2004) indicating that treatment of skin infection could offer another primary prevention strategy.

Secondary prevention is the key focus of most control strategies, due to its proven efficacy, cost-effectiveness and the difficulty with above-mentioned approaches. Secondary prophylaxis with four weekly depot intramuscular injections of benzathine penicillin significantly reduces ARF recurrence rates when compared with placebo (Padmavati et al. 1973) or with oral penicillin (Manyemba and Mayosi 2002). In Australia's Northern Territory, surveillance during 2002–2008 of Aboriginal people with ARF showed that none who received 100 % of their scheduled benzathine penicillin injections experienced a recurrent episode of ARF (Parnaby and Carapetis 2010). Secondary prophylaxis using this regimen can also reduce the severity of rheumatic valvular lesions, or can be associated with the regression of valve lesions, over time (Carapetis et al. 2005a, b; RHD Australia (ARF/RHD writing group) 2012; Lue et al. 1979). This strategy thus forms the basis of the World Health Organisation's recommendations on ARF management (WHO 2001). Guidelines on doses and durations are detailed elsewhere (RHD Australia (ARF/RHD writing group) 2012).

2.4 Contact Management

An outbreak of invasive GAS infection is currently defined as two or more cases of culture-confirmed symptomatic iGAS linked in time and place (Health Protection Agency Group A Streptococcus Working Group 2004; New South Wales Health 2015). Different approaches to the management of contacts have been advocated.

Key UK recommendations on contact management include education about symptoms of GAS infection, and antibiotic chemoprophylaxis with oral penicillin (or azithromycin in cases of penicillin allergy) for: (a) mothers and infants if either develops iGAS disease in the neonatal period; (b) close contacts if they have symptoms suggestive of localised GAS infection and (c) entire households if there are two or more cases of iGAS within a 30-day time period (Health Protection Agency Group A Streptococcus Working Group 2004). If two or more cases of invasive *S. pyogenes* occur in a residential facility, some argue that targeted or mass antibiotic chemoprophylaxis for residents and staff should be considered, in addition to ensuring monitoring, screening for infection and carriage and review of infection control practices (Milne et al. 2011).

Smith and colleagues in a 2005 review concluded that currently available evidence does not justify the routine administration of chemoprophylaxis to close contacts, arguing that the number needed to treat to avert one infection would be excessive, based on the secondary attack rate (Smith et al. 2005). Few studies have quantified the risk of iGAS (let alone other manifestations of GAS infection) among close contacts, making it difficult to determine the risk–benefit ratio of chemoprophylaxis. However, the secondary attack rate of iGAS has been the subject of ongoing surveillance in Ontario, Canada, where the population incidence of iGAS was found to be 1.5/100,000, but among close household contacts was 320/100,000 (95 % confidence interval 39–1,200/100,000), based on the occurrence of two secondary cases (Davies et al. 1996). Subsequent surveillance found a similar attack rate over ensuing years [see review, (Smith et al. 2005)]. These rates are comparable to or somewhat less than secondary attack rates among household contacts for *Neisseria meningitides*, (Purcell et al. 2004) an organism for which contact prophylaxis is routinely employed. Given the high mortality of iGAS of around 15 %, (Davies et al. 1996) the Ontario recommendations are therefore to administer antibiotic prophylaxis to close household contacts of iGAS cases (Ontario Ministry of Health 2005).

The diverse approaches to contact management, with lack of consensus on this question, (Health Protection Agency Group A Streptococcus Working Group 2004; Smith et al. 2005; Ontario Ministry of Health 2005; Prevention of Invasive Group ASIWP 2002) may reflect the fact that diverse GAS strains with differing potential for transmission, and propensity to result in invasive disease, provide for a range of experiences, and hence differing recommendations have arisen.

3 Trends in the Burden of Disease due to GAS

Here, we provide an historical perspective of the global burden of GAS infections, including important fluctuations during the last 30 years, and the changing drug susceptibility profile of *S. pyogenes*.

3.1 Long-Term Temporal Trends

As with tuberculosis and other communicable diseases associated with poverty, GAS incidence rates began to decline during the late nineteenth–early twentieth centuries in industrialised nations as living conditions improved, then continued to fall with the advent of the antibiotic era (Quinn 1989; Kaplan 1993). Historical trends in GAS rates, using scarlet fever ('scarlatina') as a proxy because of its distinctive clinical features, have been studied in attempts to shed light on modern day changes in GAS epidemiology (Katz and Morens 1992). As noted by Quinn in his comprehensive 1989 review on rheumatic and scarlet fevers: "the incidence and prevalence of these diseases have declined steadily as has mortality in the past 150–200 years. The decline predated by at least 100 years the availability of penicillin in 1943 as an effective preventive and therapeutic agent. Undoubtedly, penicillin has accelerated their general decline, but it certainly did not initiate it" (Quinn 1989). It should be noted, however, that there were very few data in Quinn's review from low- and middle-income countries, where the benefits of economic and social development have not been as great, and hence where high rates of GAS diseases appear to continue unabated. The lack of surveillance and notification processes in low-resource settings means that the bulk of the available literature on longitudinal trends in GAS disease, described in this section, relate to higher resource settings.

In a further parallel with tuberculosis rates, during the post-modern era *S. pyogenes* again began attracting attention in the 1980s when new as well as classical manifestations of GAS infection began to be reported in increasing rates from northern Europe and the USA (Veasy et al. 1987; Stevens 1995; Martin and Hoiby 1990; Stromberg et al. 1991). While the rise in tuberculosis rates at this time was attributed to public health service delivery issues and the onset of the HIV pandemic, (Bloom and Murray 1992) the mid-1980s rise in GAS rates appears more likely to have been related to organism factors, with the emergence of different GAS strains distinguishable on *emm* genotyping (Lamagni et al. 2005).

Trends in puerperal sepsis over time provide important insights into GAS transmission dynamics. Puerperal sepsis, largely attributable to group A (but also group B) streptococci became highly prevalent when 'lying in hospitals' first started to replace homebirths in Europe in the 1600s (Adriaanse et al. 2000). Deaths attributed to this cause reportedly occurred in up to 20 % of parturient women in some facilities, [see review, (Adriaanse et al. 2000)]. Although Semmelweis's evidence on the role of hand hygiene in reducing rates of puerperal sepsis from 11.4 in 1846 to 3.1 % in 1847 (Adriaanse et al. 2000) was infamously rejected by his contemporaries, later appreciation of germ theory and hygiene requirements, promulgated by Pasteur among others in the later 1800s, contributed to a decline in rates of (presumed GAS) puerperal sepsis. Rates fell to low enough levels in high-resource settings in the twentieth century to be worthy of publication when they did still sporadically occur, as demonstrated by a 1993 case report (Nathan et al. 1993). Unfortunately, since that time, a resurgence has again occurred, mirroring rates of other GAS manifestations in high-income countries (while in low-income countries, rates have remained far more constant, and high, over time). In the UK, there has been a recent increase in deaths due to maternal sepsis, particularly from *S. pyogenes*. The mortality rate from sepsis increased from 0.85 deaths per 100,000 maternities in 2003–2005 to 1.13 deaths in 2006–2008, and sepsis is now the most common cause of direct maternal death (Harper 2011). In the most recent UK surveillance report, of 29 maternal deaths from genital sepsis in 2006–2008, 13 were due to GAS; this compares with eight maternal GAS deaths in 2003–2005 and three total maternal deaths from sepsis in 2000–2002 (Harper 2011). A further complication of *S. pyogenes* puerperal sepsis is bacteraemia in the neonate, discussed in Sect. 4.3.

3.2 Epidemics

On the background of overall downward trends over time, well-recognised epidemics of GAS infections occur [see reviews, (Stevens 1995; Stevens et al. 1989)], likely to be related to shifts in the expression of virulence factors, and/or community acquisition, then waning, of immunity to given circulating strains (Katz and Morens 1992). While scarlet fever outbreaks typify GAS epidemics in the older literature, all GAS manifestations can occur in epidemic form.

Military barracks outbreaks, of various sizes and severity, occur not infrequently as mentioned above (Coburn and Young 1949; Centers for Disease Control 2002; Wasserzug et al. 2009; Wannamaker 1954; Kaplan 2009). Outbreaks of scarlet fever were observed to occur in 5 or 6-year cycles in Germany between 1972 and 1987, with outbreak strains shown to produce different quantities of erythrogenic toxin type A; this was thought to explain the severity differential observed between outbreaks (Kohler et al. 1987). Five and six-year cycles have also been observed recently in outbreaks of APSGN in northern Australia (Marshall et al. 2011). In this study, specific *emm* types (e.g. *emm*55) appeared rapidly in communities, became the dominant circulating GAS type, then disappeared after approximately 3 months; this pattern, and the ensuing periodicity of APSGN outbreaks, was hypothesised to be due to the development of type-specific immunity in the population (Marshall et al. 2011). Of note, regular cyclical outbreaks of rheumatic fever by contrast have not been observed. No such patterns are evident in the observed trends in longitudinal incidence rates from registers of ARF in Australia (Parnaby and Carapetis 2010; Brown et al. 2003) or New Zealand, (Jaine et al. 2008) and there is an absence of publications suggesting any cyclical patterns in ARF epidemiology. One-off ARF outbreaks do occur, such as the widely reported Utah outbreak of 75 cases of ARF in 1985–1986, thought to be associated with the emergence of a rheumatogenic mucoid M type 18 GAS (Veasy et al. 1987). Nevertheless, regular 5–6 yearly outbreaks, in environments of either high or low ARF incidence, are not observed as far as we are aware. A plausible hypothesis to explain the different epidemiologies of these GAS manifestations is the difference in underlying immunological pathogenesis. ARF develops after multiple exposures to rheumatogenic GAS strains, with resultant immune 'priming' that eventually leads to a clinically apparent episode of rheumatic fever, which can sometimes (especially in the case of Sydenham's chorea) occur months to years after the infective insult; thus host determinants play a contributory role, and resulting rates of ARF are relatively steady over time. APSGN, on the other hand, occurs soon after infection with a single nephritogenic GAS strain which can suddenly appear in a population, (Marshall et al. 2011) and spread with high transmissibility. Accordingly, this disease has greater outbreak potential.

The 'new' manifestations of GAS infection reported from Europe and North America in the 1980–1990s comprised toxin-mediated disease (Lamagni et al. 2005; Cone et al. 1987; Hoge et al. 1993). Toxic shock syndrome due to GAS was first described in the USA in 1987 (Cone et al. 1987); rather than being a genuinely new disease, however, this is probably the same disease as what had previously been termed 'septic scarlet fever'. Similarly, 'Streptococcal gangrene' was not a new concept (Bisno and Stevens 1996), but the occurrence of a cluster of necrotising fasciitis cases in the US Rocky Mountain region, as well as STSS, with high case fatality of 30 % (Stevens et al. 1989), led to fears of a new epidemic of 'flesh eating bacteria' (Stevens 1995). Clinical case definitions for STSS and necrotising fasciitis were then developed (Stevens 1995; The Working Group on Severe Streptococcal Infections 1993). STTS requires identification of GAS with features of shock plus $\geq$ 2 specific laboratory or clinical abnormalities; necrotising fasciitis comprises necrosis of soft tissues and fascia plus serious systemic disease (e.g. shock, disseminated intravascular coagulation or organ failure). Concerns about these outbreaks of virulent GAS strains prompted many European countries from this time onwards to make invasive *S. pyogenes* infections notifiable (Lamagni et al. 2005). The lack of reportable status in many other nations poses difficulties for accurate ascertainment of disease burden.

3.3 Trends in GAS Drug Resistance

Fortunately, *S. pyogenes* remains fully susceptible to penicillin, despite over five decades of exposure to this antibiotic. Hypothesised explanations, as summarised by Horn et al., include that GAS do not have the capacity to express β-lactamase or low-affinity penicillin-binding proteins, or that expression of either is toxic to the organism (Horn et al. 1998). Resistance does, however, readily develop to non-β-lactam antibiotics. The distribution and burden of resistance to other antibiotics is important to monitor, since alternative agents are required in the context of adverse reactions to β-lactam antibiotics, with erythromycin the recommended second-line agent for rheumatic fever prophylaxis, for instance (RHD Australia (ARF/RHD writing group) 2012). Alternative agents are also required for the treatment of mixed infections where an agent broader than penicillin is required. Furthermore, there is increasing popularity in using the lincosamide class of antibiotic for GAS treatment. Agents from this class (clindamycin or lincomycin) are utilised either as monotherapy or as an adjunct to β-lactams in the treatment of invasive GAS infections, due to their in vitro antitoxin effect (Stevens et al. 1987), their ability to improve *S. pyogenes* phagocytosis by impairing expression of surface M protein (Gemmell et al. 1981), and their good tissue penetration and long post-antibiotic effect; thus resistance to this antibiotic class is of particular interest.

Cross-resistance between lincosamides and the macrolide antibiotic class occurs if the macrolide resistance is attributable to the expression of *erm* genes, which generates the 'MLS_B' phenotype (resistant to **m**acrolides, **l**incosamides and **s**treptogramin B). Macrolide resistance in GAS isolates is now well established. Surveys of Korean school children identified rising rates of resistance among GAS isolates to erythromycin and clindamycin, from 29–10 %, respectively in 1995, to 51–34 % in 2000 (Kim and Lee 2004). Monitoring of macrolide resistance in European countries shows substantial variation, ranging from 11 % of isolates in Russia and Portugal in the 1990–2000s, to 32 % of isolates in Italy (Lamagni et al. 2005).

Erythromycin resistance appears, not unexpectedly, to track erythromycin prescribing habits, with the rise and fall in prescriptions for this antibiotic in Finland corresponding to changes in GAS resistance rates over time (Seppala et al. 1997). High utilisation of the macrolide antibiotic class, thus potentially driving macrolide $\pm$ lincosamide resistance in Streptococci, is often required in geographical locations where concurrent high GAS rates occur. In Central Australian Aboriginal populations for instance (where erythromycin remains the second-line agent for ARF secondary prophylaxis, and clindamycin is used as adjunctive treatment for toxin-mediated manifestations of iGAS), mass community azithromycin treatments for chlamydial infections (trachoma and sexually transmitted infection) are periodically instituted (Ewald et al. 2003). Indeed, rapid appearance of macrolide-resistant *Streptococcus pneumoniae* in Aboriginal children after a single mass administration of azithromycin for a community trachoma program has been documented (Leach et al. 2008). This highlights the importance of ongoing surveillance of antibiotic resistance among GAS, despite the reassuring susceptibility to penicillin.

4 The Global Burden of GAS Disease

It is estimated that $\geq$95 % of global cases of ARF, APSGN and iGAS occur in low-resource settings (World Health Organization 2005). An in-depth evaluation of the global burden of GAS disease, published as a WHO discussion paper (World Health Organization 2005) and review (Carapetis et al. 2005a, b) in 2005, provides the most comprehensive estimates to date of morbidity and mortality attributable to GAS. A subsequent review in 2008 provides updated data from Asia (Carapetis 2008). A major recommendation from these papers was for reporting systems for GAS infections, especially bacteraemia surveillance, to be established in Sub-Saharan

Africa, Pacific Island nations and Asia (World Health Organization 2005). As evident in the following section, improved capacity for case detection and notification is associated with increases in reported rates of GAS-related diseases in low-resource settings.

4.1 Acute Rheumatic Fever

ARF and consequent RHD remain as significant causes of cardiovascular disease, having particularly devastating impacts on children and young adults. There is no diagnostic test for ARF, clinical presentations can be subtle and hence overlooked, and few jurisdictions routinely notify ARF cases; therefore, accurate estimations of case numbers are difficult. Overall, ARF incidence estimates vary widely, from a low of 0.7 per 100,000 children per year in Slovenia in 1990–1991 (Cernay et al. 1993) to 21.2/100,000 in Malaysian children in 1981–1990 (Omar 1995), to 374–508/100,000 in selected Australian Aboriginal populations in the 1980–1990s (Carapetis et al. 2000; Richmond and Harris 1998).

A systematic review in 2008 aimed to summarise worldwide population-based studies of ARF incidence, but restricted the analysis to first ARF episode only (excluding recurrences) (Tibazarwa et al. 2008). Studies from 10 countries on all continents except Africa were available for review. The mean incidence rate of first attack of ARF was 19 per 100,000 per population per year (95 % confidence interval 9–30/100 000), lowest rates (≤10/100,000) being in America and Western Europe, and higher incidences (>10/100,000) documented in Eastern Europe, Middle East, Asia and Australasia (Omar 1995).

ARF incidence rates reported from each country would be expected to correlate with that country's RHD prevalence, but such correlation is not always evident; in particular, although South Africa has the highest global prevalence of RHD [estimated to harbour 50 % of the world's childhood RHD cases (Carapetis et al. 2005a, b)], ARF incidence is only reported there at 13.4/100,000, suggesting substantial underdiagnosis of ARF.

"High risk" groups have been defined as those living in communities with ARF incidence >30/100,000 per year in 5–14 year olds, or RHD all-age prevalence >2/1,000 (RHD Australia (ARF/RHD writing group) 2012). In Northern Australia, such rates continue to be reported in Indigenous populations, but register-based programs have been shown to improve case detection and management in these settings. In North Queensland, where the annual incidence of ARF is 155 per 100,000 Indigenous children aged 5–14 years, ARF reporting increased 41 % in 2004–2009 compared with the previous 5 years, attributed to better case detection and notification, achieved through the establishment of a register and appointment of a program co-ordinator (Hanna and Clark 2010). The proportion of ARF cases which were recurrences (a marker of success of secondary prophylaxis) decreased significantly during programme implementation (Hanna and Clark 2010). In the neighbouring Northern Territory of Australia, register-derived data indicate that since 2002, ARF incidence in Aboriginal children of the peak age group (5–14 years) was 150–380/100,000, and remained high in the next age bracket of 15–24 year olds at approximately 70–230/100,000 (Parnaby and Carapetis 2010). Unfortunately, the proportion of cases which are recurrences has remained high there at 12–40 %, indicating further scope for the bolstering of secondary prophylaxis programmes (Parnaby and Carapetis 2010).

A small resurgence of cases in North-eastern Italy in 2007–2008 indicates that ARF can continue to reappear in apparently low-risk (high-resource) settings; incidence among 5–15 year olds rose from the usual background rate of 2–4 to 23–27 per 100,000 (Pastore et al. 2011). The authors commented that declining antibiotic treatment of pharyngitis might have contributed (only 2/13 children with ARF had received antibiotic treatment of an antecedent pharyngitis), but the focal nature of the outbreak more likely favoured the emergence of a specific rheumatogenic strain (Pastore et al. 2011).

Consequences of ARF include temporary but potentially recurring morbidity from arthralgia/arthritis; chorea which can last up to 2–3 years; the need for 10 years or longer of secondary prophylaxis; and the most devastating outcome, established RHD. The latter occurs in 42–60 % of people with a history of prior ARF, (Carapetis et al. 2000; Rheumatic fever working party 1960) but it is unfortunately not predictable on the basis of clinical features of a given ARF episode, mandating secondary prophylaxis for each child or adolescent diagnosed with ARF. Attendant complications of RHD include the need for valve repair or replacement, anticoagulation, infective endocarditis and stroke.

4.2 Rheumatic Heart Disease

RHD rates are the best documented of GAS-related diseases, allowing for relatively confident global estimates. In keeping with the recognised socioeconomic determinants of GAS infection likelihood, highest RHD rates are reported in slum dwellers, followed by rural then urban populations (World Health Organization 2005). It is generally acknowledged that the highest national rates occur in South Africa, where RHD prevalence is around 5.7 per 1,000 in the 5–14 year population. This rate has been derived from 14 studies using clinical or echocardiographic confirmation of RHD, reviewed elsewhere (Carapetis et al. 2005a, b; World Health Organization 2005). This is followed by the rates in Pacific Island nations and Indigenous Australians and New Zealanders (Carapetis et al. 2005a, b; World Health Organization 2005; Carapetis et al. 2000). Data from the Australian Northern Territory RHD register in fact show a current RHD prevalence in Indigenous children (5–14 years) of 8.5/1,000, higher than the South African reports, although case ascertainment is likely to be greater in this than in the African setting (Parnaby and Carapetis 2010). While the peak incidence of first ARF episode is usually at around age 12, RHD develops and worsens with recurrent ARF episodes, hence peak RHD prevalence is generally reported in older age groups. The peak RHD prevalence in Indigenous people occurs in those aged 35–44 years (at a rate of 31.9/1,000 in the most recent register report), with an overall RHD prevalence of 19.4 per 1,000 (almost 2 %) (Parnaby and Carapetis 2010).

In Asia, fewer reliable estimates are available. A study from Myanmar utilising echocardiographic surveillance revealed an overall RHD prevalence of 13.2/1,000 adults, and a peak prevalence of 52.9 RHD cases per 1,000 in a subgroup of rural women aged 26–35 years (Myo Thet et al. 1992). More recently in Pakistan, population screening for RHD using physical examination and echocardiographic confirmation of the diagnosis identified 54 RHD cases among 9,430 people screened (prevalence: 5.7 per 1,000) (Rizvi et al. 2004). Over 80 % of the people identified as having RHD in this study were previously unaware of their diagnosis and were not taking secondary prophylaxis, revealing the magnitude of the problem of underdiagnosis. A 2007 study similarly found that RHD was diagnosed clinically in 8 of 3,677 Cambodian children (2.2 cases per 1,000) but notably, the case detection rate rose to 79 (21.5 cases per 1,000) when echocardiography was used (Marijon et al. 2007).

These studies show that careful research methodologies uncover many more RHD cases than would otherwise have been detected (Carapetis 2008). They also raise the issue of the role of echocardiography in RHD screening. Data using echocardiography to screen for RHD have led to a marked increase in reported RHD prevalences [see Ref. (Seckeler and Hoke 2011) and Fig. 2a and 2b]. Concerns about potential overdiagnosis of ARF/RHD with this modality exist, since it is unclear how much valvular regurgitation can be considered physiological, (Ferrieri 2002) but these specificity concerns in classifying mild cases are generally outweighed by the major apparent improvements in sensitivity (ability to detect clinically inapparent valvular disease) which echocardiography offers (RHD Australia (ARF/RHD writing group) 2012). A major difficulty is the lack of an established 'gold standard' for RHD diagnosis, which in the past was a clinical diagnosis based on cardiovascular examination including auscultation. Auscultation even by experienced clinicians can, however, miss valvular lesions in as many as 90 % of children subsequently shown to have valve pathology consistent with RHD on echocardiogram (Marijon et al. 2007). A number of studies have shown that echocardiogram increases the RHD detection rate in children by around 10 fold (Marijon et al. 2012). Wider use of echocardiography may thus impact greatly on future reported rates of RHD prevalence, with the global burden of this disease likely appear substantially elevated compared with figures derived from clinical diagnoses.

Accurate mortality estimates are more difficult than ARF/RDH incidence and prevalence estimates. In low-resource settings, where secondary prophylaxis programmes are not delivered and medical and surgical management of RHD is limited, an average of 1.5 % of RHD patients are estimated to die annually (World

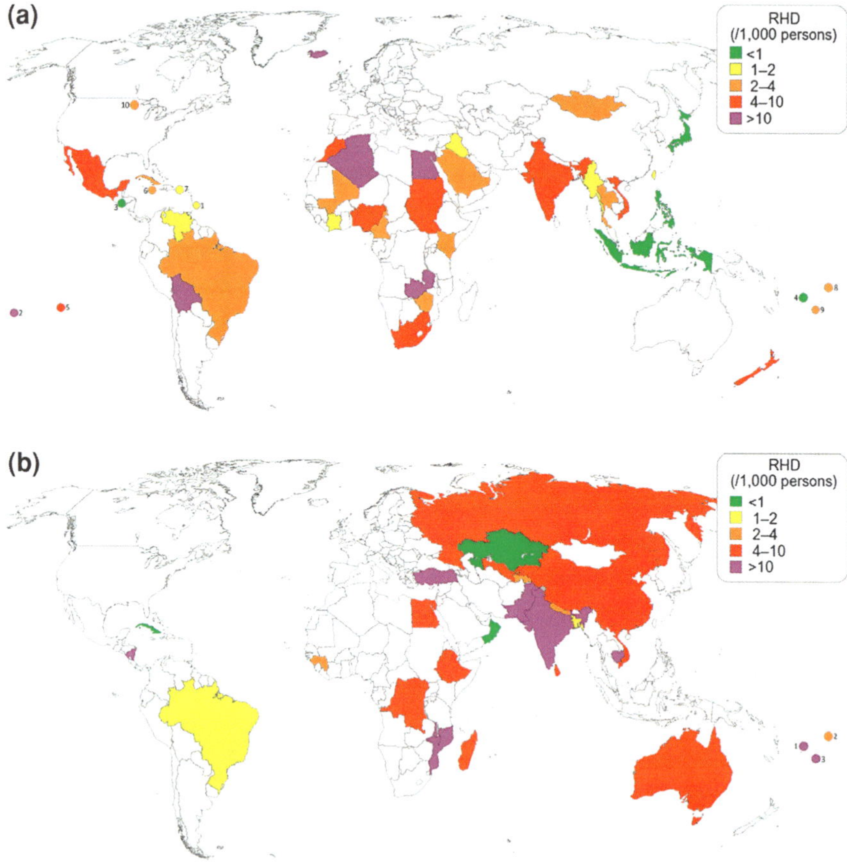

Fig. 2 **a** Map of global RHD prevalence, 1970–1990. **b** Map of global rheumatic heart disease prevalence, 1999–2011. (Reproduced from Ref. Seckeler and Hoke 2011). These maps illustrate: *1* the difficulty in accurate RHD rate ascertainment, with changes over time often reflecting different screening methods rather than real epidemiological changes; *2* the problem in assigning a single national RHD rate when large within country discrepancies exist; *3* the absence of data on RHD rates from many countries.

Health Organization 2005). In high-resource settings, where ARF rates are high in Indigenous people, access to secondary and tertiary level care may result in lower mortality rates. Conservative estimates of annual mortality attributable to RHD are between 233,000 and 300,000 deaths per year (Carapetis et al. 2005a, b; World Health Organization 2005). The mean age at death from ARF or RHD in Aboriginal Australians was previously found to be 35.7 years in a 1999 report, (Carapetis and Currie 1999) and 44 years in the same region in a 2010 study (Parnaby and Carapetis 2010). A recent study from New Zealand shows a high burden of RHD mortality, and health-care expenditure, in middle-aged people (Milne et al. 2012a, b). The mean age at RHD death was 56.4 and 58.4 years in

Māori men and women, respectively, 50.9 and 59.8 years for Pacific and considerably older at 78.2 and 80.6 years for non-Māori, non-Pacific men and women. Greatest heathcare expenditures in this study were found to be related to valve surgery (Milne et al. 2012a, b).

RHD during pregnancy can have a grave prognosis, being an important cause of maternal mortality in low-resource settings, (Sawhney et al. 2003) and of morbidity in high-resource settings (Sartain et al. 2012).

Additional complications of RHD include infective endocarditis and stroke. In a systematic review of 11 studies, RHD was the underlying cause of valve disease in 63 % of cases of infective endocarditis, and the endocarditis mortality rate in low-resource countries was 25 % (Carapetis et al. 2005a, b). Further review of eight studies of stroke from less-developed countries concluded that between 3 and 7.5 % of all strokes are directly attributable to RHD (Carapetis et al. 2005a, b).

The burden of RHD is thus of great magnitude, affecting children and young adults including pregnant women, resulting in potential lifelong morbidity and premature death, and thereby accounting for millions of disability adjusted life years (World Health Organization 2005). The World Health Organisation's forthcoming update to the Global Burden of Disease study (World Health Organisation 1990), which aims to provide complete systematic assessments of the data on all diseases and injuries, will include a review of the global burden RHD, including years of life lost (YLL) and years of life lived with disability (YLD) statistics. These data will be available soon and should provide more robust estimates of the RHD burden.

4.3 Acute Post-streptococcal Glomerulonephritis

Although APSGN is often considered as a relatively benign disease, in some populations it is associated with a 5–6-fold increased risk of chronic renal disease (White et al. 2001). The existence of multidimensional concurrent risk factors for renal disease in certain disadvantaged populations may compound the likelihood of adverse outcomes of APSGN (Hoy et al. 1998).

The median APSGN incidence in children in low-resource settings is estimated at 24.3/100,000 per year, (World Health Organization 2005) compared with approximately 6/100,000 per year in affluent settings (Lennon et al. 1988; Carapetis 1998). Similarly for other GAS manifestations, 97 % of deaths (complicating about 1 % of cases) occur in low-resource countries (World Health Organization 2005).

In Australia's Northern Territory, annual incidence rates of APSGN during the period 1992–2007 were found to be exceedingly high at 94.3 and 7.3 per 1,000 in the 0–14 and >14 year age groups, respectively (Marshall et al. 2011). High year-to-year variation in rates was noted, with three outbreaks occurring at 5–6 yearly intervals during the 16 year study period (also noted above, Sect. 3.2). Different GAS *emm* types were recovered from cases ± their contacts during each of the three outbreak years: *emm* 19.7 in 1995, *emm* 3.22 in 2000, and *emm* 55 in the 2005 outbreak. In this environment, the vast majority of APSGN cases are secondary to pyoderma rather than pharyngitis. This was reflected in APSGN cases showing seasonal variation in the northern monsoonal climate, with peaks occurring in the dry season, (Marshall et al. 2011) when pyoderma rates have also been noted to peak (McDonald et al. 2006) (see Sect. 4.1).

4.4 Invasive GAS Disease

Different definitions of 'invasive GAS disease' are in use, being either the isolation of GAS from blood and other sterile sites, or from blood alone (Lamagni et al. 2005). Invasive GAS rates again illustrate the communicable disease burden gradient between low- and high-income settings. Reported incidences of iGAS infections are 1.5–3.9 cases per 100,000 population per year in high-income countries [Canada, USA, Northern Europe; see review (World Health Organization 2005)], 6.4–10.2 per 100,000 in Australian non-Indigenous population, (Carapetis et al. 1999; Norton et al. 2004) 13.0 per 100,000 in Kenyan children (Berkley et al. 2005) and up to 82.5 per 100,000 in Australian Indigenous populations (Norton et al. 2004) (discussed in Sect. 5). In Fiji, incidence of GAS bacteraemia, in all-ages over 5 years was found to be 11.6/100,000, over 5 years was found to be 11.6/100,000, with Indigenous Fijians disproportionately affected (Steer et al. 2008). Moreover, *emm* genotyping of GAS isolates revealed that a wide diversity of *emm* types occurred, differing from types commonly identified in industrialised countries, with important implications for vaccine development (Steer et al. 2008).

The proportion of blood culture isolates which are *S. pyogenes* is highest in infants, progressively falling during the first years of life. A study drawing on data from four low-resource countries found GAS to be the third leading cause of bacteraemia in infants, accounting for 29 % (WHO Young Infants Study Group 1999). In Kenya, it is the fifth commonest cause of community bacteraemia in children <5 years, with a case fatality among all children of 25 % (Berkley et al. 2005). As GAS in neonates in low-resource countries is likely to have been acquired during birth, GAS puerperal sepsis is also likely to be occurring in these settings, but with poor documentation of rates.

Mortality rates from iGAS also remain high in adults and in affluent settings at around 15 % (Davies et al. 1996) among Swedish adults, iGAS was identified in 556 patients during 1994–1995, in whom the mortality rate was 16 %, rising to 37 % in those with STSS (Svensson et al. 2000). During the 1980–1990s period of iGAS resurgence in the USA, high mortality despite appropriate antibiotic and supportive care prompted renewed vigour in GAS research (Stevens 1995). Factors identified at this time as being associated with increased mortality were, not unexpectedly, extremes of age, hypotension or multi-organ failure (Hoge et al. 1993).

4.5 Superficial GAS Disease

The superficial GAS infections pyoderma and pharyngitis cause morbidity of their own, but more importantly, are the precursors to the autoimmune consequences of ARF and APSGN. They may also be the portal of entry for iGAS or toxin-mediated disease.

Variations in pyoderma prevalence are chiefly related to accessibility to appropriate housing and hygiene, but also show seasonal variation as noted in Sect. 4.2, being more common in dry than wet seasons in monsoonal climates (McDonald et al. 2006). Communities with high scabies rates also have elevated pyoderma prevalence (Andrews et al. 2009). In previous extensive literature reviews, (Carapetis et al. 2005a, b) pyoderma prevalence was found to range from 1 to 20 % among children in less-developed countries, but was as high as 40–90 % in surveys of Pacific (Steer et al. 1999) and Indigenous Australian children (Shelby-James et al. 2002).

Cellulitis and erysipelas are universally common causes of healthcare utilisation. Unlike many other manifestations of GAS infection, the occurrence of these conditions increases with age. The incidence of cellulitis and erysipelas reported from Minnesota in 2007 was 200 cases per 100,000 patient-years (McNamara et al. 2007). At San Francisco General Hospital, skin and soft tissue infections, including cellulitis, were the leading cause of admission for medical or surgical treatment in one study (Centers for Disease Control and Prevention (CDC) 2001). Almost all erysipelas, and approximately 85 % of cellulitis episodes, are attributable to *S. pyogenes*, the remainder being mostly due to *Staphylococcus aureus*; microbiological confirmation, however, is uncommon unless there is associated bacteraemia (Bisno and Stevens 1996; Hook et al. 1986). Long-term morbidity from these conditions is rare, but can include skin discolouration or scarring and damage to underlying vascular and lymphatic structures, increasing the risk of subsequent recurrences which can become a major burden for affected patients. Recurrent cellulitis occasionally necessitates long-term penicillin prophylaxis, although the benefit of this, as demonstrated in a randomised trial, is only slight (Sjoblom et al. 1993). Recurrent episodes of streptococcal lower limb cellulitis as a complication of saphenous vein harvest for coronary artery bypass grafting were first described by Baddour and Bisno in 1982 (Baddour and Bisno 1982). This problematic complication is an important reason that such surgery is no longer advocated where alternatives, such as minimally invasive vein harvest or radial artery harvest, can be used instead (Mahmood 2006).

GAS pharyngitis remains a nuisance and a common cause of presentation to general practitioners in high-income settings. Infection rates are best determined from studies using serological testing, to avoid the aforementioned problem of misclassification of pharyngeal carriage as infection. Previously it has been estimated, based on a number of population studies, that in high-resource settings, around 15 % of school children and 4–10 % of adults suffer an episode of symptomatic GAS pharyngitis annually, and in low-resource settings, rates may be 5–10 times higher (World Health Organization 2005; Danchin et al. 2004; Nandi et al. 2001; Majeed 1993).

5 Within-Country Variation: Disparities Between Different Populations

Between country variations described above overlook important within-country inequities in GAS rates, reflecting socioeconomic differences between populations within a given nation. These particularly affect Indigenous versus non-Indigenous and migrant versus non-migrant populations.

In Australia, nationwide GAS infection rates are generally low, while rates in rural and remote-dwelling Australian Aboriginal people are among the world's highest. The burden of superficial GAS infection in children in these communities, predominantly skin rather than pharynx, is phenomenally high (McDonald et al. 2006). Pyoderma is so common as to be considered as a normal fact of life by many affected individuals and healthcare staff, which greatly impairs treatment-seeking behaviour and prescribing habits. As noted above, a stark differential is evident in iGAS incidence in Australian Indigenous versus non-Indigenous populations with rates in these groups of 32.2–82.5 and 6.4–10.2/100,000, respectively (Carapetis et al. 1999; Norton et al. 2004). A hospital-based study in northern Australia during the years 1993–1995 found APSGN rates of 239/100,000 in Aboriginal children <15 years, compared with 6/100,000 in non-Aboriginal children (Carapetis 1998). More recently, the gap unfortunately remains, with comparative APSGN rates in children aged 0–14 years reported as 94.3 and 2.3/1,000 in Aboriginal and non-Aboriginal children, respectively (Marshall et al. 2011). Data from rheumatic fever/RHD registers in Australia indicate ARF rates persisting at levels of 186–375 cases per 100,000 per year in Aboriginal school age children, and a RHD prevalence of over 2 % (Carapetis 2009). A report from a Central Australian Aboriginal community in 1990 noted extremely worrying ARF rates of 815 per 100,000 school-aged children (Brennan and Patel 1990). By contrast, ARF is now almost unheard of in Australian-born, non-Indigenous children (Noonan et al. 2012). In the Northern Territory, Indigenous people account for 93 % of those with RHD (Parnaby and Carapetis 2010).

In New Zealand, current ARF incidence rates in Indigenous Māori and Pacific Islander children are comparable to rate estimates for non-Māori in the 1920s (Milne et al. 2012a, b). In a recent hospital-based ARF study, incidence rates for 5–14 year-old Māori children were 40.2 per 100,000, Pacific Islanders 81.2 per 100,00 and non-Māori/non-Pacific 2.1 per 100,000 (Milne et al. 2012a, b). Another New Zealand study reported rate ratios for ARF risk of 10.0 for Māori and 20.7 for Pacific peoples, compared with other New Zealand ethnic groups (Jaine et al. 2008).

In the USA, Native Americans also experience higher GAS infection rates than non-Native peoples. In a retrospective survey in Arizona, USA, 1985–1990, the annual age-adjusted incidence of iGAS was calculated to be 4.3 per 100,000 overall, but 46.0 per 100,000 among Native Americans (Hoge et al. 1993).

In societies which are affluent overall, migrant groups often experience lower socioeconomic status than non-migrants, and may travel more frequently to environments where GAS infection pressures are greater. People of Polynesian (Pacific Islander) ethnicity appear to have especially notable rates of ARF and other GAS-related disease, when surveyed in their countries of origin (Steer et al. 1999) and after migration to high-income countries (Jaine et al. 2008; Milne et al. 2012a, b; Kurahara et al. 2006; Smith et al. 2011). Polynesians living in Hawaii were found to be at higher risk of developing ARF (odds ratio 4.8), after controlling for socioeconomic factors (Kurahara et al. 2006). In a small study from Western Sydney, Australia, 36 % of ARF cases occurred among Pacific Islander children (Smith et al. 2011). Similarly, in a new urban Australian multi-centre study of 151 children with ARF, the children's ethnicity was noted to be Indigenous Australian (131), non-Indigenous Australian (10), Pacific Islanders (8), African (1) or unknown (1) (Noonan et al. 2012). While the Indigenous or Pacific Islander overrepresentation was thus again noted, ongoing cases in the affluent white population still occur, and the low level of diagnostic suspicion in this group can contribute to diagnostic delay (Noonan et al. 2012). A specific identifiable genetic predisposition to the development of ARF after GAS infection in the Pacific Islander ethnic group remains unproven (Carapetis et al. 2000; 2005a, b; Bryant et al. 2009). These poor health statistics for a communicable disease affecting disadvantaged populations pose important challenges to the healthcare systems of these nations.

6 Conclusions

The global burden of GAS infections is predominantly borne by least affluent people in resource-poor nations, with children and young people disproportionally affected. Arguably the most important but perhaps most challenging control strategy, therefore comprises improvements in societal socioeconomic status. The difficulty in reducing GAS infection rates without major gains in the social determinants of health is illustrated by the failure to achieve the goal of GAS infection reductions among Aboriginal Australians, even though concerted efforts in secondary and tertiary care have been made over decades, and good medical resources are available. Improvements in GAS control could be gained through the implementation of recognised evidence-based prevention and control strategies, such as secondary prophylaxis for ARF, but to be successful, this strategy requires major commitment of resources and personnel. This lack of progress makes a strong argument to accelerate development and availability of an effective GAS vaccine (Steer et al. 2009a, b). Better epidemiological reporting methods are required for GAS-related diseases globally, to enable the appropriate targeting of control strategies and assess the success of interventions. Comprehensive reporting systems would require both laboratory notification of GAS isolated from sterile sites, and clinician notification of ARF and APSGN, which lack specific diagnostic tests (except where renal biopsy is diagnostic of APSGN).

Continued improvements in the understanding of host and organism factors accounting for regional differences in the rates and manifestation of GAS infection will be able to improve the focus of control strategies, including better recognition and targeting of populations and individuals at risk. Greater knowledge of disease pathogenesis will help guide treatment approaches, especially in the field of adjunctive therapies. The long sought-after possibility of a GAS vaccine remains a key research focus. There is important scope for further antimicrobial treatment trials to address optimal clinical management. These issues will each be addressed in subsequent chapters. The major over-representation of socioeconomically disadvantaged people means that prevention and treatment strategies need to be feasible, accessible and affordable in low-resource settings.

Acknowledgment APR is supported by National Health and Medical Research Council (Australia) Fellowship 1016567.

References

Adriaanse AH, Pel M, Bleker OP (2000) Semmelweis: the combat against puerperal fever. Eur J Obstet Gynecol Reprod Biol 90:153–158

Andrews RM, McCarthy J, Carapetis JR, Currie BJ (2009) Skin disorders, including pyoderma, scabies, and tinea infections. Pediatr Clin North Am 56:1421–1440

Baddour LM, Bisno AL (1982) Recurrent cellulitis after saphenous venectomy for coronary bypass surgery. Ann Intern Med 97:493–496

Berkley JA, Lowe BS, Mwangi I, Williams T, Bauni E, Mwarumba S et al (2005) Bacteremia among children admitted to a rural hospital in Kenya. N Engl J Med 352:39–47

Bisno AL, Stevens DL (1996) Streptococcal infections of skin and soft tissues. N Engl J Med 334:240–245

Bisno AL, Stevens DL (2000) Streptococcus pyogenes (including Streptococcal toxic shock syndrome and necrotising fasciitis). In: Mandell GL, Bennett JE, Dolin R (eds) Mandell, Douglas and Bennett's principles and practice of infectious diseases, 5th edn. Churchill Livingstone, Philadelphia, pp 2101–2117

Bloom BR, Murray CJ (1992) Tuberculosis: commentary on a reemergent killer. Science 257:1055–1064

Brennan RE, Patel MS (1990) Acute rheumatic fever and rheumatic heart disease in a rural central Australian aboriginal community. Med J Aust 153:335, 338–339

Brown A, Purton L, Schaeffer G, Wheaton G (2003) Central Australian Rheumatic Heart Disease Control Program. Central Australian RHD Steering Committee. Northern Territory Disease Control Bulletin 10:1–8

Bryant PA, Robins-Browne R, Carapetis JR, Curtis N (2009) Some of the people, some of the time: susceptibility to acute rheumatic fever. Circulation 119:742–753

Carapetis JR (1998) Ending the heartache: the epidemiology and control of acute rheumatic fever and rheumatic heart disease in the top end of the Northern Territory, University of Sydney

Carapetis JR (2008) Rheumatic heart disease in Asia. Circulation 118:2748–2753

Carapetis JR (2009) Epidemiology of group A streptococcus in Australia. Microbiol Australia 30:171–176

Carapetis JR, Currie BJ (1999) Mortality due to acute rheumatic fever and rheumatic heart disease in the Northern territory: a preventable cause of death in aboriginal people. Aust N Z J Public Health 23:159–163
Carapetis JR, Walker AM, Hibble M, Sriprakash KS, Currie BJ (1999) Clinical and epidemiological features of group A streptococcal bacteraemia in a region with hyperendemic superficial streptococcal infection. Epidemiol Infect 122:59–65
Carapetis JR, Currie BJ, Mathews JD (2000) Cumulative incidence of rheumatic fever in an endemic region: a guide to the susceptibility of the population? Epidemiol Infect 124:239–244
Carapetis JR, Steer AC, Mulholland EK, Weber M (2005a) The global burden of group A streptococcal diseases. Lancet Infect Dis 5:685–694
Carapetis JR, McDonald M, Wilson NJ (2005b) Acute rheumatic fever. Lancet 366:155–168
Centers for Disease Control, Prevention (1999) Nosocomial group A streptococcal infections associated with asymptomatic health-care workers–Maryland and California, 1997. MMWR Morb Mortal Wkly Rep. 48:163-166
Centers for Disease Control and Prevention (CDC) (2001) Soft tissue infections among injection drug users—San Francisco, California, 1996–2000. MMWR Morb Mortal Wkly Rep 50(19):381–384. http://www.ncbi.nlm.nih.gov/pubmed/11465906
Centers for Disease Control, Prevention (2002) Outbreak of group A streptococcal pneumonia among Marine Corps recruits, 1 Nov–20 Dec, California. MMWR Morb Mortal Wkly Rep. 2003, 52:106–109
Cernay J, Rusnak M, Raisova A (1993) Incidence of rheumatic fever in Slovakia during the last 20 years. Ceskoslovenska Pediatrie 48:79–83
Coburn AF, Young DC (1949) The epidemiology of hemolytic streptotoccus during World War II in the United States Navy. Williams and Wilkins, Baltimore
Cone LA, Woodard DR, Schlievert PM, Tomory GS (1987) Clinical and bacteriologic observations of a toxic shock-like syndrome due to Streptococcus pyogenes. N Engl J Med 317:146–149
Danchin MH, Rogers S, Selvaraj G, Kelpie L, Rankin P, Vorich R et al (2004) The burden of group A streptococcal pharyngitis in Melbourne families. Indian J Med Res 119(Suppl): 144–147
Davies HD, McGeer A, Schwartz B, Green K, Cann D, Simor AE et al (1996) Invasive group A streptococcal infections in Ontario, Canada. Ontario Group A Streptococcal study group. N Engl J Med 335:547–554
Del Mar CB, Glasziou PP, Spinks AB (2006) Antibiotics for sore throat. Cochrane database of systematic reviews. :CD000023
Ewald DP, Hall GV, Franks CC (2003) An evaluation of a SAFE-style trachoma control program in Central Australia. Med J Aust 178:65–68
Ferrieri P (2002) Proceedings of the Jones Criteria workshop. Circulation 106:2521–2523
Gemmell CG, Peterson PK, Schmeling D, Kim Y, Mathews J, Wannamaker L et al (1981) Potentiation of opsonization and phagocytosis of Streptococcus pyogenes following growth in the presence of clindamycin. J Clin Invest 67:1249–1256
Hanna JN, Clark MF (2010) Acute rheumatic fever in Indigenous people in North Queensland: some good news at last? Med J Aust 192:581–584
Harper A for the Centre for Maternal and Child Enquiries (2011) Sepsis (Chapter 7). In: Saving mothers' lives. Reviewing maternal deaths to make motherhood safer, 2006–2008. BJOG 118 (Supp 1):85–96
Health Protection Agency Group A Streptococcus Working Group (2004) Interim UK guidelines for management of close community contacts of invasive group A streptococcal disease. Commun Dis Public Health 7:354–361
Heggie AD, Jacobs MR, Linz PE, Han DP, Kaplan EL, Boxerbaum B (1992) Prevalence and characteristics of pharyngeal group A beta-hemolytic streptococci in US Navy recruits receiving benzathine penicillin prophylaxis. J Infect Dis 166:1006–1013
Hoge CW, Schwartz B, Talkington DF, Breiman RF, MacNeill EM, Englender SJ (1993) The changing epidemiology of invasive group A streptococcal infections and the emergence of

streptococcal toxic shock-like syndrome: a retrospective population-based study. JAMA 269:384–389

Hook EW 3rd, Hooton TM, Horton CA, Coyle MB, Ramsey PG, Turck M (1986) Microbiologic evaluation of cutaneous cellulitis in adults. Arch Intern Med 146:295–297

Horn DL, Zabriskie JB, Austrian R, Cleary PP, Ferretti JJ, Fischetti VA et al (1998) Why have group A streptococci remained susceptible to penicillin? Report on a symposium. Clin Infect Dis 26:1341–1345

Hoy WE, Mathews JD, McCredie DA, Pugsley DJ, Hayhurst BG, Rees M et al (1998) The multidimensional nature of renal disease: rates and associations of albuminuria in an Australian Aboriginal community. Kidney Int 54:1296–1304

Jaine R, Baker M, Venugopal K (2008) Epidemiology of acute rheumatic fever in New Zealand 1996–2005. J Paediatr Child Health 44:564–571

Kaplan EL (1993) T. Duckett Jones Memorial Lecture. Global assessment of rheumatic fever and rheumatic heart disease at the close of the century. Influences and dynamics of populations and pathogens: a failure to realize prevention? Circulation 88:1964–1972

Kaplan EL (2009) Some things never change! Clin Infect Dis 48:1220–1222

Kaplan EL, Top FH Jr, Dudding BA, Wannamaker LW (1971) Diagnosis of streptococcal pharyngitis: differentiation of active infection from the carrier state in the symptomatic child. J Infect Dis 123:490–501

Katz AR, Morens DM (1992) Severe streptococcal infections in historical perspective. Clin Infect Dis 14:298–307

Kim S, Lee NY (2004) Epidemiology and antibiotic resistance of group A streptococci isolated from healthy schoolchildren in Korea. J Antimicrob Chemother 54:447–450

Kohler W, Gerlach D, Knoll H (1987) Streptococcal outbreaks and erythrogenic toxin type A. Zentralbl Bakteriol Mikrobiol Hyg A 266:104–115

Kurahara DK, Grandinetti A, Galario J, Reddy DV, Tokuda A, Langan S et al (2006) Ethnic differences for developing rheumatic fever in a low-income group living in Hawaii. Ethn Dis 16:357–361

Lamagni TL, Efstratiou A, Vuopio-Varkila J, Jasir A, Schalen C, Strep E (2005) The epidemiology of severe Streptococcus pyogenes associated disease in Europe. Euro Surveill 10:179–184

Leach AJ, Shelby-James TM, Mayo M, Gratten M, Laming AC, Currie BJ et al (1997) A prospective study of the impact of community-based azithromycin treatment of trachoma on carriage and resistance of Streptococcus pneumoniae. Clin Infect Dis 24:356–362

Lehmann D, Tennant MT, Silva DT, McAullay D, Lannigan F, Coates H et al (2003) Benefits of swimming pools in two remote Aboriginal communities in Western Australia: intervention study. BMJ 327:415–419

Lennon D, Martin D, Wong E, Taylor LR (1988) Longitudinal study of post streptococcal disease in Auckland; rheumatic fever, glomerulonephritis, epidemiology and M typing 1981–86. N Z Med J 101:396–398

Lennon D, Stewart J, Farrell E, Palmer A, Mason H (2009) School-based prevention of acute rheumatic fever: a group randomized trial in New Zealand. Pediatr Infect Dis J 28:787–794

Lue HC, Chen CL, Wei H, Okuni M, Mabilangan LM, Dharmasakti D et al (1979) The natural history of rheumatic fever and rheumatic heart disease in the Orient. Jpn Heart J 20:237–252

Mahmood Z, Al Benna S, Nkere U, Murday A (2006) Decreased morbidity following long saphenous vein harvesting using a minimally invasive technique: a randomised controlled trial comparing two techniques for long saphenous vein harvest. J Cardiothorac Surg 1:15

Majeed HA, Al-Doussary L, Moussa MM, Yusuf AR, Suliman AH (1993) Office diagnosis and management of group A streptococcal pharyngitis employing the rapid antigen detecting test. A 1-year prospective study of reliability and cost in primary care centres. Ann Trop Paediatr 13:65–72

Manyemba J, Mayosi BM (2002) Penicillin for secondary prevention of rheumatic fever. Cochrane database of systematic reviews :CD002227

Marijon E, Ou P, Celermajer DS, Ferreira B, Mocumbi AO, Jani D et al (2007) Prevalence of rheumatic heart disease detected by echocardiographic screening. N Engl J Med 357:470–476
Marijon E, Mirabel M, Celermajer DS, Jouven X (2012) Rheumatic heart disease. Lancet 379:953–964
Marshall CS, Cheng AC, Markey PG, Towers RJ, Richardson LJ, Fagan PK et al (2011) Acute post-streptococcal glomerulonephritis in the Northern Territory of Australia: a review of 16 years data and comparison with the literature. Am J Trop Med Hyg 85:703–710
Martin PR, Hoiby EA (1990) Streptococcal serogroup A epidemic in Norway 1987–1988. Scand J Infect Dis 22:421–429
McDonald M, Currie BJ, Carapetis JR (2004) Acute rheumatic fever: a chink in the chain that links the heart to the throat? Lancet Infect Dis 4:240–245
McDonald MI, Towers RJ, Andrews RM, Benger N, Currie BJ, Carapetis JR (2006) Low rates of streptococcal pharyngitis and high rates of pyoderma in Australian aboriginal communities where acute rheumatic fever is hyperendemic. Clin Infect Dis 43:683–689
McNamara DR, Tleyjeh IM, Berbari EF, Lahr BD, Martinez JW, Mirzoyev SA et al (2007) Incidence of lower-extremity cellulitis: a population-based study in Olmsted county. Minnesota. Mayo Clin Proc 82:817–821
Milne LM, Lamagni T, Efstratiou A, Foley C, Gilman J, Lilley M et al (2011) Streptococcus pyogenes cluster in a care home in England April to June 2010. Euro Surveill 16:20021
Milne RJ, Lennon D, Stewart JM, Vander Hoorn S, Scuffham PA (2012a). Mortality and hospitalisation costs of rheumatic fever and rheumatic heart disease in New Zealand. J Paediatr Child Health 48:692–697
Milne RJ, Lennon DR, Stewart JM, Vander Hoorn S, Scuffham PA (2012b) Incidence of acute rheumatic fever in New Zealand children and youth. J Paediatr Child Health 48:685–691
Morita JY, Kahn E, Thompson T, Laclaire L, Beall B, Gherardi G et al (2000) Impact of azithromycin on oropharyngeal carriage of group A Streptococcus and nasopharyngeal carriage of macrolide-resistant Streptococcus pneumoniae. Pediatr Infect Dis J 19:41–46
Myo Thet H, Thein N, Nyan T, May Mon K (1992) Prevalence of cardiovascular diseases in rural area of Hmawbi and urban Yangon city. Asia Pac J Public Health 6:188–194
Nandi S, Kumar R, Ray P, Vohra H, Ganguly NK (2001) Group A streptococcal sore throat in a periurban population of northern India: a one-year prospective study. Bull World Health Organ 79:528–533
Nathan L, Peters MT, Ahmed AM, Leveno KJ (1993) The return of life-threatening puerperal sepsis caused by group A streptococci. Am J Obstet Gynecol 169:571–572
New South Wales Health (2012) Invasive group A Streptococcus—Control guidelines for NSW Public Health Units 2015. [cited October 2012].http://www.health.nsw.gov.au/factsheets/guideline/maternal_sepsis.html
Noonan S, Zurynski YA, Currie BJ, McDonald M, Wheaton G, Nissen M, et al. (2012) A National Prospective Surveillance Study of Acute Rheumatic Fever in Australian Children. Pediatr Infect Dis J [Epub ahead of print]
Norton R, Smith HV, Wood N, Siegbrecht E, Ross A, Ketheesan N (2004) Invasive group A streptococcal disease in North Queensland (1996–2001). Indian J Med Res 119(Suppl): 148–151
Omar A (1995) Pattern of acute rheumatic fever in a local teaching hospital. Medical J Malaysia 50:125–130
Ontario Ministry of Health (2005) Guidelines for management of contacts of cases of invasive group A streptococcal disease (GAS) including streptococcal toxic shock syndrome (STSS) and necrotising fasciitis (NF) [cited October 2012]. http://microbiology.mtsinai.on.ca/protocols/pdf/k5c.pdf
Padmavati S, Sharma KB, Jayaram O (1973) Epidemiology and prophylaxis of rheumatic fever in Delhi–a five year follow-up. Singapore Med J 14:457–461
Parnaby MG, Carapetis JR (2010) Rheumatic fever in indigenous Australian children. J Paediatr Child Health 46:527–533

Pastore S, De Cunto A, Benettoni A, Berton E, Taddio A, Lepore L (2011) The resurgence of rheumatic fever in a developed country area: the role of echocardiography. Rheumatology 50:396–400

Prevention of Invasive Group ASIWP (2002) Prevention of invasive group A streptococcal disease among household contacts of case patients and among postpartum and postsurgical patients: recommendations from the Centers for Disease Control and Prevention. Clin Infect Dis 35:950–959

Purcell B, Samuelsson S, Hahne SJ, Ehrhard I, Heuberger S, Camaroni I et al (2004) Effectiveness of antibiotics in preventing meningococcal disease after a case: systematic review. BMJ 328:1339

Quinn RW (1989) Comprehensive review of morbidity and mortality trends for rheumatic fever, streptococcal disease, and scarlet fever: the decline of rheumatic fever. Rev Infect Dis 11:928–953

RHD Australia (ARF/RHD writing group) (2012) The Australian guideline for prevention, diagnosis and management of acute rheumatic fever and rheumatic heart disease (2nd edition). National Heart Foundation of Australia and the Cardiac Society of Australia and New Zealand [cited Oct 2012]. http://www.rhdaustralia.org.au/sites/default/files/guideline_0.pdf

Rheumatic fever working party (1960) The evolution of rheumatic heart disease in children. Five-year report of a co-operative clinical trial of A.C.T.H. cortisone, and aspirin. Circulation 22:503–515

Richmond P, Harris L (1998) Rheumatic fever in the Kimberley region of Western Australia. J Trop Pediatr 44:148–152

Rizvi SF, Khan MA, Kundi A, Marsh DR, Samad A, Pasha O (2004) Status of rheumatic heart disease in rural Pakistan. Heart 90:394–399

Sartain JB, Anderson NL, Barry JJ, Boyd PT, Howat PW (2012) Rheumatic heart disease in pregnancy—cardiac and obstetric outcomes. Intern Med J [epub ahead of print]

Sawhney H, Aggarwal N, Suri V, Vasishta K, Sharma Y, Grover A (2003) Maternal and perinatal outcome in rheumatic heart disease. Int J Gynaecol Obstet 80:9–14

Seckeler MD, Hoke TR (2011) The worldwide epidemiology of acute rheumatic fever and rheumatic heart disease. Clin Epidemiol 3:67–84

Seppala H, Klaukka T, Vuopio-Varkila J, Muotiala A, Helenius H, Lager K et al (1997) The effect of changes in the consumption of macrolide antibiotics on erythromycin resistance in group A streptococci in Finland. Finnish Study Group for Antimicrobial Resistance. N Engl J Med 337:441–446

Shelby-James TM, Leach AJ, Carapetis JR, Currie BJ, Mathews JD (2002) Impact of single dose azithromycin on group A streptococci in the upper respiratory tract and skin of Aboriginal children. Pediatr Infect Dis J 21:375–380

Sjoblom AC, Eriksson B, Jorup-Ronstrom C, Karkkonen K, Lindqvist M (1993) Antibiotic prophylaxis in recurrent erysipelas. Infection 21:390–393

Smith A, Lamagni TL, Oliver I, Efstratiou A, George RC, Stuart JM (2005) Invasive group A streptococcal disease: Should close contacts routinely receive antibiotic prophylaxis? The Lancet Infect Dis 5:494–500

Smith MT, Lester-Smith D, Zurynski Y, Noonan S, Carapetis JR, Elliott EJ (2011) Persistence of acute rheumatic fever in a tertiary children's hospital. J Paediatr Child Health 47:198–203

Steer AC, Adams J, Carlin J, Nolan T, Shann F (1999) Rheumatic heart disease in school children in Samoa. Arch Dis Child 81:372

Steer AC, Jenney AJ, Oppedisano F, Batzloff MR, Hartas J, Passmore J et al (2008) High burden of invasive beta-haemolytic streptococcal infections in Fiji. Epidemiol Infect 136:621–627

Steer AC, Batzloff MR, Mulholland K, Carapetis JR (2009a) Group A streptococcal vaccines: facts versus fantasy. Curr Opin Infect Dis 22:544–552

Steer AC, Law I, Matatolu L, Beall BW, Carapetis JR (2009b) Global emm type distribution of group A streptococci: systematic review and implications for vaccine development. Lancet Infect Dis 9:611–616

Stevens DL (1995) Streptococcal toxic-shock syndrome: spectrum of disease, pathogenesis, and new concepts in treatment. Emerg Infect Dis 1:69–78
Stevens DL, Maier KA, Mitten JE (1987) Effect of antibiotics on toxin production and viability of Clostridium perfringens. Antimicrob Agents Chemother 31:213–218
Stevens DL, Tanner MH, Winship J, Swarts R, Ries KM, Schlievert PM et al (1989) Severe group A streptococcal infections associated with a toxic shock-like syndrome and scarlet fever toxin A. N Engl J Med 321:1–7
Stromberg A, Romanus V, Burman LG (1991) Outbreak of group A streptococcal bacteremia in Sweden: an epidemiologic and clinical study. J Infect Dis 164:595–598
Svensson N, Oberg S, Henriques B, Holm S, Kallenius G, Romanus V et al (2000) Invasive group A streptococcal infections in Sweden in 1994 and 1995: epidemiology and clinical spectrum. Scand J Infect Dis 32:609–614
Tanz RR, Shulman ST, Barthel MJ, Willert C, Yogev R (1985) Penicillin plus rifampin eradicates pharyngeal carriage of group A streptococci. J Pediatrics 106:876–880
The WHO Young Infants Study Group (1999) Bacterial etiology of serious infections in young infants in developing countries: Results of a multicenter study. Pediatr Infect Dis J 18:S17–S22
The Working Group on Severe Streptococcal Infections (1993) Defining the group A streptococcal toxic shock syndrome: Rationale and consensus definition. The Working Group on Severe Streptococcal Infections. JAMA 269:390–391
Tibazarwa KB, Volmink JA, Mayosi BM (2008) Incidence of acute rheumatic fever in the world: a systematic review of population-based studies. Heart 94:1534–1540
Veasy LG, Wiedmeier SE, Orsmond GS, Ruttenberg HD, Boucek MM, Roth SJ et al (1987) Resurgence of acute rheumatic fever in the intermountain area of the United States. N Engl J Med 316:421–427
Wannamaker LW (1954) The epidemiology of streptococcal infections. Columbia University Press, New York
Wasserzug O, Valinsky L, Klement E, Bar-Zeev Y, Davidovitch N, Orr N et al (2009) A cluster of ecthyma outbreaks caused by a single clone of invasive and highly infective Streptococcus pyogenes. Clin Infect Dis 48:1213–1219
White AV, Hoy WE, McCredie DA (2001) Childhood post-streptococcal glomerulonephritis as a risk factor for chronic renal disease in later life. Med J Aust 174:492–496
WHO Expert consultation on rheumatic fever and rheumatic heart disease (2001) Rheumatic fever and rheumatic heart disease: report of a WHO Expert Consultation, Geneva, 29 Oct–1 Nov. WHO Technical Report Series 923. 2004 [cited October 2012]. http://whqlibdoc.who.int/trs/WHO_TRS_923.pdf
Williams AW, Gurley CR, Sobel E, Castelda M (1932) Milk-borne Septic Sore Throat and Scarlet Fever and Their Relation to Beta Hemolytic Streptococci. J Bacteriol 23:241–258
World Health Organisation (1990) Global Burden of Disease [cited Oct 2012]. http://www.who.int/topics/global_burden_of_disease/en/
World Health Organisation (2011) Global Tuberculosis Control: WHO Report. WHO Library Cataloguing-in-Publication Data. 2011 [cited Oct 2012]. http://www.who.int/tb/publications/global_report/2011/gtbr11_full.pdf
World Health Organization (2005) The Current Evidence for the Burden of Group A Streptococcal Diseases. Discussion papers on Child Health. Department of Child and Adolescent Health and Development, WHO. [cited Oct 2012]. http://whqlibdoc.who.int/hq/2005/WHO_FCH_CAH_05.07.pdf
Yagupsky P, Landau D, Beck A, Dagan R (1995) Carriage of Streptococcus pyogenes among infants and toddlers attending day-care facilities in closed communities in southern Israel. Eur J Clin Microbiol Infect Dis 14:54–58

Molecular Markers for the Study of Streptococcal Epidemiology

David J. McMillan, Martina L. Sanderson-Smith, Pierre Robert Smeesters and Kadaba S. Sriprakash

Abstract Diseases caused by *Streptococcus pyogenes* (Group A streptococcus, GAS) range from superficial infections such as pharyngitis and impetigo to potentially fatal rheumatic heart disease and invasive disease. Studies spanning emm-typing surveillance to population genomics are providing new insights into the epidemiology, pathogenesis, and biology of this organism. Such studies have demonstrated the differences that exist in the epidemiology of streptococcal disease between developing and developed nations. In developing nations, where streptococcal disease is endemic, the diversity of GAS emm-types circulating is much greater than that found in developed nations. An association between emm-type and disease, as observed in developed countries is also lacking. Intriguingly, comparative genetic studies suggest that emm-type is not always a good predictor of the evolutionary relatedness of geographically distant isolates. A view of GAS as a highly dynamic organism, in possession of a core set of virulence genes that contribute to host niche specialization and common pathogenic processes, augmented

D. J. McMillan (✉) · K. S. Sriprakash
Bacterial Pathogenesis Laboratory, Queensland Institute of Medical Research, Herston, QLD 4006, Australia
e-mail: David.mcmillan@qimr.edu.au

M. L. Sanderson-Smith
Illawarra Health and Medical Research Institute and School of Biological Sciences, University of Wollongong, Wollongong, NSW 2522, Australia

P. R. Smeesters
Murdoch Children Research Institute, Melbourne, Australia

P. R. Smeesters
Laboratoire de Génétique et Physiologie Bactérienne, Institut de Biologie et de Médecine Moléculaires, Université Libre de Bruxelles, Bruxelles, Belgium

Current Topics in Microbiology and Immunology (2013) 368: 29–48
DOI: 10.1007/82_2012_278

Published Online: 24 November 2012

by accessory genes that change the relative virulence of specific lineages is emerging. Our inability to definitively identify genetic factors that contribute to specific disease outcome underscores the complex nature of streptococcal diseases.

Contents

1 Introduction

Diseases associated with infection by *Streptococcus pyogenes* (group A streptococcus, GAS) have been so prevalent in history that many have been given easily recognizable common names (e.g., strep throat, scarlet fever, school sores, childbed fever). While the incidence of most diseases has declined markedly in countries with high per capita income, countries and regions of low income continue to suffer a high burden of these and other GAS associated diseases. Infection by GAS has been estimated to result in half a million deaths each year (Carapetis et al. 2005), and thereby position GAS as one of the top ten bacterial killers of humans. The majority of these deaths follow the development of rheumatic heart disease (RHD) and occur in developing nations. In more affluent countries, with better access to healthcare and antibiotic treatment, the prevalence of RHD is much lower, the majority of deaths attributed to the clinical manifestations associated with streptococcal invasive disease.

The burden of GAS-related diseases, differences in the geographic occurrences of these diseases, environment and the preferred body sites colonized by the organism, and accompanying changes in disease outcomes have all impelled continuing molecular and epidemiological studies of this organism. The aims of these studies have been to develop a greater understanding of molecular differences that underpin relative differences in virulence of discrete GAS lineages, identify plausible microbiological based rationale for differences in the epidemiology of GAS in different geographic locations, and predict the effectiveness of putative vaccine candidates (Smeesters et al. 2009; Steer et al. 2009d). Almost all GAS epidemiological studies use the M-protein or its genetic counterpart, the *emm*-gene as the basis for discriminating between GAS lineages. Here, we describe how these studies, the analysis of other genetic markers and genomics has changed our view of the epidemiology and biology of this organism.

2 *emm*-Gene-Based Molecular Epidemiology

Rebecca Lancefield first reported serotypic diversity in GAS more than 80 years ago. Unlike other bacterial species in which diversity was based on variation in capsular antigens, the serotypic variation that Lancefield observed was based on variation in the N-terminal region of the surface exposed M-protein (Cunningham 2000; Fischetti 1989). Subsequent work in the 1950s showed that the presence of type specific antibodies in animal and human serum was responsible for immunity against the homologous emm-type but did not protect against heterologous emm-types (Lancefield 1962). Thus, M-protein-based serotype diversity was given a functional credence, and became the basis of the GAS typing scheme (Lancefield 1962).

Serotype-based M-typing has given way to nucleotide-based procedures (emm-typing) that targets the nucleotide sequence corresponding to the hypervariable amino terminal region of the M-protein. More than 200 different emm-sequence types have now been reported (Beall et al. 1996; Facklam et al. 2002). Although undergoing several minor revision since it's introduction, the basic premise of emm-typing remains the same (http://www.cdc.gov/ncidod/biotech/strep/strepblast.htm). In both its serological and nucleotide-based incarnations, emm-typing has been extensively used to examine both geographic strain distribution and disease association. With the large number of emm-type-based surveillance studies carried out in developing countries over the past two decades it has become clear that the epidemiology of GAS differs between developing and developed regions. Excellent systematic reviews summarizing the global molecular epidemiology of GAS have been published separately by Smeesters and Steer in 2009 (Smeesters et al. 2009; Steer et al. 2009d). In low-income regions where streptococcal infections and disease are endemic, the diversity of circulating emm-types is high. This has been demonstrated globally in distinct area such as India, Fiji, Ethiopia, and Brazil (Abdissa et al. 2006; Dey et al. 2005; Smeesters et al. 2006, 2008, 2010a; Steer et al. 2009e; Tewodros and Kronvall 2005). These studies also demonstrate that no one emm-type is dominant in these regions. Steers review reported the greatest diversity of emm-types to be found in Pacific Regions and Africa. Figure 1, depicting the emm-type diversity as determined using Simpson Reciprocal Index, clearly demonstrates the differences in diversity between countries. The pool of emm-types recovered in separate studies in low-income regions also differs (Smeesters et al. 2009; Steer et al. 2009d). For example, only one third of the emm-types recovered in two Ethiopian studies (Abdissa et al. 2006; Tewodros and Kronvall 2005) were described in a separate Fijian study (Steer et al. 2009e). Taken together the data appear to fit a model where the number of circulating emm-types is high, and subject to temporal flux. Few longitudinal studies, examining the rate of emm-type replacement have been

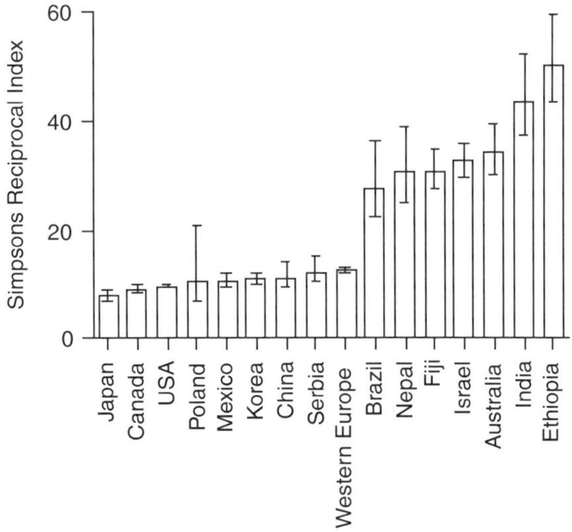

Fig. 1 Diversity of circulating emm-types in 15 countries and Western Europe. Diversity was calculated and presented as Simpsons Reciprocal Index with 95 % confidence intervals. (Adapted from Expert Review of Vaccines, December 2009, Vol. 8, No. 12, Pages 1705–1720 with permission of Expert Reviews Ltd)

carried out in endemic regions (Steer et al. 2009b). There is also lack of surveillance and prospective surveys in some areas of the world where the burden of streptococcal disease is high (Carapetis et al. 2005). This is particularly true for Africa, South America, and some locations in Asia (Smeesters et al. 2009; Steer et al. 2009d).

In developed countries, fewer emm-types appear to circulate, with even fewer dominant. Steer et al. (Steer et al. 2009d) reported that 25 emm-types accounted for more 90 % of all isolates recovered in developed nations, with 146 different emm-types accounting for the remaining 10 %. Recovery of these predominant emm-types in multiple studies also suggests that are consistently present in the wider population (McNeil et al. 2005; Shulman et al. 2009; Smeesters et al. 2006). Longitudinal studies carried out in smaller geographic region show that different emm-types are recovered at different time points (Shulman et al. 2009), demonstrating that local temporal flux in circulating emm-types occurs.

Molecular techniques, such as vir-typing (Gardiner et al. 1995) and emm pattern-typing (Hollingshead et al. 1993) represent alternate methods to categorize GAS based on variation in the emm-gene and surrounding DNA. Emm pattern typing categorizes emm-types into three distinct patterns based on chromosomal architecture (pattern types A-C, D, and E). The patterns are determined by the presence and arrangement of *emm* and *emm*-like genes, as determined by specific PCR reactions (Hollingshead et al. 1993). emm-type correlates well with specific emm pattern type (McGregor et al. 2004b). In general emm pattern-type correlates with host colonization site. Pattern A-C strains are usually associated with throat colonization, pattern D strains mainly recovered from superficial skin infection, while pattern E represents a "generalist" group associated with both tissue sites (Bessen and Lizano 2010). Similar to emm-typing, exceptions to the emm pattern type/tissue tropism have been reported, especially in low-income populations

(Bessen and Lizano 2010). Only about 20 % of the 223 known emm-types belong to pattern A-C type. These emm-types predominantly circulate in high-income countries. The remaining 80 % of emm-types are equally distributed between the pattern D and E groups, and are mostly found in developing countries. Of note, emm-types belonging to emm pattern A-C, which contain the so-called 'rheumatogenic' and 'invasive' M-types have been most extensively studied in relation to disease outcome (Erdem et al. 2005; Fischetti 1989; Smeesters et al. 2010b).

Certain emm-types, belonging to pattern A-C, are predominantly recovered from the throat rather than skin in developing countries (Shulman et al. 2004). As rheumatic fever (RF) is considered to follow an episode of pharyngitis, these types have also come to be known as 'rheumatogenic' emm-types (Bisno et al. 2003; Martin and Barbadora 2006; Stollerman 2001). Other emm-types are more commonly associated with skin infection and glomerulonephritis, and have come to be known as 'nephritogenic' emm-types. A third group of emm-types (e.g., emm1, emm12, emm3, and emm18) associated with the outbreak of severe invasive disease in the USA and Europe over the past two decades are known as 'invasive' emm-types (O'Loughlin et al. 2007; Vlaminckx et al. 2007). Interestingly recent epidemiological studies carried out in geographic regions where RHD and streptococcal infections are endemic have failed to recover significant numbers of these rheumatogenic emm-types (Bessen et al. 2000; Steer et al. 2009c). In fact, epidemiological studies in tropical regions, where the incidence of RF, but the incidence of pharyngitis is low is challenging the dogma that there is an association between GAS, throat infection, and RF (Bessen et al. 2000; Parks et al. 2012). Unique emm-types have also been implicated in RF in Hawaii (Erdem et al. 2007). Similarly, 'invasive' emm-types while present are not common in low income countries (Steer et al. 2009a).

Emm-typing has served the scientific community well for several decades, and is still relevant to tracking disease outbreaks. However, the differences in emm-type diversity and disease associations have created challenges for the streptococcal community when attempting to extrapolate data and interpretation of results from developed nations to developing nations. It should be acknowledged that comparisons of emm-type prevalence in different locations can be complicated by the different methods used to acquire isolates. Some studies represent single timepoint surveillance, whereas others analyses isolates collected over time (Shulman et al. 2009). Isolates recovered during epidemic outbreaks are also likely to differ from those recovered during non-disease related surveillance in the same location. Nevertheless, the data from different studies conducted in endemic and non-endemic regions collectively suggest differences in diversity observed between these groups are real.

Emm-type is also assumed to be marker for evolutionary relatedness, and by extension, similarity in pathogenic potential of isolates. While this is undoubtedly true for isoaltes collected during clonal outbreaks, it is less so for geographically or temporally unrelated strains. As described in more detail below, evidence for lateral gene transfer (LGT) and recombination of DNA, including the emm-gene is strong. The section of the emm-gene used for emm-typing encodes the region of the corresponding protein that may be targeted by the host immune system, and is therefore likely to be under strong immune selection pressure. Replacement of this region, through recombination involving part or all of the emm-gene is an efficient method for escaping the host immune response (Panchaud et al. 2009; Whatmore and Kehoe 1994). In countries where streptococcal disease is endemic, and emm-type diversity is high, the conditions for LGT of the emm-gene is much greater than non-endemic countries.

3 Non-emm-Gene Virulence Factors and GAS Epidemiology

GAS pathogenesis depends on the coordinated regulation of a complex repertoire of virulence factors. While host factors undoubtedly contribute to variation in susceptibility to GAS, the frequent association of specific emm-types with disease in developed countries lends itself strongly to the argument that specific bacterial characteristics are associated with defined epidemiologies. The most comprehensive body of work examining association between non-*emm* genetic factors and GAS disease has occurred in the area of serious invasive disease. However, despite many years of research, the identification of a defined subset of invasive disease related virulence factors remains elusive.

There is increasing evidence that in populations where GAS is endemic, despite the diversity of emm-types, there is conservation of certain virulence factors linked to tissue tropism. While emm pattern typing has been useful for classification of GAS into these three patterns, several elegant epidemiological studies have led to the identification of a specific subset of genes which may confer GAS with a tissue specific phenotype (Fig. 2) (Bessen et al. 2005; Kalia and Bessen 2004; Kratovac et al. 2007). Linkages to tissue-specificity are particularly strong for genes encoding the colonization factors serum opacity factor (*sof*), collagen-binding protein *(cpa),* and fibronectin binding proteins *(prtF1)* and *(fbaA)*, which have a defined pattern of presence or absence in strains associated with distinct tissue sites (Kratovac et al. 2007). There is also significant historical evidence to implicate these colonization factors in tissue tropism, and most recently, comparative genomic hybridization analysis of 96 GAS isolates revealed that a major defining factor of tissue tropism in skin versus throat isolates is the presence or absence of genes encoding fibronectin binding proteins (Bessen et al. 2011).

Historically, SOF has provided the basis for an alternative GAS typing scheme. Streptococci of differing emm-types produce distinct variants of SOF, and type specific antibodies can be used to inhibit serum opacification by SOF variants (Maxted et al. 1973). There is a strong correlation between emm-type and SOF phenotype, and SOF antisera has been used as a method of GAS typing in the past. A comprehensive study of GAS isolates in 2000 showed that SOF is a useful predictor of emm-type, and that discordance between *sof* and *emm* is rare. However, GAS strains of the same emm-type do not always contain the same *sof* allele, and strains with diverse emm-types can have highly similar SOF genes (Beall et al.

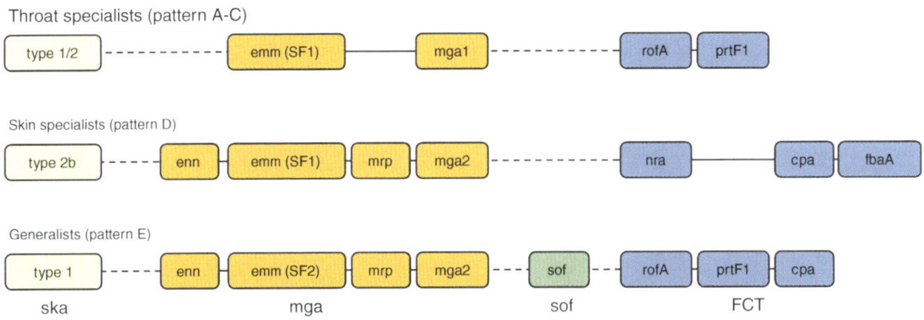

Fig. 2 Typical virulence gene repertoires of throat (pattern A-C), skin (pattern D) and generalist (pattern E) GAS isolates. Allelic variants of genes located in the streptokinase (ska), mga, serum opacity factor (sof) and FCT loci are shown. SF indicates emm gene subfamily as defined by Hollingshead et al. (1994)

2000; Johnson et al. 2006; Sakota et al. 2006). No correlation has been found between various *sof* alleles and different disease states, or geographical locations. A comparison of published studies from around the world shows that most reported *sof/emm* combinations are distributed globally (Dhanda et al. 2011; Goodfellow et al. 2000; Johnson et al. 2006; Sakota et al. 2006).

This lack of geographical segregation may, in part, be explained by the linkage between *sof* and the "generalist" tissue phenotype, that is, isolates which are equally associated with skin and pharyngeal infection, which represent approximately 50 % of GAS isolates globally (Bessen and Lizano 2010). Typically, pharyngeal specialists are opacity factor negative, and harbor a *prtF1* allele (Courtney and Pownall 2010; Cunningham 2000; Kratovac et al. 2007). In contrast, skin specialists are more likely to be endowed with *fbaA*. In a study of GAS isolates from the Northern territory of Australia, where skin infection predominates, 92 % of isolates were found to be *fbA* positive (Ramachandran et al. 2004). Similarly, *cpa* is more frequently associated with skin trophic isolates than strains which would be regarded as throat specialists (Kratovac et al. 2007). The differential distribution of these genes between skin versus throat versus generalist strains largely parallels the architecture of the overall *mga* regulon.

Additionally, a number of genes appear to have evolved into discrete phylogenetic lineages associated with tissue site preference. The regulatory gene *mga* is present in the GAS genome in one of two allelic forms, *mga1*, which is predominantly associated with throat specialist strains, and *mga2*, which appears to be found in all emm pattern D and E strains. These allelic forms are mutually exclusive (Bessen et al. 2005). Similarly, *rofA/nra* encodes as transcriptional regulator which regulates expression of pillus structural genes (Kreikemeyer et al. 2003). *rofA* is associated with emm pattern A-C and E strains (throat specialists and generalists), skin specialists typically harbor *nra* (Bessen et al. 2005).

The plasminogen activator, streptokinase, has been found in all GAS strains screened to date. Streptokinase is comprised of three distinct domains, α, β, and γ. There is significant variability within the β domain of *ska*, and 3 distinct *ska* alleles have been described—type 1, type 2a, and 2b (Kalia and Bessen 2004). There is strong genetic linkage between the gene encoding the plasminogen binding M protein (PAM) and type 2b *ska* alleles, which appear to be almost exclusively coinherited (Kalia and Bessen 2004; McArthur et al. 2008). Isolates harboring this allele are typically regarded as "skin trophic". In contrast, cluster 2a alleles are almost exclusively associated with isolates that would be regarded as throat-trophic based on emm pattern A-C, while type 1 *ska* alleles are found in both emm pattern A-C and E backgrounds (Kalia and Bessen 2004).

These preferences for tissue site, and distribution of virulence factors, are not absolute. In geographical regions where GAS infection is endemic, the demarcation line between emm-type and tissue site preference is less well defined. Population surveillance of GAS isolates in Nepal found that 19 % of isolates associated with skin infection display a genetic architecture usually associated with pharyngeal specialists (Sakota et al. 2006), while in an Ethiopian study, 28 % of isolates associated with tonsillitis harbored an emm pattern D, or skin specialist genomic arrangement (Tewodros and Kronvall 2005). Similarly, GAS with an emm pattern D, or skin trophic chromosomal arrangement harboring *ska* type 1 alleles have been reported (Kalia and Bessen 2004). This underscores the high degree of genetic diversity among GAS isolates. In areas where GAS is endemic, recombination between skin and throat specialists appears to be common (Bessen et al. 2011; Kalia et al. 2002). Given these findings, it will remain important to monitor the relationships between virulence factor distribution and tissue tropism in epidemiological studies.

Significant intra-emm-type genetic variation, particularly with respect to regulatory gene profile, phage content, and toxin profile have been observed in the dominant emm-types associated with invasive disease in developed countries. Perhaps the most striking example of this is the reported difference in transcription profile between invasive and colonizing GAS strains of the same emm-type in outbreaks of GAS infection (Johnson et al. 1992; Marcon et al. 1988; Sumby et al. 2006). Expression microarray analysis of M1 isolates from the United States, Canada, and Finland revealed the existence of distinct non-invasive and invasive transcriptome profiles within a group of clinical M1 strains (Sumby et al. 2006). These transcriptome profiles have since been linked to mutations in the control of virulence regulatory sensor kinase (*covRS*; alternatively designated *csrRS*), which is responsible for the regulation of approximately 10 % of the GAS genome (Graham et al. 2002; Sumby et al. 2006). Acquisition of mutations in *covRS* which inactivate the ability of this system to negatively regulate the GAS genome during infection results in upregulation of capsule, loss of SpeB expression and increased disease severity in animal models of infection (Walker et al. 2007). Acquisition of these mutations appears to be dependent on the expression of specific virulence factor genes, including *emm*, *hasA*, and the phage encoded *sda1* (Cole et al. 2011). The presence of mutations in *covRS* associated with clinical invasive GAS isolates have been extensively reported in studies of isolates from developed nations, and these mutations are not restricted to emm1 GAS. A recent study of isolates from Canada that compared the genome sequences of GAS serotype M3 isolates from human pharyngitis cases and from human invasive disease reported that *covS* mutations occur with a higher frequency in invasive-disease isolates than in pharyngeal isolates (Shea et al. 2011). Several studies report *covS* mutations in GAS isolates recovered from patients with severe STSS (Ato et al. 2008; Ikebe et al. 2010). Similarly, a Hawaiian emm81.0 GAS isolate from the throat that had spread to the bloodstream was shown to have acquired a mutation in *covS* (Garcia et al. 2010).

While these mutations can be linked to "hypervirulence", the association is not absolute. In a retrospective study of GAS clinical isolates from Chile, 77/110 isolates were found to have SNPs in the *covRS* regulon, irrespective of their site of isolation. However, only two of these isolates were found to express the high levels of capsule associated with hypervirulence as a result of mutations in *covS* (Wozniak et al. 2012). In contrast, screening of isolates from the Northern Territory of Australia, where skin infection predominates, found a much lower frequency of *covRS* mutation. Of the invasive ($n = 12$) and superficial ($n = 13$) Northern Territory isolates screened, only one was found to contain a *covS* mutation (Maamary et al. 2010). Whether this lower frequency of mutation is common in regions such as the Northern Territory, where emm-types are diverse, and skin infection is endemic, is unclear due to a lack of data on the *covS* status of clinical isolates from these and other similar regions.

Studies of other virulence markers have been used to track the origins of outbreak strains, and may prove useful in characterizing newly emergent strains of GAS. Exotoxin profile has been used as a means of differentiating GAS isolates in a number of studies. To date, 11 distinct superantigens have been identified. Three of these are chromosomally encoded (SpeG, SpeJ, and SMEZ), while the remaining eight (SpeA, SpeC, SpeH, SpeI, SpeJ, SpeK, SpeL, SpeM, and SSA) are located on temperate phage. Exchange of mobile genetic elements (MGE) such as phage contributes significantly to the genetic diversity seen within the GAS species, and is thought to be involved in the emergence of highly successful virulent clones (Lintges et al. 2010; Maamary et al. 2012; Maripuu et al. 2008). Superantigen profiling may therefore prove useful as a marker for the presence and transfer of prophages that encode additional genes playing a role in GAS pathogenesis or tissue tropism.

A high level of allelic variation within certain superantigen genes has been reported (Bianco et al. 2006; Talkington et al. 1993), and this, combined with the fact that studies of superantigen distribution from different groups often use different primer sets, confounds the comparison of superantigen profiling by different groups. However, a 2012 study of the distribution of superantigens amongst 480 diverse GAS isolates from Portugal has reported the use, for the first time, of a defined primer set enabling amplification of all reported superantigen alleles to date (Friaes et al. 2012). Using this technique, 11 different superantigen profiles were identified. Interestingly, this study found that none of the individual superantigen genes could be linked with a specific emm-type, but rather, superantigen profile shows a strong association with emm-type. Superantigen profile could not be accurately predicated using alternative typing methods (other than PFGE subtyping), leading to the conclusion that superantigen profile may be a useful predictor of emm-type, but that superantigen profile cannot be inferred from emm-type (Friaz et al. 2012). This assertion is supported by previous findings from numerous studies undertaken in various countries suggesting a link between emm-type and superantigen profile (Commons et al. 2008; Le Hello et al. 2010; Schmitz et al. 2003; Vlaminckx et al. 2003).

The reported circulation of superantigens within the streptococcal population varies temporally. Screening of GAS isolates associated with a reemergence of invasive disease in Denmark between 1999 and 2002 found that the frequency of emm1 isolates harboring *speA* decreased from 94 % in 1999 to 71 % in 2002, while the emm1-specific prevalence of *speC* increased from 25 to 53 % over the same period (Ekelund et al. 2005). Similarly, a comprehensive study of GAS isolates from Melbourne, Australia indicates that strains before the mid-1980s do not typically harbor *speK*, which is found in the genome of a high proportion of contemporary emm3 isolates (Commons et al. 2008). This highlights the potential usefulness of superantigen profiling to monitor the emergence of highly successful GAS clones during epidemics. There is evidence to suggest that superantigen prevalence in the GAS population differs geographically (Commons et al. 2008; Proft et al. 2003). However, cross-study comparison is, as stated previously, confounded by the use of different primer sets between studies.

4 Multilocus Sequence Typing

As virulence factors may be under strong selection pressure, and as is the case with superantigens, be present on mobile genetic elements (MGEs), they are not an ideal target for determination of evolutionary relationships between GAS strains. Multilocus Sequence Typing (MLST) is nucleotide-based method for investigating relationships among bacteria of the same species that targets seven selection neutral house-keeping genes. The genes used in the GAS MLST scheme (http://spyogenes.mlst.net/) are glucose kinase (*gki*), glutamine transport ATP-binding protein (*gtr*), glutamate racemase (*murI*), DNA mismatch repair (*mutS*), transketolase (*recP*), xanthine phosphoribosyltransferase (*xpt*), and acetoacetyl-CoA thiolase (*yql*). The combination of seven allelic variants is used to denote a specific multilocus sequence type (ST). More than 600 STs are listed on the *S. pyogenes* MLST website, indicating that ST is more discriminatory than emm-type.

A strong linkage between emm-type and ST has been reported for isolates from endemic countries (Enright et al. 2001). However a subsequent study by McGregor et al. (McGregor et al. 2004a) examining STs of isolates recovered from the Indigenous population of a remote Australian island (representing a low income region) reported a weaker relationship between emm-type and ST combinations present in isolates from the island and emm-type/ST combinations from other populations. The data can be interpreted as providing strong circumstantial evidence for LGT of the *emm*-gene in regions where GAS diversity is high. Further complicating interpretation of MLST data was the observation of LGT involving housekeeping present on the 'core' genome (McGregor et al. 2004b). The proportion of emm-types associated with distant genetic backgrounds, as determined by MLST, were found to be much high for skin specialists (pattern D strains) and generalists (pattern E strains) than throat specialists (pattern A-C), suggesting that recombination events in latter occur at a lower frequency (Bessen et al. 2008).

5 Genomic Studies of Group A Streptococcus

It is becoming clear that emm-typing provides limited value when attempting to identify genetic factors linked to niche specialization of specific disease outcomes. Other approaches, examining the presence or absence of multiple genetic factors has been one pathway taken in attempt to provide a more thorough analysis of streptococcal diseases (Bessen et al. 2011; McMillan et al. 2006). Genomic studies are also proving to be pivotal in understanding the level of diversity that exists in the streptococcal population. The first M1 GAS genome was published in 2001 (Ferretti et al. 2001), with genome from other emm-types following soon after (Beres et al. 2002; Green et al. 2005; Smoot et al. 2002). Comparative analyses of these genomes clearly demonstrated the importance of MGEs, LGT, and recombination in generating diversity between emm-types. The majority of GAS genomes possess multiple bacteriophage, which in turn often carry virulence genes predicted to change the virulence of a lineage. Integrative conjugative elements (ICE) and remnant MGEs are also common (Green et al. 2005; Maamary et al. 2012; Smoot et al. 2002; Sumby et al. 2005).

However, it has been the intra-emm-type genomic studies utilizing multiple isolates with defined clinical histories that are providing the greatest insight into how strains with altered capacity to cause disease may evolve (Ben Zakour et al. 2012; Fittipaldi et al. 2012; Maamary et al. 2012; Sumby et al. 2005; Tse et al. 2012). These studies suggest that the acquisition of new MGEs, resulting in the elaboration of novel virulence gene repertoire is an important facet underlying the changes in virulence (Maamary et al. 2012; Sumby et al. 2005). Genomic analysis of the recent invasive disease outbreak associated with GAS emm59 provides an illustrative example of the value of population-based genomics. Prior to 2005, emm59 was not considered to be an emm-type associated with major outbreaks of invasive disease. However, in the second half of the decade, more than 500 invasive disease cases associated with this emm-type were reported in Canada (Tyrrell et al. 2010). Comparison of the core genome of outbreak isolates with historical and temporally distinct clones found the outbreak isolates to form a genetically distinct group (Fittipaldi et al. 2012). However, differences in the core genome between outbreak and non-outbreak isolates were remarkably small. While differences in MGE content between the outbreak strains and non-outbreak strain were apparent, as were difference in biological properties, the identity of putative genes within the MGEs that contribute to the relative differences in virulence could not be identified.

The first genomic analysis of GAS emm12 isolates involved in a scarlet fever outbreak were also recently reported (Tse et al. 2012). Unlike the emm59 outbreak, this outbreak appeared to be multiclonal. A novel ICE and bacteriophage were found to be distributed throughout the different clonal lineages involved in the outbreak. The latter contains genes encoding superantigens, and were hypothesized to be one of the reason for the observed increase in virulence.

6 The Role of Other *β*-Hemolytic Bacteria in 'GAS-Associated Diseases

Streptococcus dysgalactiae subsp. *equisimilis* (SDSE) is generally considered as a commensal organism with occasional potential to cause diseases in humans. Over the past few decades there have been several reports of SDSE causing diseases that are normally attributed to GAS. The list of diseases includes tonsillitis, skin infections, post streptococcus arthritis, pleuropneumonia, meningitis, endocarditis, puerperal septicaemia, necrotizing fasciitis, toxic shock syndrome (Jensen and Kilian 2012).

However, a possible link between SDSE and RF/RHD is still deemed as circumstantial, and controversial, with a general recalcitrance among the medical community against the view that *β*-hemolytic streptococci other than GAS (for example SDSE) could also be associated with RF/RHD. A major problem in definitively demonstrating a link between SDSE and RF in clinical settings is the long lag between infection and RF, and possible requirement for multiple episodes of infection, both of which call for well-controlled follow up studies. Such studies are near impossible as SDSE and GAS do cocolonize at the same tissue sites. Animal models of RHD are also in their infancy (Lymbury et al. 2003).

Despite the resistance to the accommodation of alternative views, several observations have strongly supported possible involvement of SDSE in the pathogenesis of RF/RHD. We and others (Bramhachari et al. 2010; McDonald et al. 2007) have observed that the GAS throat isolation rate in some communities is not commensurate with the RF/RHD burden. However, in these same population SDSE isolation rates from the throats are high. Indeed SDSE, but not GAS, has been recovered from an Indigenous Australian child after recurrent severe pharyngitis which was followed by RF in this patient (Davies et al. 2005) suggesting a role for SDSE in the pathogenesis of RF. All SDSE strains also express the M protein which may elicit heart-tissue cross-reactive autoantibodies (Haidan et al. 2000). In this study purified F(ab')2 fragments from the Indigenous Australians reacted with surface protein extracts of SDSE, but not with those of GAS skin isolates (Haidan et al. 2000). RF/RHD associated GAS M types are also reported to possess collagen binding motifs in their M proteins suggesting a possible role for this interaction in the pathogenesis of RF/RHD (Dinkla et al. 2003) Interestingly, similar motifs were also seen in some SDSE M proteins (Dinkla et al. 2007). Collectively, these above observations strongly suggest SDSE possessed many of the same characteristics as GAS that is linked to the pathogenesis of RF/RHD.

Phylogenically SDSE and GAS are closely related to each other (Lefebure et al. 2012). While there are conspicuous absences of some virulence genes in SDSE, up to half of GAS genes encoding virulence factors or surface associated proteins could be present in SDSE (Davies et al. 2007). Intra and Interspecies LGTs in GAS and SDSE have been long recognized (Sriprakash and Hartas 1996; Towers et al. 2004) with phages and ICEs contributing significantly to the population structure of these streptococci. Clonal relationships revealed that recombination far outweigh mutation in generating clonal variants. While both replacement and additive changes occur through LGT, characteristics that influence host-pathogen interactions are likely to be due to additive LGT (Choi et al. 2012).

Interestingly, cross-species LGTs between GAS and SDSE seem to be predominantly unidirectional; from the former to the latter (Choi et al. 2012). This seems to be true even for selective-neutral housekeeping genes (McMillan et al. 2010). However, although the same emmSTs were found in different clonal complexes of GAS and SDSE suggesting occurrence of LGT of the emm-gene within the species, cross-species replacement of this gene was rarely observed. Likewise, SDSE with group A carbohydrate was found rarely. All SDSE isolates are speB-negative (Bramhachari et al. 2010) and no instance of additive LGT for this locus was ever reported to our knowledge. These observations, and mechanisms such as general resistance to phage mediated changes in SDSE due to clustered regularly interspaced short palindromic repeats (Shimomura et al. 2011) suggest that despite extensive ongoing LGTs, species demarcation between SDSE and GAS is unlikely to blur. However, with extensive additive LGTs into SDSE and sharing of many virulence characteristics the epidemiology of GAS diseases may be blurred in regions of high endemicity.

7 Conclusion

In a relatively short time, studies of the molecular epidemiology of GAS have progressed from the study of single genes, to population-based genomic comparisons. These more comprehensive studies have led to a view of GAS as highly dynamic organism. In endemic regions, where the number of circulating emm-types is high, no one lineage is associated with disease. SDSE is also abundant in these regions. In non-endemic countries, a small pool of emm-types that fluctuates at the local level is present. Ongoing LGT and recombination occasionally give rise to new lineages whose novel genetic repertoire may increase their virulence and disease causing potential. Mutation of core genes is a second mechanism that can also change the virulence properties of streptococci. Our inability to identify or validate the role of genes is specific disease outcomes, highlights the complex nature of streptococcal virulence, and suggests that different genetic repertoires may be responsible for the same disease in different lineages. Changes in accessory virulence gene repertoire occur against a background of core virulence genes that are associated with niche specialization and core pathogenic processes. For a complete understanding of streptococcal functional genomics of large global collections of streptococci will be essential.

References

Abdissa A, Asrat D, Kronvall G, Shittu B, Achiko D, Zeidan M, Yamuah LK, Aseffa A (2006) High diversity of group A streptococcal emm types among healthy schoolchildren in Ethiopia. Clin Infect Dis 42:1362–1367

Ato M, Ikebe T, Kawabata H, Takemori T, Watanabe H (2008) Incompetence of neutrophils to invasive group A streptococcus is attributed to induction of plural virulence factors by dysfunction of a regulator. PLoS One 3:e3455

Beall B, Facklam R, Thompson T (1996) Sequencing emm-specific PCR products for routine and accurate typing of group A streptococci. J Clin Microbiol 34:953–958

Beall B, Gherardi G, Lovgren M, Facklam RR, Forwick BA, Tyrrell GJ (2000) emm and sof gene sequence variation in relation to serological typing of opacity-factor-positive group A streptococci. Microbiology 146(Pt 5):1195–1209

Ben Zakour NL, Venturini C, Beatson SA, Walker MJ (2012) Analysis of a Streptococcus pyogenes puerperal sepsis cluster by use of whole-genome sequencing. J Clin Microbiol 50:2224–2228

Beres SB, Sylva GL, Barbian KD, Lei B, Hoff JS, Mammarella ND, Liu MY, Smoot JC, Porcella SF, Parkins LD, Campbell DS, Smith TM, McCormick JK, Leung DY, Schlievert PM, Musser JM (2002) Genome sequence of a serotype M3 strain of group A Streptococcus: phage-encoded toxins, the high-virulence phenotype, and clone emergence. Proc Natl Acad Sci U S A 99:10078–10083

Bessen DE, Lizano S (2010) Tissue tropisms in group A streptococcal infections. Future Microbiol 5:623–638

Bessen DE, Carapetis JR, Beall B, Katz R, Hibble M, Currie BJ, Collingridge T, Izzo MW, Scaramuzzino DA, Sriprakash KS (2000) Contrasting molecular epidemiology of group A streptococci causing tropical and nontropical infections of the skin and throat. J Infect Dis 182:1109–1116

Bessen DE, Manoharan A, Luo F, Wertz JE, Robinson DA (2005) Evolution of transcription regulatory genes is linked to niche specialization in the bacterial pathogen Streptococcus pyogenes. J Bacteriol 187:4163–4172

Bessen DE, McGregor KF, Whatmore AM (2008) Relationships between emm and multilocus sequence types within a global collection of Streptococcus pyogenes. BMC Microbiol 8:59

Bessen DE, Kumar N, Hall GS, Riley DR, Luo F, Lizano S, Ford CN, McShan WM, Nguyen SV, Dunning Hotopp JC, Tettelin H (2011) Whole-genome association study on tissue tropism phenotypes in group A Streptococcus. J Bacteriol 193:6651–6663

Bianco S, Allice T, Zucca M, Savoia D (2006) Survey of phenotypic and genetic features of streptococcus pyogenes strains isolated in Northwest Italy. Curr Microbiol 52:33–39

Bisno AL, Brito MO, Collins CM (2003) Molecular basis of group A streptococcal virulence. Lancet Infect Dis 3:191–200

Bramhachari PV, Kaul SY, McMillan DJ, Shaila MS, Karmarkar MG, Sriprakash KS (2010) Disease burden due to Streptococcus dysgalactiae subsp. equisimilis (group G and C streptococcus) is higher than that due to Streptococcus pyogenes among Mumbai school children. J Med Microbiol 59:220–223

Carapetis JR, Steer AC, Mulholland EK, Weber M (2005) The global burden of group A streptococcal diseases. Lancet Infect Dis 5:685–694

Choi SC, Rasmussen MD, Hubisz MJ, Gronau I, Stanhope MJ, Siepel A (2012) Replacing and additive horizontal gene transfer in Streptococcus. Mol Biol Evol 29:3309–3320
Cole JN, Barnett TC, Nizet V, Walker MJ (2011) Molecular insight into invasive group A streptococcal disease. Nat Rev Microbiol 9:724–736
Commons R, Rogers S, Gooding T, Danchin M, Carapetis J, Robins-Browne R, Curtis N (2008) Superantigen genes in group A streptococcal isolates and their relationship with emm types. J Med Microbiol 57:1238–1246
Courtney HS, Pownall HJ (2010) The structure and function of serum opacity factor: a unique streptococcal virulence determinant that targets high-density lipoproteins. J Biomed Biotechnol 2010:956071
Cunningham MW (2000) Pathogenesis of group A streptococcal infections. Clin Microbiol Rev 13:470–511
Davies MR, Tran TN, McMillan DJ, Gardiner DL, Currie BJ, Sriprakash KS (2005) Inter-species genetic movement may blur the epidemiology of streptococcal disease in endemic regions. Microbes Infect 7:1128–1138
Davies MR, McMillan DJ, Beiko RG, Barroso V, Geffers R, Sriprakash KS, Chhatwal GS (2007) Virulence profiling of Streptococcus dysgalactiae subspecies equisimilis isolated from infected humans reveals 2 distinct genetic lineages that do not segregate with their phenotypes or propensity to cause diseases. Clin Infect Dis 44:1442–1454
Dey N, McMillan DJ, Yarwood PJ, Joshi RM, Kumar R, Good MF, Sriprakash KS, Vohra H (2005) High diversity of group A Streptococcal emm types in an Indian community: the need to tailor multivalent vaccines. Clin Infect Dis 40:46–51
Dhanda V, Vohra H, Kumar R (2011) Virulence potential of Group A streptococci isolated from throat cultures of children from north India. Indian J Med Res 133:674–680
Dinkla K, Rohde M, Jansen WT, Kaplan EL, Chhatwal GS, Talay SR (2003) Rheumatic fever-associated Streptococcus pyogenes isolates aggregate collagen. J Clin Invest. 111:1905–1912
Dinkla K, Nitsche-Schmitz DP, Barroso V, Reissmann S, Johansson HM, Frick IM, Rohde M, Chhatwal GS (2007) Identification of a streptococcal octapeptide motif involved in acute rheumatic fever. J Biol Chem 282:18686–18693
Ekelund K, Skinhoj P, Madsen J, Konradsen HB (2005) Reemergence of emm1 and a changed superantigen profile for group A streptococci causing invasive infections: results from a nationwide study. J Clin Microbiol 43:1789–1796
Enright MC, Spratt BG, Kalia A, Cross JH, Bessen DE (2001) Multilocus sequence typing of Streptococcus pyogenes and the relationships between emm type and clone. Infect Immun 69:2416–2427
Erdem G, Ford JM, Kanenaka RY, Abe L, Yamaga K, Effler PV (2005) Molecular epidemiologic comparison of 2 unusual clusters of group a streptococcal necrotizing fasciitis in Hawaii. Clin Infect Dis 40:1851–1854
Erdem G, Mizumoto C, Esaki D, Reddy V, Kurahara D, Yamaga K, Abe L, Johnson D, Yamamoto K, Kaplan EL (2007) Group A streptococcal isolates temporally associated with acute rheumatic fever in Hawaii: differences from the continental United States. Clin Infect Dis (an official publication of the Infectious Diseases Society of America) 45:e20–e24
Facklam RF, Martin DR, Lovgren M, Johnson DR, Efstratiou A, Thompson TA, Gowan S, Kriz P, Tyrrell GJ, Kaplan E, Beall B (2002) Extension of the lancefield classification for group A streptococci by addition of 22 new M protein gene sequence types from clinical isolates: emm103 to emm124. Clin Infect Dis 34:28–38
Ferretti JJ, McShan WM, Ajdic D, Savic DJ, Savic G, Lyon K, Primeaux C, Sezate S, Suvorov AN, Kenton S, Lai HS, Lin SP, Qian Y, Jia HG, Najar FZ, Ren Q, Zhu H, Song L, White J, Yuan X, Clifton SW, Roe BA, McLaughlin R (2001) Complete genome sequence of an M1 strain of Streptococcus pyogenes. Proc Natl Acad Sci U S A 98:4658–4663
Fischetti VA (1989) Streptococcal M protein: molecular design and biological behavior. Clin Microbiol Rev 2:285–314
Fittipaldi N, Beres SB, Olsen RJ, Kapur V, Shea PR, Watkins ME, Cantu CC, Laucirica DR, Jenkins L, Flores AR, Lovgren M, Ardanuy C, Linares J, Low DE, Tyrrell GJ, Musser JM

(2012) Full-genome dissection of an epidemic of severe invasive disease caused by a hypervirulent, recently emerged clone of group A Streptococcus. Am J Pathol 180:1522–1534

Friaes A, Pinto FR, Silva-Costa C, Ramirez M, and Melo-Cristino J (2012) Superantigen gene complement of Streptococcus pyogenes-relationship with other typing methods and short-term stability. Eur J Clin Microbiol Infect Dis [Epub ahead of print]

Garcia AF, Abe LM, Erdem G, Cortez CL, Kurahara D, Yamaga K (2010) An insert in the covS gene distinguishes a pharyngeal and a blood isolate of Streptococcus pyogenes found in the same individual. Microbiology 156:3085–3095

Gardiner D, Hartas J, Currie B, Mathews JD, Kemp DJ, Sriprakash KS (1995) Vir typing: a long-PCR typing method for group A streptococci. PCR Methods Appl. 4:288–293

Goodfellow AM, Hibble M, Talay SR, Kreikemeyer B, Currie BJ, Sriprakash KS, Chhatwal GS (2000) Distribution and antigenicity of fibronectin binding proteins (SfbI and SfbII) of Streptococcus pyogenes clinical isolates from the northern territory. Australia. J Clin Microbiol 38:389–392

Graham MR, Smoot LM, Migliaccio CA, Virtaneva K, Sturdevant DE, Porcella SF, Federle MJ, Adams GJ, Scott JR, Musser JM (2002) Virulence control in group A Streptococcus by a two-component gene regulatory system: global expression profiling and in vivo infection modeling. Proc Natl Acad Sci U S A. 99:13855–13860

Green NM, Zhang S, Porcella SF, Nagiec MJ, Barbian KD, Beres SB, LeFebvre RB, Musser JM (2005) Genome sequence of a serotype M28 strain of group a streptococcus: potential new insights into puerperal sepsis and bacterial disease specificity. J Infect Dis 192:760–770

Haidan A, Talay SR, Rohde M, Sriprakash KS, Currie BJ, Chhatwal GS (2000) Pharyngeal carriage of group C and group G streptococci and acute rheumatic fever in an Aboriginal population. Lancet 356:1167–1169

Hollingshead SK, Readdy TL, Yung DL, Bessen DE (1993) Structural heterogeneity of the emm gene cluster in group A streptococci. Mol Microbiol 8:707–717

Hollingshead SK, Arnold J, Readdy TL, Bessen DE (1994) Molecular evolution of a multigene family in group A streptococci. Mol Biol Evol 11:208–219

Ikebe T, Ato M, Matsumura T, Hasegawa H, Sata T, Kobayashi K, Watanabe H (2010) Highly frequent mutations in negative regulators of multiple virulence genes in group A streptococcal toxic shock syndrome isolates. PLoS Pathog 6:e1000832

Jensen A, Kilian M (2012) Delineation of Streptococcus dysgalactiae, its subspecies, and its clinical and phylogenetic relationship to Streptococcus pyogenes. J Clin Microbiol 50:113–126

Johnson DR, Stevens DL, Kaplan EL (1992) Epidemiologic analysis of group A streptococcal serotypes associated with severe systemic infections, rheumatic fever, or uncomplicated pharyngitis. J Infect Dis 166:374–382

Johnson DR, Kaplan EL, VanGheem A, Facklam RR, Beall B (2006) Characterization of group A streptococci (Streptococcus pyogenes): correlation of M-protein and emm-gene type with T-protein agglutination pattern and serum opacity factor. J Med Microbiol 55:157–164

Kalia A, Bessen DE (2004) Natural selection and evolution of streptococcal virulence genes involved in tissue-specific adaptations. J Bacteriol 186:110–121

Kalia A, Spratt BG, Enright MC, Bessen DE (2002) Influence of recombination and niche separation on the population genetic structure of the pathogen Streptococcus pyogenes. Infect Immun 70:1971–1983

Kratovac Z, Manoharan A, Luo F, Lizano S, Bessen DE (2007) Population genetics and linkage analysis of loci within the FCT region of Streptococcus pyogenes. J Bacteriol 189:1299–1310

Kreikemeyer B, McIver KS, Podbielski A (2003) Virulence factor regulation and regulatory networks in Streptococcus pyogenes and their impact on pathogen-host interactions. Trends Microbiol 11:224–232

Lancefield RC (1962) Current knowledge of type-specific M antigens of group A streptococci. J Immunol 89:307–313

Le Hello S, Doloy A, Baumann F, Roques N, Coudene P, Rouchon B, Lacassin F, Bouvet A (2010) Clinical and microbial characteristics of invasive Streptococcus pyogenes disease in

New Caledonia, a region in Oceania with a high incidence of acute rheumatic fever. J Clin Microbiol 48:526–530

Lefebure T, Richards VP, Lang P, Pavinski-Bitar P, Stanhope MJ (2012) Gene repertoire evolution of Streptococcus pyogenes inferred from phylogenomic analysis with Streptococcus canis and Streptococcus dysgalactiae. PLoS One 7:e37607

Lintges M, van der Linden M, Hilgers RD, Arlt S, Al-Lahham A, Reinert RR, Plucken S, Rink L (2010) Superantigen genes are more important than the emm type for the invasiveness of group A Streptococcus infection. J Infect Dis 202:20–28

Lymbury RS, Olive C, Powell KA, Good MF, Hirst RG, LaBrooy JT, Ketheesan N (2003) Induction of autoimmune valvulitis in Lewis rats following immunization with peptides from the conserved region of group A streptococcal M protein. J Autoimmun 20:211–217

Maamary PG, Sanderson-Smith ML, Aziz RK, Hollands A, Cole JN, McKay FC, McArthur JD, Kirk JK, Cork AJ, Keefe RJ, Kansal RG, Sun H, Taylor WL, Chhatwal GS, Ginsburg D, Nizet V, Kotb M, Walker MJ (2010) Parameters governing invasive disease propensity of non-M1 serotype group A streptococci. J Innate Immun 2:596–606

Maamary PG, Ben Zakour NL, Cole JN, Hollands A, Aziz RK, Barnett TC, Cork AJ, Henningham A, Sanderson-Smith M, McArthur JD, Venturini C, Gillen CM, Kirk JK, Johnson DR, Taylor WL, Kaplan EL, Kotb M, Nizet V, Beatson SA, and Walker MJ (2012) Tracing the evolutionary history of the pandemic group A streptococcal M1T1 clone. FASEB J 26:4674–4684

Marcon MJ, Hribar MM, Hosier DM, Powell DA, Brady MT, Hamoudi AC, Kaplan EL (1988) Occurrence of mucoid M-18 Streptococcus pyogenes in a central Ohio pediatric population. J Clin Microbiol 26:1539–1542

Maripuu L, Eriksson A, Norgren M (2008) Superantigen gene profile diversity among clinical group A streptococcal isolates. FEMS Immunol Med Microbiol 54:236–244

Martin JM, Barbadora KA (2006) Continued high caseload of rheumatic fever in western Pennsylvania: possible rheumatogenic emm types of streptococcus pyogenes. J pediatr 149:58–63

Maxted WR, Widdowson JP, Fraser CA (1973) Antibody to streptococcal opacity factor in human sera. J Hyg (Lond). 71:35–42

McArthur JD, McKay FC, Ramachandran V, Shyam P, Cork AJ, Sanderson-Smith ML, Cole JN, Ringdahl U, Sjobring U, Ranson M, Walker MJ (2008) Allelic variants of streptokinase from Streptococcus pyogenes display functional differences in plasminogen activation. FASEB J 22:3146–3153

McDonald M, Towers RJ, Andrews RM, Carapetis JR, Currie BJ (2007) Epidemiology of Streptococcus dysgalactiae subsp. equisimilis in tropical communities. Northern Austral Emerg Infect Dis. 13:1694–1700

McGregor KF, Bilek N, Bennett A, Kalia A, Beall B, Carapetis JR, Currie BJ, Sriprakash KS, Spratt BG, Bessen DE (2004a) Group A streptococci from a remote community have novel multilocus genotypes but share emm types and housekeeping alleles with isolates from worldwide sources. J Infect Dis 189:717–723

McGregor KF, Spratt BG, Kalia A, Bennett A, Bilek N, Beall B, Bessen DE (2004b) Multilocus sequence typing of Streptococcus pyogenes representing most known emm types and distinctions among subpopulation genetic structures. J Bacteriol 186:4285–4294

McMillan DJ, Beiko RG, Geffers R, Buer J, Schouls LM, Vlaminckx BJ, Wannet WJ, Sriprakash KS, Chhatwal GS (2006) Genes for the majority of group a streptococcal virulence factors and extracellular surface proteins do not confer an increased propensity to cause invasive disease. Clin Infect Dis 43:884–891

McMillan DJ, Bessen DE, Pinho M, Ford C, Hall GS, Melo-Cristino J, Ramirez M (2010) Population genetics of Streptococcus dysgalactiae subspecies equisimilis reveals widely dispersed clones and extensive recombination. PLoS One 5:e11741

McNeil SA, Halperin SA, Langley JM, Smith B, Warren A, Sharratt GP, Baxendale DM, Reddish MA, Hu MC, Stroop SD, Linden J, Fries LF, Vink PE, Dale JB (2005) Safety and immunogenicity of 26-valent group a streptococcus vaccine in healthy adult volunteers. Clin Infect Dis 41:1114–1122

O'Loughlin RE, Roberson A, Cieslak PR, Lynfield R, Gershman K, Craig A, Albanese BA, Farley MM, Barrett NL, Spina NL, Beall B, Harrison LH, Reingold A, Van Beneden C (2007) The epidemiology of invasive group A streptococcal infection and potential vaccine implications: United States, 2000–2004. Clin Infect Dis 45:853–862

Panchaud A, Guy L, Collyn F, Haenni M, Nakata M, Podbielski A, Moreillon P, Roten CA (2009) M-protein and other intrinsic virulence factors of Streptococcus pyogenes are encoded on an ancient pathogenicity island. BMC Genomics 10:198

Parks T, Smeesters PR, Steer AC (2012) Streptococcal skin infection and rheumatic heart disease. Curr Opin Infect Dis 25:145–153

Proft T, Webb PD, Handley V, Fraser JD (2003) Two novel superantigens found in both group A and group C Streptococcus. Infect Immun 71:1361–1369

Ramachandran V, McArthur JD, Behm CE, Gutzeit C, Dowton M, Fagan PK, Towers R, Currie B, Sriprakash KS, Walker MJ (2004) Two distinct genotypes of prtF2, encoding a fibronectin binding protein, and evolution of the gene family in Streptococcus pyogenes. J Bacteriol 186:7601–7609

Sakota V, Fry AM, Lietman TM, Facklam RR, Li Z, Beall B (2006) Genetically diverse group A streptococci from children in far-western Nepal share high genetic relatedness with isolates from other countries. J Clin Microbiol 44:2160–2166

Schmitz FJ, Beyer A, Charpentier E, Normark BH, Schade M, Fluit AC, Hafner D, Novak R (2003) Toxin-gene profile heterogeneity among endemic invasive European group A streptococcal isolates. J Infect Dis 188:1578–1586

Shea PR, Beres SB, Flores AR, Ewbank AL, Gonzalez-Lugo JH, Martagon-Rosado AJ, Martinez-Gutierrez JC, Rehman HA, Serrano-Gonzalez M, Fittipaldi N, Ayers SD, Webb P, Willey BM, Low DE, Musser JM (2011) Distinct signatures of diversifying selection revealed by genome analysis of respiratory tract and invasive bacterial populations. Proc Natl Acad Sci U S A 108:5039–5044

Shimomura Y, Okumura K, Murayama SY, Yagi J, Ubukata K, Kirikae T, Miyoshi-Akiyama T (2011) Complete genome sequencing and analysis of a Lancefield group G Streptococcus dysgalactiae subsp. equisimilis strain causing streptococcal toxic shock syndrome (STSS). BMC Genomics 12:17

Shulman ST, Tanz RR, Kabat W, Kabat K, Cederlund E, Patel D, Li Z, Sakota V, Dale JB, Beall B (2004) Group A streptococcal pharyngitis serotype surveillance in North America, 2000–2002. Clin Infect Dis 39:325–332

Shulman ST, Tanz RR, Dale JB, Beall B, Kabat W, Kabat K, Cederlund E, Patel D, Rippe J, Li Z, Sakota V (2009) Seven-year surveillance of North American pediatric group a streptococcal pharyngitis isolates. Clin Infect Dis 49:78–84

Smeesters PR, Vergison A, Campos D, de Aguiar E, Miendje Deyi VY, Van Melderen L (2006) Differences between Belgian and Brazilian group A Streptococcus epidemiologic landscape. PLoS One 1:e10

Smeesters PR, Mardulyn P, Vergison A, Leplae R, Van Melderen L (2008) Genetic diversity of Group A Streptococcus M protein: implications for typing and vaccine development. Vaccine 26:5835–5842

Smeesters PR, McMillan DJ, Sriprakash KS, Georgousakis MM (2009) Differences among group A streptococcus epidemiological landscapes: consequences for M protein-based vaccines? Expert Rev Vaccines 8:1705–1720

Smeesters PR, Dramaix M, Van Melderen L (2010a) The emm-type diversity does not always reflect the M protein genetic diversity—is there a case for designer vaccine against GAS. Vaccine 28:883–885

Smeesters PR, McMillan DJ, Sriprakash KS (2010b) The streptococcal M protein: a highly versatile molecule. Trends Microbiol 18:275–282

Smoot JC, Barbian KD, Van Gompel JJ, Smoot LM, Chaussee MS, Sylva GL, Sturdevant DE, Ricklefs SM, Porcella SF, Parkins LD, Beres SB, Campbell DS, Smith TM, Zhang Q, Kapur V, Daly JA, Veasy LG, Musser JM (2002) Genome sequence and comparative microarray analysis of serotype M18 group A Streptococcus strains associated with acute rheumatic fever outbreaks. Proc Natl Acad Sci U S A 99:4668–4673

Sriprakash KS, Hartas J (1996) Lateral genetic transfers between group A and G streptococci for M-like genes are ongoing. Microb Pathog 20:275–285

Steer AC, Jenney A, Kado J, Good MF, Batzloff M, Waqatakirewa L, Mullholland EK, Carapetis JR (2009a) Prospective surveillance of invasive group a streptococcal disease, Fiji, 2005–2007. Emerg Infect Dis 15:216–222

Steer AC, Jenney AW, Kado J, Good MF, Batzloff M, Magor G, Ritika R, Mulholland KE, Carapetis JR (2009b) Prospective surveillance of streptococcal sore throat in a tropical country. Pediatr Infect Dis J 28:477–482

Steer AC, Kado J, Jenney AW, Batzloff M, Waqatakirewa L, Mulholland EK, Carapetis JR (2009c) Acute rheumatic fever and rheumatic heart disease in Fiji: prospective surveillance, 2005–2007. Med J Aust 190:133–135

Steer AC, Law I, Matatolu L, Beall BW, Carapetis JR (2009d) Global emm type distribution of group A streptococci: systematic review and implications for vaccine development. Lancet Infect Dis 9:611–616

Steer AC, Magor G, Jenney AW, Kado J, Good MF, McMillan D, Batzloff M, Carapetis JR (2009e) emm and C-repeat region molecular typing of beta-hemolytic streptococci in a tropical country: implications for vaccine development. J Clin Microbiol 47(8):2502–2509

Stollerman GH (2001) Rheumatic fever in the 21st century. Clin Infect Dis 33:806–814

Sumby P, Porcella SF, Madrigal AG, Barbian KD, Virtaneva K, Ricklefs SM, Sturdevant DE, Graham MR, Vuopio-Varkila J, Hoe NP, Musser JM (2005) Evolutionary origin and emergence of a highly successful clone of serotype M1 group a Streptococcus involved multiple horizontal gene transfer events. J Infect Dis 192:771–782

Sumby P, Whitney AR, Graviss EA, DeLeo FR, Musser JM (2006) Genome-wide analysis of group a streptococci reveals a mutation that modulates global phenotype and disease specificity. PLoS Pathog 2:e5

Talkington DF, Schwartz B, Black CM, Todd JK, Elliott J, Breiman RF, Facklam RR (1993) Association of phenotypic and genotypic characteristics of invasive Streptococcus pyogenes isolates with clinical components of streptococcal toxic shock syndrome. Infect Immun 61:3369–3374

Tewodros W, Kronvall G (2005) M protein gene (emm type) analysis of group A beta-hemolytic streptococci from Ethiopia reveals unique patterns. J Clin Microbiol 43:4369–4376

Towers RJ, Gal D, McMillan D, Sriprakash KS, Currie BJ, Walker MJ, Chhatwal GS, Fagan PK (2004) Fibronectin-binding protein gene recombination and horizontal transfer between group A and G streptococci. J Clin Microbiol 42:5357–5361

Tse H, Bao JY, Davies MR, Maamary P, Tsoi HW, Tong AH, Ho TC, Lin CH, Gillen CM, Barnett TC, Chen JH, Lee M, Yam WC, Wong CK, Ong CL, Chan YW, Wu CW, Ng T, Lim WW, Tsang TH, Tse CW, Dougan G, Walker MJ, Lok S, Yuen KY (2012) Molecular characterization of the 2011 Hong Kong scarlet fever outbreak. J Infect Dis 206:341–351

Tyrrell GJ, Lovgren M, St Jean T, Hoang L, Patrick DM, Horsman G, Van Caeseele P, Sieswerda LE, McGeer A, Laurence RA, Bourgault AM, Low DE (2010) Epidemic of group A Streptococcus M/emm59 causing invasive disease in Canada. Clin Infect Dis 51:1290–1297

Vlaminckx BJ, Mascini EM, Schellekens J, Schouls LM, Paauw A, Fluit AC, Novak R, Verhoef J, Schmitz FJ (2003) Site-specific manifestations of invasive group a streptococcal disease: type distribution and corresponding patterns of virulence determinants. J Clin Microbiol 41:4941–4949

Vlaminckx BJ, Schuren FH, Montijn RC, Caspers MP, Fluit AC, Wannet WJ, Schouls LM, Verhoef J, Jansen WT (2007) Determination of the relationship between group A streptococcal genome content, M type, and toxic shock syndrome by a mixed genome microarray. Infect Immun 75:2603–2611

Walker MJ, Hollands A, Sanderson-Smith ML, Cole JN, Kirk JK, Henningham A, McArthur JD, Dinkla K, Aziz RK, Kansal RG, Simpson AJ, Buchanan JT, Chhatwal GS, Kotb M, Nizet V (2007) DNase Sda1 provides selection pressure for a switch to invasive group A streptococcal infection. Nat Med 13:981–985

Whatmore AM, Kehoe MA (1994) Horizontal gene transfer in the evolution of group A streptococcal emm-like genes: gene mosaics and variation in Vir regulons. Mol Microbiol 11:363–374

Wozniak A, Rojas P, Rodriguez C, Undabarrena A, Garate C, Riedel I, Roman JC, Kalergis AM, Garcia P (2012) M-protein gene-type distribution and hyaluronic acid capsule in group A Streptococcus clinical isolates in Chile: association of emm gene markers with csrR alleles. Epidemiol Infect 140:1286–1295

Epidemiology and Pathogenicity of Zoonotic Streptococci

Marcus Fulde and Peter Valentin-Weigand

Abstract Zoonotic infections caused by *Streptococcus* spp. have been neglected in spite of the fact that frequency and severity of outbreaks increased dramatically in recent years. This may be due to non-identification since respective species are often not considered in human medical diagnostic procedures. On the other hand, an expanding human population concomitant with an increasing demand for food and the increased number of companion animals favour conditions for host species adaptation of animal streptococci. This review aims to give an overview on streptococcal zoonoses with focus on epidemiology and pathogenicity of four major zoonotic species, *Streptococcus canis*, *Streptococcus equi* sub. *zooepidemicus*, *Streptococcus iniae* and *Streptococcus suis*.

Contents

M. Fulde
Medizinische Hochschule Honnover, Institut für Medizinische Mikrobiologie und Krankenhaushygiene, Carl-Neuberg-Str. 1 30625 Hannover, Germany
e-mail: fulde.marcus@mh-hannover.de

P. Valentin-Weigand (✉)
Institut für Mikrobiologie, Zentrum für Infektionsmedizin,
Stiftung Tierärztliche Hochschule Hannover,
Bischofsholer Damm 15, 30173 Hannover, Germany
e-mail: peter.valentin@tiho-hannover.de

Current Topics in Microbiology and Immunology (2013) 368: 49–81
DOI: 10.1007/82_2012_277

Published Online: 29 November 2012

1 Introduction

Infections caused by *Streptococcus* spp. have received little attention in zoonosis research. Some zoonotic species, e.g. *S. suis*, are considered to be neglected due to non-identification during routine diagnostic procedures in human medicine. On the other hand, since streptococci *per se* are classified as facultative pathogens, almost all endothermic and many poikilothermic animal species as well as humans harbour at least one streptococcal species as commensal on skin and/or mucosal surfaces (Facklam 2002). Poor hygiene, secondary infections or underlying immunodeficiency, often results in streptococcosis of the natural host, whereas true host–species jumps seem to be rather rare events. Usually, zoonotic infections caused by streptococci do not result in notified epidemics or pandemics, but they can lead to severe and life-threatening diseases of individuals (Abbott et al. 2010; Baiano and Barnes 2009; Galperine et al. 2007; Lam et al. 2007; Lun et al. 2007; Wertheim et al. 2009). On the other hand, outbreaks have been reported which were related to highly virulent zoonotic strains, as exemplified by the *S. suis* outbreak in China 2005 or the *S. zooepidemicus* outbreak in Brazil 1997/98 (see below). Fortunately, such outbreaks have not been caused by human-to-human transmission, but are limited to infections from contaminated food or close contact with infected animals (Balter et al. 2000; Chen et al. 2007; Tang et al. 2006). This suggests that these zoonotic streptococci are not well adapted to the human host.

The wide distribution of streptococcal species and the incomplete characterisation of clinical isolates to species-level render a predication about the zoonotic potential of certain species difficult (Facklam 2002). Furthermore, Lancefield typing is not sufficient to discriminate zoonotic streptococci. For example, *S. agalactiae* (GBS) isolates leading to neonatal sepsis and meningitis harbour a different geno- and phenotype than GBS strains causing mastitis in cattle or infections in fish (Dogan et al. 2005; Martinez et al. 2000; Oliveira et al. 2006; Pereira et al. 2010; Sukhnanand et al. 2005). Similarly, a different genotype is discussed for *S. pneumoniae* serotype 3 strains causing pneumonia in race horses

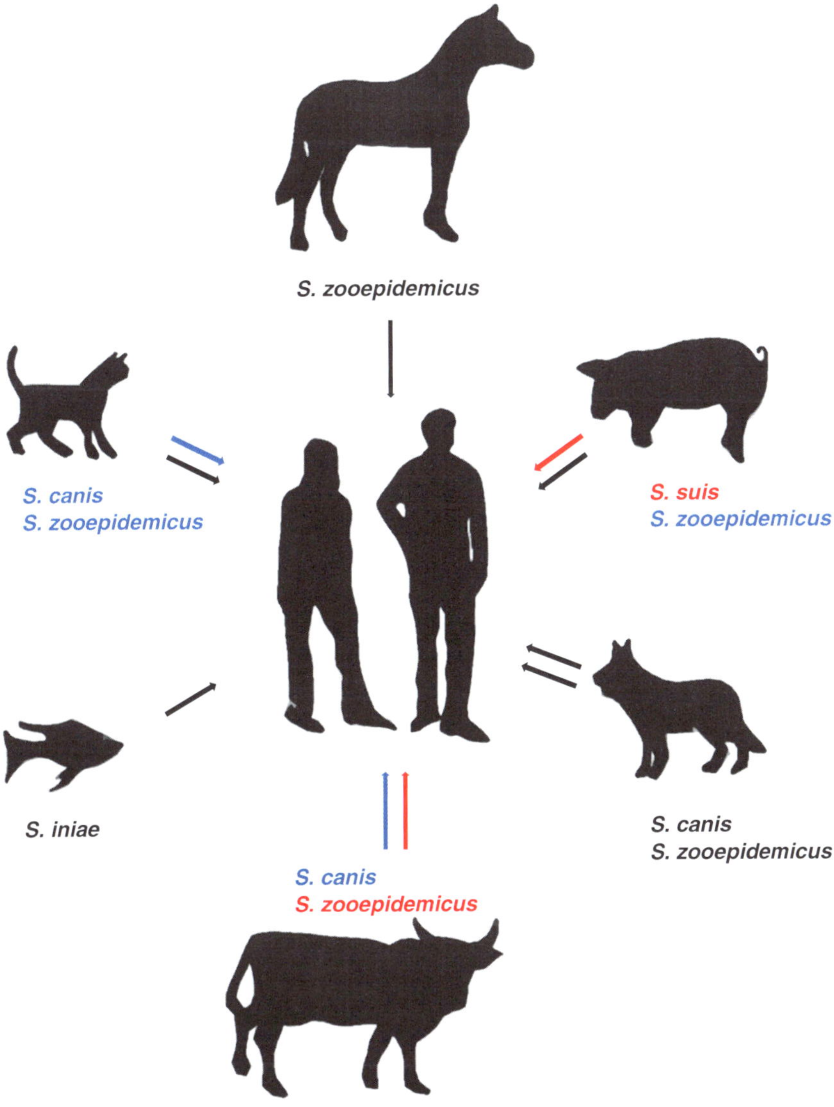

Fig. 1 Schematic figure representing host–pathogen relations of zoonotic streptococci. *Black arrows* indicate the transmission to one individual, whereas red arrows illustrate the origin of outbreaks. An identification of zoonotic species in animals without a proven transmission to humans is coloured *blue*

as compared to those from humans and strains isolated from a variety of rodents (Blunden et al. 1994; Van der Linden 2009).

Considering the complexity of zoonotic and potentially zoonotic streptococcal infections, we will focus in this review on the proven zoonotic species *S. canis*, *S. equi* sub. *zooepidemicus* (*S. zooepidemicus*), *S. iniae* and *S. suis* with emphasis on epidemiology and pathogenicity. A scheme illustrating the natural hosts of these species and their transmission to humans is depicted in Fig. 1.

2 *Streptococcus canis*

2.1 General Features

S. canis was first described taxonomically by Devriese and co-workers in 1986, but was known as a causative agent of mastitis in cows and a variety of different clinical pictures in dogs and cats much earlier (Devriese et al. 1986). Phenotypically, *S. canis* is characterised as large-colony-forming, ß-haemolytic group G streptococcus (GGS) of animal origin distinguishable from other GGS, such as *S. dysgalactiae* subsp. *equisimilis* (*S. equisimilis*, large-colony-forming, ß-haemolytic GGS derived from human) and members of the *S. anginosus* group (minute ß-haemolytic colonies derived from humans) (Efstratiou et al. 1994; Devriese et al. 1986). Biochemical identification of human GGS is based on positive α- and β- galactosidase activity, negative hyaluronidase- and β- glucoronidase activity as well as on negative acid production from trehalose catabolism (Efstratiou et al. 1994; Devriese et al. 1986). The composition of the cell wall peptidoglycan represents a remarkable rare phenotype with its main constituents being lysine, threonine and glycine (Devriese et al. 1986). Interestingly, the inability of *S. canis* to lyse human fibrin represents another factor for distinguishing human and animal GGS, as already shown by Butaye (1956). However, this feature might be also important for pathogenesis of *S. canis* infections (as described in Sect. 2.3).

In addition to the phenotypic properties, early genetic studies based on DNA hybridisation and 16S rRNA sequencing revealed that *S. canis* is a member of the large group of pyogenic streptococci (Kawamura et al. 1995; Devriese et al. 1986). This was confirmed in detailed phylogenetic studies by Jensen and Kilian (2012) and Lefebure et al. (2012). Both groups found that human specific *S. pyogenes* (GAS) and *S. dysgalactiae*, comprising GGS from human origin (*S. equisimilis*) and bovine group C streptococcus (GCS, *S. dysgalactiae* sub. *dysgalactiae*), represent the closest relatives of *S. canis*.

Several groups identified genetic markers for species-specific detection of *S. canis*, such as specific sequences of genes coding for the superoxide dismutase *sod*A (Poyart et al. 1998; Whatmore et al. 2001), the DNA mismatch repair protein *mut*S (Whatmore et al. 2001), the CAMP factor (*cfg*) (Hassan et al. 2003) and the RNase P (*rnp*B) (Tapp et al. 2003). Hassan et al. (2003) developed a PCR based on species-specific sequences of the 16S rRNA gene and the 16S-23S rRNA gene intergenic spacer region. Nevertheless, routine diagnostic of streptococci is usually limited to determination of the Lancefield group which might result in an under-representation of *S. canis* in data banks (Lam et al. 2007).

2.2 Epidemiology and Zoonotic Relevance

S. canis represents a constituent of the resident microflora of domestic carnivores colonising skin and mucosae of urinary, gastrointestinal and reproductive tract mainly in dogs and cats (Devriese et al. 1986). Lyskova et al. (2007) reported a carrier rate of 18 % for dogs and 12.7 % for cats, respectively, in an epidemiological study of 926 samples from 324 animals. The importance of *S. canis* as an infectious agent in dogs was clearly demonstrated by Lamm et al. (2010) who showed in a retrospective study of 393 samples from streptococcal infections in dogs that 22.4 % were caused by *S. canis*. In addition to domestic carnivores as primary host species, Hamilton and Stark (1970) reported cases of bovine mastitis with animal GGS as the causative agent, suggesting that the bovine udder might be a further niche of this facultative pathogen. In a few cases, *S. canis* was isolated from rats, minks, mice, rabbit, foxes and horses (Corning et al. 1991; Iglauer et al. 1991).

Severe and life-threatening diseases in dogs with β-haemolytic streptococci as the causative agent were reported since the 1930s of the last century (Hare and Frye 1938). Today, it is known that the spectrum of clinical symptoms resulting from *S. canis* infections in dogs is highly divers. It ranges from mild and superficial pyogenic infections of the skin and mucosae of the respiratory and urogenital tract to more severe clinical pictures such as pneumonia, mastitis, abortion, cellulitis and septicaemia of newborns (Biberstein et al. 1980; DeWinter and Prescott 1999; Kruger et al. 2010). In addition, in the mid-1990s, several reports were published describing identification of *S. canis* as the causative agent of streptococcal toxic shock-like syndrome (STSS) and Necrotising Fasciitis (NF) in dogs in Southern Canada and USA (Miller et al. 1996, Prescott et al. 1995). In addition to the clinical pictures described in dogs, there are some reports on sporadic cases of *S. canis* associated contagious lymphadenitis, arthritis and myositis in cats (Iglauer et al. 1991; Swindle et al. 1980; Tillman et al. 1982). Nevertheless, *S. canis* infections in cats seem to be more often associated with severe outbreaks presenting as STSS, NF, sepsis and meningitis reaching mortality rates of up to 30 %. It is very likely that this phenomenon is due to the limited space in catteries and shelters resulting in close contact of the animals and easier transmission. Interestingly, a recent study by Kruger et al. (2010) proposed a clonal origin of *S. canis* strains causing STSS and NF in cats, which seems to be contradictory to the scenario described in dogs (DeWinter et al. 1999).

In contrast to the broad spectrum of clinical pictures and tissue sites involved in *S. canis* infections of domestic carnivores, in cattle infection may lead to subclinical mastitis characterised by elevated somatic cell counts (SCCs) (Chaffer et al. 2005; Hassan et al. 2005). Although the carrier rate of *S. canis* in the udder of cows is low ($\sim$1 %, Hamilton and Stark 1970; McDonald and McDonald 1976), and outbreaks of mastitis are rare events, morbidity might reach >30 %. Tikofsky and Zadoks (2005) described a scenario for an outbreak of *S. canis* mastitis in a dairy herd. A cat suffering from chronic sinusitis caused by *S. canis* was the most likely source as evaluated by bacterial culture and ribotyping. The authors stated that dissemination of the bacterium from cow to cow was a result of an insufficient hygiene management including contaminated udder and insufficient post-milking teat disinfection. Nevertheless, this example shows the ability of *S. canis* to spread across species barriers.

Cases of human infections caused by *S. canis* are rare, but may be underestimated due to insufficient diagnostic practice limited to Lancefield typing (Lam et al. 2007). It is believed that transmission to humans occurs mainly via direct contact or as a result of animal bites. However, in most cases, aetiology of *S. canis* infection is unknown (Galperine et al. 2007). Two cases of septicaemia due to dog bite and close vicinity to pet dogs were reported by Takeda et al. (2001) and Bert and Lambert-Zechovsky (1997), respectively. Interestingly, detection of *S. canis* as the causative agent of ulcers in dog owners increased during the last years (Lam et al. 2007), suggesting a possible emerging role of this zoonosis in the near future.

2.3 Virulence and Pathogenesis

Despite its role as an emerging infective agent in animals and humans, knowledge on pathogenesis and virulence factors of *S. canis* remains elusive. However, similarities in disease establishment and progression, clinical pictures, pathology and outcome of infections with *S. pyogenes* infections suggest the existence of comparable virulence mechanisms, although data from animal infections are lacking. An initial attempt to prove this hypothesis was reported by DeWinter et al. (1999). Using Southern Blot analyses, the genomes of 15 *S. canis* isolates from cases of STSS and NF in dogs were screened for GAS-associated virulence factor genes, such as *emm*, *ska*, *spe*A, *spe*B, *spe*C, *mf, ssa*, *scp*, *has*A and *slo*. Interestingly, only two respective genes were detected, *slo* and *emm* (DeWinter et al. 1999). In another study, Igwe et al. (2003) identified a new allele of the GAS superantigen (SAg) smeZ. SAgs are phage-encoded virulence factors extensively studied in GAS. Their toxoid nature is based on high affinities for major histocompatibility complex II (MHC class II) molecules and T cell receptors on professional antigen presenting cells. Upon activation, a cytokine "storm" comprising pro-inflammatory cytokines (TNF-α and IL-1β) as well as T cell mediators (IL-2) are induced leading to severe clinical and pathological changes, such as fever, hypotension, shock and multi-organ failure (Igwe et al. 2003; Proft and Fraser 2007) which resembles the clinical symptoms of STSS in *S. canis* infected dogs. The identification of putative *emm* genes in *S. canis* genomes was surprising, since *S. canis*, like other animal specific and zoonotic pathogens, is characterised as *emm* negative according to the established PCR test for *emm* typing described by the Center for Disease Control (CDC, Atlanta, USA). Probably, due to poor genome information, it took 12 years until a second publication verified the existence of an M-like protein in *S. canis* SCM (Fulde et al. 2011a). Detailed biochemical studies characterised SCM as an alpha-helical protein with a high probability to form dimers on the bacterial surface. SCM specifically binds to plasminogen (PLG) with high specificity and affinity. Accordingly, numerous other M proteins identified in human-specific GAS and *S. equisimilis*, as well as M-like proteins of animal specific (e.g. *S. equi* sub *equi*) and zoonotic species, e.g. *S. zooepidemicus* and *S. iniae* (Baiano et al. 2008, 2009; Bergmann et al. 2011; Meehan et al. 2001; Nitsche-Schmitz et al. 2007; Nitsche et al. 2006; Smeesters et al. 2008; Timoney et al. 1997, 2010), bind extracellular matrix (ECM) and serum proteins. However, in contrast the well-known PLG-binding M Protein PAM on the surface of GAS (Wistedt et al. 1995), SCM interacts with the C-terminal part of PLG, named mini-PLG (mPLG). Only few studies have been published on the interaction of pathogens with mPLG (Ljungh 2000; Rojas et al. 2008; Ullberg et al. 1992) and SCM displays the first receptor proven so far. The role of mPLG in pathogenesis is unclear but an involvement in endothelial cell migration during wound repair is discussed (Hayashi et al. 2009). Since wounds display the main entry site for S. canis, it may be speculated that interaction with mPLG in wound fluids represents a yet unknown pathogenicity mechanism.

The immobilisation of the zymogen plasminogen on the surface of different streptococcal species and its subsequent activation to the broad spectrum serine protease plasmin by the main streptococcal PLG-activator (PA) streptokinase (SKA) is a common and well-known mechanism for tissue destruction and dissemination of invasive isolates within the host (Sun et al. 2004). However, as already described by Butaye (1956), *S. canis* does not possess any fibrinolytic activity in human serum. This inability is not restricted to a certain host as described for *S. zooepidemicus* (Fulde et al. 2011a; Schroeder et al. 1999) and supports results by DeWinter et al. (1999) who postulated that the genomes of 15 different *S. canis* strains lack a gene coding for streptokinase. Nevertheless, surface-bound plasminogen could be activated by host-derived PA, such as urokinase (uPA) which enables *S. canis* to degrade fibrin matrices and disseminate through semi-synthetic thrombi. An illustration of fibrin degradation by S. canis is shown in Fig. 2.

In addition to the interaction of *S. canis* with plasminogen, an association with albumin, IgG and fibrinogen was reported by Laemmler and co-workers (Lammler et al. 1988). Although the authors did not comment on the responsible bacterial binding proteins or possible virulence mechanisms, it is conceivable that in analogy to pyogenic streptococci, binding to ECM and serum proteins is a prerequisite for establishing *S. canis* infections. For example, since M protein-mediated binding to fibrinogen confers anti-phagocytic properties, immobilisation of albumin leads to an inactivation of the antibacterial peptide MIG/CXCL9 (Egesten et al. 2011). Furthermore, the interaction of streptococci with IgG occurs mainly in a "non-immunogenic" way by binding to the conserved F_c-fragment (Lewis et al. 2008) which seems to be similar in zoonotic *S. canis* (own unpublished data).

Surface exposed metabolic enzymes with a secondary function (moonlighting enzymes) are well-known players in streptococcal disease progression (Pancholi and Chhatwal 2003). Hitzmann et al. (2012) recently reported the existence of the Arginine Deiminase System (ADS), an energy providing pathway catabolising arginine, on the surface of zoonotic *S. canis*. Its possible role in pathogenesis was

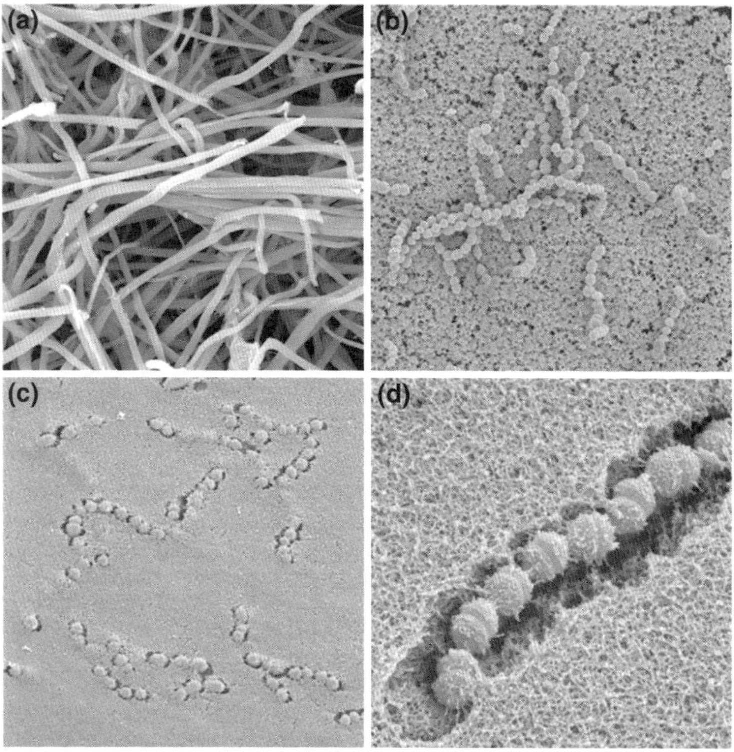

Fig. 2 Fibrin degradation by zoonotic *S. canis* **a** Fibrin bundles form a dense network in semi-synthetic thrombi. **b** PLG-coated *S. canis* strains strongly attach to fibrin bundles. **c, d** Immobilisation of proteolytic plasmin activity on the bacterial surface leads to fibrinolytic phenotype of *S. canis*. (EM pictures provided by Manfred Rohde, HZI Braunschweig, Germany).

described earlier for the zoonotic pathogen *S. suis* by Gruening et al. (2006), and therefore is described in Sect. 5.3.

3 *Streptococcus equi* sub. *zooepidemicus* (*Streptococcus zooepidemicus*)

3.1 General Features

S. zooepidemicus belongs to the large group of pyogenic streptococci with a ß-haemolytic, large-colony-forming phenotype comparable to other members such as GAS, S. *equisimilis* and *S. canis*, respectively (Facklam 2002; Efstratiou et al. 1994; Barnham et al. 1987a). *S. zooepidemicus* harbours the Lancefield group C antigen, and is, therefore, serologically closely related to *S. equi* and *S. equisimilis*, respectively. Differentiation of these species is done biochemically based on species-specific fermentation patterns of carbon sources. For example, *S. equisimilis* is able to utilize trehalose, whereas *S. equi* and *S. zooepidemicus* lack this ability. Instead, *S. zooepidemicus* ferments lactose and sorbitol in contrast to *S. equisimilis* and *S. equi*, respectively (Bannister et al. 1985).

Recent phylogenetic studies shed more light on the inter-species relationships in the pyogenic group of streptococci. These studies revealed that *S. equisimilis* is closely related to human pathogenic GAS forming a common clade together with the zoonotic GGS *S. canis* (Jensen and Kilian 2012), whereas *S. equi* and *S. zooepidemicus* share a genome identity of more than 98 % (Timoney 2004). Moreover, it is widely accepted that *S. equi* has evolved from an ancestral strain of *S. zooepidemicus* (Webb et al. 2008). Holden et al. (2009) compared the genomes of a highly virulent *S. equi* isolate with a *S. zooepidemicus* strain from a healthy thoroughbred racehorse. As expected, both genomes were highly identical comprising approximately 2,000 open reading frames (ORFs). However, the authors conclude from their studies that gain and loss of genes or their functions, respectively, might have forced *S. equi* to evolve from a mucosal colonizer of different animals (*S. zooepidemicus*) to a highly specialised pathogen of horses as sole host (Holden et al. 2009).

The close relationship between *S. equi* and *S. zooepidemicus* complicates the development of a discrimination scheme other than time-consuming biochemical analysis (Webb et al. 2012). However, discrimination of both species is important since strangles, the main disease caused by *S. equi*, is a highly contagious infection of the upper respiratory tract constituting a major infectious disease threat in horses (Waller and Jolley 2007). The availability of the genome sequences will hopefully allow to identify new genetic markers for differentiation of both subspecies. Very recently, Webb et al. (2012) presented a quantitative triplex PCR with improved discrimination, a promising step towards eradication of strangles.

3.2 Epidemiology and Zoonotic Relevance

Although *S. zooepidemicus* is considered as an opportunistic pathogen in a large variety of mammalian species including cats, rodents, minks, monkeys and seals (Ryu et al. 2011; Akineden et al. 2005, 2007; Matz-Rensing et al. 2009; Blum et al. 2010; Britton and Davies 2010; Literak and Mraz 1991), the majority of cases were reported from domestic animals such as horses, dogs, pigs and ruminants, respectively. In horses, *S. zooepidemicus* represents a commensal of mucosae of the upper respiratory and the lower genital tract (Barquero et al. 2010; Priestnall and Erles 2011). As an opportunistic pathogen it can cause secondary infections after primary virus infection, as well as after predisposition by stress or tissue injuries (Timoney 2004; Ryu et al. 2011). Nevertheless, in several epidemiological studies *S. zooepidemicus* was found to constitute a major causative agent of purulent infections in horses and foals (Ryu et al. 2011; Panchaud et al. 2010; Erol et al. 2012; Gaede et al. 2010; Clark et al. 2008). Typically, such infections lead to severe respiratory diseases with sudden onset dyspnoea and haemorrhagic nasal discharge accompanied by pyrexia, coughing, leucocytosis and neutrophilia (Priestnall and Erles 2011; Oikawa et al. 1994, 1995). Furthermore, dissemination of *S. zooepidemicus* in inner organs, such as liver, lung, brain and kidney as well were joints (Timoney 2004) often results in clinical pictures constituting as neonatal septicaemia, pyogenic abscesses, ulcerative keratitis and endometritis concomitant with reproductive failures (Timoney 2004; Priestnall and Erles 2011; Wada 2012).

Notably, *S. zooepidemicus* is emerging as cause of severe and life-threatening diseases in dogs (Priestnall and Erles 2011; Pesavento et al. 2008; Byun et al. 2009). In contrast to infections in horses, emerging dog infections frequently occur as outbreaks (Priestnall and Erles 2011). Clinical pictures are similar to those obtained in horses. Several studies reported that fatal outbreaks exhibit as haemorrhagic pneumonia and septicaemia, respectively, with a rapid disease progression (Priestnall and Erles 2011; Pesavento et al. 2008; Byun et al. 2009; Kim et al. 2007a; Sundberg et al. 1981; Garnett et al. 1982) often ending fatal within 24–48 h. In milder cases, typical symptoms are pyrexia, coughing, anorexia and tachypnoea (see also an excellent review by Priestnall and Erles (2011)).

In addition to *S. suis*, *S. zooepidemicus* is a major pathogen in swine in Asia. Outbreaks were reported from Sichuan, China (1975) and Indonesia (1994) with fatal cases of more than 300,000 pigs in 2 weeks. Typical clinical symptoms include arthritis, diarrhoea, bronchopneumonia, endocarditis and meningitis (Fan et al. 2008; Feng et al. 2010; Soedarmanto et al. 1996). Interestingly, during the large outbreak in Indonesia, a single, highly virulent *S. zooepidemicus* clone spread to a monkey population and led to severe disease similar as described for pigs (Soedarmanto et al. 1996).

S. zooepidemicus-induced severe and deep tissue infections have been observed in ruminants, such as goat, sheep, cattle, lama, camels and alpaca (Aubry et al. 2000; Barnham et al. 1987a; Hewson and Cebra 2001; Jones et al. 2009; Las et al. 2002; Younan et al. 2005; Pisoni et al. 2009). However, the majority of reports focussed on its role as a causative agent of mastitis.

Human outbreaks have been associated with unpasteurised milk and its products. For example, two outbreaks occurred in 2003 in Gran Canaria and Finland, respectively (Kuusi et al. 2006; Bordes-Benitez et al. 2006). In both cases, cheese produced from raw milk was the source of infection. People suffered septicaemia, aortic aneurism, pneumonia, meningitis and septic arthritis with partially fatal outcome. A special complication with regard to *S. zooepidemicus* infections is the development of post-streptococcal glomerulonephritis (PSGN) (Francis et al. 1993; Barnham et al. 1983, 1987b). One unusual large outbreak occurred from 1997 to 1998 in Nova Serrana, a rural province in Brazil (Balter et al. 2000). Again cheese from contaminated milk was the source of infection. In total, 253 cases of acute nephritis were reported of which three had a fatal progression and seven required dialysis. Extensive epidemiological studies including whole genome sequencing of the respective outbreak strain were applied. However, bacterial factors promoting the establishment of PSGN remains unknown (Beres et al. 2008; Sesso and Pinto 2005; Nicholson et al. 2000). The transmission of *S. zooepidemicus* from livestock products other than milk or cheese, respectively, is discussed but not yet clearly proven. For example, Yuen et al. (1990) characterised *S. zooepidemicus* strains from Hong Kong isolated from infected human and pig individuals and found major genotypical and phenotypical similarities. The fact that *S. zooepidemicus* represents one of the major pathogens in swine in Asia and that Asian people frequently consume undercooked pork strongly suggests that transmission from pigs to human occurs (Yuen et al. 1990).

Transmission of *S. zooepidemicus* from companion animals to humans is frequently described in the literature. Most of the cases originated in horses, probably due to the wide distribution of *S. zooepidemicus* as a commensal in this species (Minces et al. 2011). Nevertheless, one case of transmission from dog to human was described (Abbott et al. 2010). In this case, the likely mode of transmission was via wound infection or aerosols rather than ingestion. Typical clinical symptoms in humans range from purulent abscesses of skin and mucosae to severe illnesses, such as endocarditis, endophthalmitis, septic arthritis, pneumonia and meningitis. Clinical outcome depends on the mode of transmission, the immune-competence of the patient and the repertoire of virulence factors of the pathogen.

3.3 Virulence and Pathogenesis

S. zooepidemicus is believed to be the evolutionary ancestor of *S. equi* and both species share an overall genome identity of more than 98 %. But there is much less known about virulence factors in *S. zooepidemicus* as compared to *S. equi*. Comparative genome analysis by Holden et al. (2009) identified phage-associated genes, such as sAGs, a phospholipase A_2 toxin and an integrative conjugative element harbouring an iron acquisition system with similarities to a pathogenicity island of *Yersinia pestis* (Holden et al. 2009). Furthermore, the identification of genes possibly encoding pili in the genome of *S. zooepidemicus*, which could facilitate adherence and colonisation in horses, might reflect the evolution from a generalist to a specialist.

S. zooepidemicus possesses a hyaluronic capsule synthesised by genome products of the *has* operon (Blank et al. 2008; Paillot et al. 2010). Hyaluronic capsule is one of the most important virulence factors of streptococci facilitating anti-phagocytic properties. Furthermore, Wibawan et al. (1999) demonstrated that capsule positively correlates with the ability to adhere to HeLa cells. Whether or not this mechanism resembles the interaction between GAS and CD44 remains to be proven (Schrager et al. 1998). Similar to *S. canis*, *S. zooepidemicus* is also characterised as *emm*-negative according to the guidelines of the CDC. Nevertheless, Timoney et al. (1995) were able to identify a M-like protein (SZP) which protects mice against a lethal infectious dose of *S. zooepidemicus* but not against *S. equi* (Timoney et al. 1995). Similar to SCM, SZP is assumed to form an alpha–

helical structure but it does not possess A, B and C repeats characteristic for M proteins of human pathogenic GAS and *S. equisimilis* (Smeesters et al. 2008). SZP harbours variable- and hyper-variable regions in the N-terminal part of the mature protein (Walker and Timoney 1998). In accordance to other M and M-like proteins, SZP binds to fibrinogen (Timoney et al. 1997). Thus, an anti-phagocytic activity is plausible. However, to the best of our knowledge, a direct correlation was not yet reported. Recently, Hong-Jie et al. (2009) found that *szp*-deficient *S. zooepidemicus* mutant strains were >1,000-fold less virulent in an intramuscular mouse model (Hong-Jie et al. 2009). An interesting explanation was given by Ma et al. (2012). They found that *S. zooepidemicus* recruits thioredoxin (TRX) via SZP to the bacterial surface which, in turn, inhibits *inter alia* the deposition of complement factor 3, and thereby protects against phagocytic killing.

Binding to ECM- and plasma proteins is a crucial step in streptococcal pathogenesis. For example, as extensively studied for GAS, the interaction with fibronectin (FN) leads to a significant increase in adherence and invasion in epithelial and endothelial cells (Schwarz-Linek et al. 2006), and thus might represent a prerequisite for colonisation of and dissemination in the host. FN binding and subsequent attachment to host cells is well-known for *S. zooepidemicus* isolates (Valentin-Weigand et al. 1988). During the years, two proteins (FNZ and SFS) were identified with affinities to FN (Lindmark and Guss 1999; Lindmark et al. 1996). Although at least FNZ seems to induce a protective immunity in a mouse infection model, its role in pathogenesis remains elusive (Flock et al. 2006). ZAG is a surface-associated proteinous receptor for IgG, albumin and the plasma proteinase inhibitor α_2-macroglobulin (α_2M) (Jonsson et al. 1995). Potential benefits resulting from a recruitment of IgG and albumin, respectively, are already discussed in the *S. canis* section. Interestingly, Valentin-Weigand et al. (1990) showed an anti-phagocytic effect of surface bound α_2M on equine group C streptococci in vitro. The in vivo relevance, however, has not yet been demonstrated. In addition to the ability to bind (and inactivate) immunoglobulins, *S. zooepidemicus* possess two endopeptidases (IdeZ and IdeZ2) in its proteome which specifically cleave IgG from a variety of different species (Hulting et al. 2009; Lannergard and Guss 2006). Homologues in GAS constitute important virulence factors essential for circumventing antimicrobial properties (Pawel-Rammingen 2012; Akesson et al. 2006; Pawel-Rammingen et al. 2002). Furthermore, factors known to be important for establishing streptococcal infections were also reported in *S. zooepidemicus*. For example, the cytolysin streptolysin S (SLS), responsible for the β-haemolytic phenotype of pyogenic streptococci, was identified in the genome of *S. zooepidemicus* (Flanagan et al. 1998). SLS has been extensively studied and characterised as a virulence factor in GAS and human GGS. Thus, it may also be important for disseminating infections of other streptococci. Very recently, Paillot et al. (2010) identified the genetic determinants of three novel sAG- encoding genes in the genome of a *S. zooepidemicus* strain (Paillot et al. 2010). This was an interesting finding since *S. zooepidemicus*, in contrast to its closest relative *S. equi*, was classified as sAG negative, a likely prerequisite for its commensal lifestyle (Paillot et al. 2010).

4 *Streptococcus iniae*

4.1 General Features

S. iniae is a major fish pathogen first isolated in 1976 from a freshwater dolphin. It is widely spread geographically, mainly in North America, Middle East and the Asia–Pacific region. *S. iniae* infections lead to meningoencephalitis and other pathologies which result in high economic losses in aquaculture due to high rates of morbidity and mortality (Locke et al. 2007a). Death might occur without accompanying signs or specific clinical symptoms related to central nervous system (CNS) dysfunction, including loss of orientation and erratic swimming (Agnew and Barnes 2007).

S. iniae is also considered as an emerging agent of zoonotic infections, most of which are associated with processing of fresh fish (Baiano and Barnes 2009). In humans, infection often leads to bacteraemic cellulitis which may be followed by other invasive forms of diseases, such as endocarditis, meningitis, arthritis or sepsis.

S. iniae is genetically closely related to group B streptococci as revealed from 16S rRNA sequencing. It is not assigned to any Lancefield group and shows ß-haemolysis on blood agar (noteably, highest expression was observed with blood from freshwater dolphin (Pier et al. 1978)). Isolates show some variation in colony morphology, e. g. isolates from Asian patients are more mucoid (Lau et al. 2006). Virulent strains express an exopolysaccharide capsule. The capsule operon genes show homology to those of other streptococci, including *S. agalactiae*, *S. suis* and *S. thermophilus* (Lowe et al. 2007). *S. iniae* isolates from diseased fish and humans showed less genetic diversity than isolates from healthy fish in Canada (Facklam et al. 2005; Fuller et al. 2001). Two distinct serotypes of *S. iniae* have been identified based on their reaction with rainbow trout antisera. Serotypes also differ in their biochemical properties. Serotype 1 strains are positive for reaction with arginine dihydrolase and ribose, whereas serotype 2 strains are negative for both (as reviewed by Agnew and Barnes (2007)). Serotype 2 emerged after a vaccination program in Israel and includes strains producing and releasing large amounts of extracellular polysaccharide (Eyngor et al. 2008). Importantly, serotype 2, but not serotype I strains have been reported to survive in piscine phagocytes and induce their apoptosis (Zlotkin et al. 2003), probably due to the differences in the extent of capsule coverage of the bacterial surface.

4.2 Epidemiology and Zoonotic Relevance

S. iniae is an emerging invasive fish pathogen responsible for major economic losses in warm water finfish aquaculture worldwide (Agnew and Barnes 2007; Shoemaker et al. 2001). Diseases caused by *S. iniae* infection have been described in at least 27 fish species, including trout (Eldar and Ghittino 1999), tilapia (Shoemaker et al. 2001), barramundi (Bromage et al. 1999) and hybrind striped bass (Shoemaker et al. 2001) (see also an excellent recent review by Agnew and Barnes (2007)). Carrier fish are considered to play an important role in fish-to-fish transmission. Currently, it is not known whether *S. iniae* has its origin in wild fish and has been distributed by ocean current and movement of fish, or whether its amplification in aquaculture has led to distribution into neighbouring fish populations (Agnew and Barnes 2007).

S. iniae is considered as an emerging zoonotic agent as noted at the International Conference on Emerging Infectious Diseases in 2000. In humans, it mainly causes bacterial cellulitis as a consequence of soft tissue injuries during preparation of fresh fish. However, severe complications have been described, including arthritis, meningitis, endocarditis and osteomyelitis (Sun et al. 2007; Weinstein et al. 1997). Notably, human infections by ingestion of contaminated fish have not been reported. Human cases are rare and sporadic, but continue to be reported and are likely to increase in the future due to enhanced surveillance. Epidemiological data showed that mainly elderly people with Asian origin are affected. The racial predominance may be related to a cultural preference for fresh whole fish in cooking, as suggested by Finkelstein and Oren (2011).

Importantly, it has to be mentioned that *S. iniae* is not included in commercial and clinical databases and diagnostic kits. Thus, it is very likely that *S. iniae* infections in humans are underreported due to non-identification.

4.3 Virulence and Pathogenesis

S. iniae harbours a number of virulence factors with high homology to respective factors of *S. pyogenes*. Important examples include the M-like protein SiM (Locke et al. 2008), the β-haemolysin S (Locke et al. 2007b) and an α-enolase (Kim et al. 2007b). For some of these factors, the interaction with the host has been described. As expected from the homology, these functions are reminiscent of the pathogenesis of *S. pyogenes*. Nevertheless, the mechanisms underlying the adaptation of *S. iniae* to fish as preferred host are not known. This preference is very prominent, as *S. iniae* is, for example, able to multiply in naive trout blood but hardly survives in naive human blood (Zlotkin et al. 2003). Furthermore, *S. iniae* rapidly translocates through trout skin epithelial cell monolayers covered with mucus (Eyngor et al. 2008).

A number of virulence factors of *S. iniae* have been confirmed in experimental infections of white striped bass (recently reviewed by Baiano and Barnes (2009)). The capsule and the M-like protein SiM (genes: *sim*A and *sim*B) protect *S. iniae* against opsonophagocytosis (Locke et al. 2007a, 2008). The phosphoglucomutase (gene: *pgm*) is also crucial for virulence since it is involved in cell wall morphology, surface capsule expression and resistance to cationic antimicrobial peptides (Buchanan et al. 2005). Furthermore, *S. iniae* expresses a cytolysin homologous to SLS from *S. pyogenes*. The *sag* operon is involved in SLS formation and a *sag*A mutant is highly attenuated in virulence (Locke et al. 2007b).

In general, *S. iniae* resembles a typical blood pathogen that disseminates to different organs similar to some other invasive streptococcal species, such as group A and B streptococci. However, the pathogenesis of diseases caused by *S. iniae* is only partially understood. Most likely, the first step is adherence and colonisation of tissues followed by translocation into the blood circulation, similar to other invasive streptococci. The mechanism(s) of invasion into deeper tissue are not known. However, studies using epithelial cells of rainbow trout epithelial cells showed that *S. iniae* invaded the cells and persisted intracellular for short time periods. Invasion was followed by rapid translocation and dissemination in the fish (Eyngor et al. 2007). Evasion of defense mechanisms in the blood then allows a generalised bacteraemia, which is required to establish disease. A hallmark of *S. iniae* pathogenicity is its ability to survive phagocytic killing, as reflected by its rapid dissemination to systemic sites in the blood circulation system. Therefore, the interaction between *S. iniae* and phagocytes is considered to be crucial for the pathogenesis. A virulent *S. iniae* strain survived within pronephros macrophages in vitro (Zlotkin et al. 2003). Furthermore, monocytes carried approximately 70 % of the bacteria present in the blood of diseased fish. Based on these results, Zlotkin et al. (2003) postulated in a Trojan horse theory that *S. iniae* might enter the CNS by "hijacking" migrating monocytes/macrophages. Induction of apoptosis in macrophages was also suggested as an effective mechanism of *S. iniae* to prevent priming of the immune system (Zlotkin et al. 2003).

The virulence properties of *S. iniae* and its invasive nature as a pathogen make it an excellent model pathogen to study pathogenicity mechanisms of human streptococci, such as group A and B streptococci and pneumococci. Experimental infections have been reported in a number of different species, e. g. zebrafish (Miller and Neely 2004). Recently, we described an intraperitoneal *S. iniae* infection model for tilapia. This infection model includes for the first time pathohistological screening of brains and eyes as read out parameter and a novel multiplex PCR for confirmation of the virulence genotype of the challenge strain (Baums et al. 2012). In the future such models shall allow comparative analyses of different streptococci with respect to host–pathogen-interactions at different infection stages.

5 *Streptococcus suis*

5.1 General Features

S. suis is a major porcine and zoonotic pathogen occurring worldwide. It can cause severe diseases in pigs, e.g., sepsis, meningitis, arthritis and pneumonia resulting in enormous economical losses in the swine industry (Staats et al. 1997). In humans, meningitis is the most common presentation, though *S. suis* can also be associated with other systemic complications such as sepsis, pneumonia and, as observed for the first time in two outbreaks in China, STSS (Gottschalk et al. 2007; Tang et al. 2006).

S. suis shares antigenic features with group D streptococci, but it is genetically distinct and serogrouping is not used for its identification. It produces α- or β-haemolysis on sheep and horse blood agar, respectively. Based on its capsular polysaccharides, 35 serotypes have been described to date. However, strains of serotypes 32 and 34 have been suggested to be more closely related to *S. orisratti* (Hill et al. 2005). Distribution of serotypes differs between geographical regions. Most European strains associated with disease belong to serotypes 1, 2, 7, 9 and 14 (Allgaier et al. 2001; Silva et al. 2006; Wisselink et al. 2000). Worldwide, serotype 2 is most prevalent. In recent years, serotype 9 has emerged as common pig isolate in Germany and The Netherlands, serotype 7 is highly prevalent in Scandinavia and serotypes 1 and 14 in the United Kingdom (Baums and Valentin-Weigand 2009; Perch et al. 1983; Tian et al. 2004; Wisselink et al. 2000). In contrast, in Canada and the USA serotypes 2, 1/2, and 3 are most frequently associated with disease (Messier et al. 2008). A specific sequence type (ST), ST7, which appeared to have evolved from the highly pathogenic ST1 type of a serotype 2 strain, was found to be responsible for human outbreaks in China and directly associated with the STSS (Feng et al. 2010; Ye et al. 2006, 2009). The ST7 carries a putative pathogenicity island (designated 89 K), possibly involved in development of STSS (Chen et al. 2007; Zhao et al. 2011).

5.2 Epidemiology and Zoonotic Relevance

Pigs and wild boars are considered as natural reservoirs of *S. suis* (Baums et al. 2007; Clifton-Hadley and Alexander 1980). Different mucosal surfaces might be colonised by *S. suis*. In weaning piglets, *S. suis* is an abundant coloniser of the upper respiratory and alimentary tract (Baele et al. 2001; Lowe et al. 2011; O'Sullivan et al. 2011; Su et al. 2008a, b). Tonsillar carrier rates up to 100 % have been reported (Arends et al. 1984; Clifton-Hadley 1983; Mwaniki et al. 1994a, b), persisting even after treatment with penicillin (Staats et al. 1997). The sow may harbour *S. suis* also in the genital tract. Healthy carriers of virulent *S. suis* strains are the main source of infection, and thus play an important role in the epidemiology of *S. suis* diseases in pigs and humans (Arends et al. 1984; Ngo et al. 2011). Faeces, dust, water and feed may be secondary sources of infection (Staats et al. 1997). Vectors such as flies (Enright et al. 1987) and mice (Williams et al. 1988) can play a role in disease transmission.

Pigs of any age can be infected with *S. suis*, but susceptibility generally decreases with age following weaning. Outbreaks are usually due to the introduction of a carrier into the herd. Within a carrier herd, outbreaks occur especially in young animals disposed to stress conditions. As *S. suis* is a facultative pathogen, different biotic and abiotic factors such as virus infections, corrosive gases and crowding are thought to promote *S. suis* diseases in modern swine production.

Several modes of transmission between animals in a herd have been suggested. Piglets born to sows with genital infection may acquire the infection vertically (Amass et al. 1997). Transmission can also occur by nose-to-nose contact (Berthelot-Herault et al. 2001), or by infection through the navel, genital or alimentary tract (Staats et al. 1997). After infection, colonised pigs will usually harbour the bacteria in their tonsils. Some animals will remain healthy carriers, whilst other will sooner or later develop clinical signs (Gottschalk and Segura 2000). Morbidity rarely exceeds 5 %, although it can reach more than 50 % in cases of poor hygiene and concurrent disease (Staats et al. 1997). With appropriate treatment mortality is usually low (ca. 5 %), but can be up to 20 % in untreated herds.

The infection is recognised as a zoonosis associated with meningitis, septicemia and endocarditis in humans for a long time (Arends and Zanen 1988). Occupational exposure to pigs and pork is worldwide the most important risk factor for *S. suis* infections (Fittipaldi et al. 2012). Mostly, human infections appear as sporadic cases in persons exposed to pigs or pig products, with an incidence of 3/100,000 (almost 1,500 times higher than among persons not involved in pig industry) (Arends and Zanen 1988). It seems that contamination is through direct contact with infected pigs or meat products and infection via conjunctiva or skin lesions (Arends and Zanen 1988). In addition, human infections have been associated with contact with wild pigs (Rosenkranz et al 2003).

Zoonotic infections have been reported mainly in countries with intensive swine production. In a study by Smith et al. (2008), 73 swine-exposed and 67 non-swine-exposed US adults were tested for antibodies to *S. suis* serotype 2. Results suggested that human infection with *S. suis* occurs more frequently than currently documented. Furthermore, *S. suis* pathotypes isolated from human infections were frequently detected in pigs, and recent outbreaks of porcine infections paralleled those in humans. Even though reported numbers of human infections are relatively low, there is a potential risk that *S. suis* may constitute a more important public health problem than currently recognised. This is underlined by the outbreak in China (in 2005), in which totally unexpected severe clinical presentations of streptococcal toxic shock-like syndrome due to *S. suis* infection was observed in infected individuals (Yu et al. 2006). The *S. suis*-induced toxic shock was associated with high mortality rates. This new outcome of *S. suis* infection in humans might reflect the evolutionary emergence of more virulent strains. Accordingly, the outbreak in China was due to the emergence of a serotype 2 strain that contained a pathogenicity island designated 98 K. Thus, the potential that *S. suis* may suddenly emerge as disease threat by acquiring new capacities for establishing infection, e.g. by altering the human immune response, should not be underestimated (Yu et al. 2006; Gottschalk et al. 2010; Holden et al. 2009).

Outbreaks of human infections such as described mainly occur in low-income countries with intensive pig production, which is largely due to very close contact of humans to pigs and little awareness of the disease within the population at risk. In countries with increasingly intensive pig farming, like China, Thailand or

Vietnam, the risk for humans of acquiring *S. suis* infection is unknown, since it is not a notifiable disease and has been largely neglected by routine laboratory diagnostics. On the other hand, the two largest published case series of human *S. suis* infections are from these regions and account for more than 50 % of all reported cases (Takeuchi et al. 2012; Wertheim et al. 2009). Notably, in Vietnam, *S. suis* is the most important cause of bacterial meningitis in adults (Mai et al. 2008). This is related to eating "high risk" dishes, such as undercooked blood and intestine (Nghia et al. 2008) suggesting that oral infection is an additional infection route to be considered. Therefore, zoonotic infections by *S. suis* are considered a significant, yet unrecognised burden in large parts of Southeast Asia. The urgent need for research on this neglected zoonotic pathogen is underlined by the emergence of strains which are highly virulent for humans and by the spread of antibiotic resistance within the *S. suis* population (Palmieri et al. 2011).

5.3 Virulence and Pathogenesis

Various virulence or virulence-associated factors of *S. suis* serotype 2 have been identified in the last years, among which the capsule is the only proven essential virulence factor protecting the pathogen against phagocytosis (Smith et al. 1999). Some other factors have been shown to contribute to survival in the host and virulence in experimental infections of mice and piglets, respectively. Recently, several reviews have been published on this topic, e.g. by Baums and Valentin-Weigand (2009) and Fittipaldi et al. (2012).

A number of surface associated and secreted proteins of *S. suis* serotype 2 exhibit the same or very similar functions as homologous factors of other pathogenic streptococci. Important examples are peptidoglycan polysaccharide deacetylase, opacity factor of *S. suis*, fibronectin- and fibrinogen-binding protein of *S. suis*, enolase and suilysin (Baums et al. 2006; de Greeff et al. 2002; Esgleas et al. 2008; Fittipaldi et al. 2008a; Jacobs et al. 1994). On the other hand, surface associated or secreted factor with unique functions for *S. suis* have not yet been described. Furthermore, though many functional assays were carried out with cells of porcine origin, clear evidence for functional adaptation to pigs as the main host is lacking for *S. suis*. Most recently, we have identified a porcine IGM specific protease of *S. suis* which might well represent such a functional adaptation mechanism (Seele et al., manuscript in revision).

The molecular mechanisms of pathogenesis of *S. suis* infections are only partially understood. Most studies have addressed *S. suis* meningitis in mouse and pig models. Major steps in pathogenesis resemble those of other invasive streptococcal infections, such as adherence to and colonisation of mucosal surface(s), invasion into deeper tissue and translocation in the bloodstream. A hallmark of *S. suis* pathogenicity is its ability to disseminate in the blood circulation and to maintain a bacteraemia for certain time. This is considered to be crucial to cause meningitis. Thus, major mechanisms in pathogenesis of *S. suis* infections are those involved in

(i) invasion of *S. suis* through the epithelial cell barriers, (ii) evasion of killing by complement and phagocytosis and (iii) invasion into the cerebrospinal fluid (CSF) or other target sites.

Bacterial factors, such as the FN and fibrinogen binding protein (FBPS) (de Greeff et al. 2002) or the cell wall component lipoteichonic acid (LTA) mediate adherence of bacteria to target cells for initial colonisation (Fittipaldi et al. 2008b; Vanier et al. 2007). Another bacterial mechanism involved in colonisation might be the formation of a biofilm which probably enhances bacterial resistance to innate and adaptive host defence mechanisms and treatment with antibiotics (Bonifait et al. 2008). To get access into deeper tissues bacteria might invade the respiratory epithelium either intra- or intercellularly. Suilysin, the haemolysin of *S. suis*, is discussed to play a role in interaction of *S. suis* with epithelial cells and disruption of these cells due to its cytolytic function (reviewed by Fittipaldi et al. (2012). Furthermore, the capsule is assumed to be involved in host-epithelial cell interaction. Since its main function is protection against phagocytosis after entering the bloodstream (Benga et al. 2008; Fittipaldi et al. 2012; Chabot-Roy et al. 2006; Charland et al. 1998; Segura and Gottschalk 2002), it has been proposed that the capsule is downregulated during colonisation of the mucosal epithelium to allow adherence and invasion of the bacterium to overcome this first barrier (Gottschalk and Segura 2000; Okamoto et al. 2004; Willenborg et al. 2011). Accordingly, unencapsulated *S. suis* strains showed higher adhesion and invasion rates, indicating a negative correlation between encapsulation and interaction with host cells (Benga et al. 2004; Gottschalk et al. 1991). A possible explanation for this phenotype is the masking effect of the capsule (Vanier et al. 2007; Lalonde et al. 2000; Tenenbaum et al. 2009).

A topic which has not yet received much attention in research on *S. suis* pathogenicity is the metabolic adaptation of the pathogen and its relation to virulence gene regulation in different host niches. During infection, *S. suis* encounters various (stress) conditions, e. g. low pH or low availability of nutrients and oxygen. At present very little is known about the mechanisms of streptococcal survival in the host under such conditions. A recently identified catabolic enzyme system, the Arginine Deiminase System (ADS), may play an important role in adaptation to low pH, such as in host cell phagolysosomes (Benga et al. 2004; Fulde et al. 2011; Gruening et al. 2006; Winterhoff et al. 2002). The ADS per se is a secondary enzymatic system catabolising arginine, and concomitantly produces citrulline, ATP, CO_2 and ammonia. The system is strictly regulated by a variety of different environmental stimuli, e.g. carbon starvation, oxygen content and substrate availability. It is plausible to assume that *S. suis* is able to alkalize its environment by synthesis of ammonia during arginine catabolism, and thereby protects itself against acidic damage. This mechanism may allow *S. suis* to reside in acidified phagolysosomes, thus representing an advantage in establishing invasive infections as proposed recently (Benga et al. 2004; Fulde et al. 2011; Gruening et al. 2006). Further hints towards a link between metabolic adaptation and regulation of virulence traits have been found in a recent paper on the role of glucose and the regulator catabolite control protein A (CcpA) in capsule expression and virulence of *S. suis* (Willenborg et al. 2011).

A major first line of host defense against *S. suis* is phagocytic killing by neutrophils. Accordingly, infiltrations with large number of neutrophils are typically found in respective lesions, such as in meningitis. In the absence of opsonising antibodies virulent *S. suis* serotype 2 strains are not efficiently killed by porcine neutrophils, suggesting that the pathogen has evolved strategies to evade intra- and extracellular antimicrobial activity of neutrophils. Several virulence-associated factors have been identified, among which the most intensively studies is the bacterial capsule (see above). Other factors possibly involved include D-alanylation of lipoteichoic acid and suilysin, as reviewed recently by Baums and Valentin-Weigand (2009) and Fittipaldi et al. (2012). Future studies will have to dissect the yet unknown molecular steps involved in *S. suis* interactions with neutrophils.

For induction of meningitis *S. suis* has to reach the CSF. It has been shown that *S. suis* adheres to and invades into brain microvascular endothelial cells (BMEC) and porcine choroid plexus epithelial cells (PCPEC), the main components of the blood–brain barrier (BBB) (Tenenbaum et al. 2005, 2009; Benga et al. 2005; Charland et al. 2000; Vanier et al. 2004). An increased tight junction permeability and loss of barrier function was proposed to be associated with cytotoxic effects of suilysin (Charland et al. 2000; Vanier et al. 2004). Furthermore, *S. suis* can stimulate the production of pro-inflammatory cytokines, such as interleukin-6 (IL-6), IL-8 and monocyte chemotactic protein-1 (MCP-1) by BMEC, which in turn may alter the BBB permeability (Vadeboncoeur et al. 2003). However, Tenenbaum et al. (2009) described the entry of *S. suis* into the CNS as a transcellular translocation without destruction of CPEC lining of the BBB (Tenenbaum et al. 2009).

A hallmark of pathogenicity (and immune control) of *S. suis* is its ability to induce inflammation, which is characterised by massive infiltrations of neutrophils into lesions (see above). A number of *S. suis* components, both cell-wall associated and secreted factors, have been shown to induce release of pro-inflammatory cytokines which in turn may either help to control acute infection or contribute to immunopathology (Segura et al. 1999, 2002, 2006). The interaction of *S. suis* with innate and adaptive immune mechanisms of the host is crucial for both pathogenesis and immune control of *S. suis* infections. Nevertheless, this topic is out of the scope of this article, and thus will not be addressed in further detail here.

Taken together, at present, we still have only a very scattered picture of the molecular host–pathogen interactions that contribute to pathogenesis of *S. suis* infections. Furthermore, it is unclear which of the virulence-associated factors and mechanisms identified in mouse and pig play similar roles in zoonotic infections. The experience during the Chinese outbreak 2005 (see above) has shown that highly virulent strains/clones can evolve that are equally pathogenic for pigs and humans. This emphasizes the urgent need for future studies on the evolution, epidemiology and pathogenicity of such highly virulent strains.

6 Conclusions

For a long time, streptococcal species other than GAS and *S. pneumoniae* were considered solely as animal specific pathogens. Zoonotic infections were neglected, since they were assumed to occur only very rarely and accidentally in human individuals. The emergence of two large severe outbreaks in human population caused by *S. suis* and *S. zooepidemicus*, respectively, increased awareness of public health authorities. A rapidly growing human population concomitant with an increased demand for food as well as the increase in companion animals with very close daily contact within families will most likely foster evolution of zoonotic streptococci. Thus, the future task will be to develop strategies against zoonotic streptococcosis by improving diagnosis and control of infections in pets, livestock and humans. The poor current knowledge of the evolution of human pathogenic strains and of the mechanisms of host-specific adaptation urgently demands in-depth information about the possible modes of transmission, emergence of antibiotic resistance and prevalence of zoonotic streptococci. This requires that experts from different geographical regions and scientific backgrounds combine their efforts in a synergistic matter to follow the one medicine concept also in research on zoonotic streptococci.

Acknowledgments Our studies are supported by grants from the Deutsche Forschungsgemeinschaft (SFB587, DFG-SPP 1316), Bonn, Germany and Deutsches Zentrum für Infektionsforschung (DZIF).

References

Abbott Y, Acke E, Khan S, Muldoon EG, Markey BK, Pinilla M, Leonard FC, Steward K, Waller A (2010) Zoonotic transmission of *Streptococcus equi subsp. zooepidemicus* from a dog to a handler. J Med Microbiol 59:120–123

Agnew W, Barnes AC (2007) *Streptococcus iniae*: an aquatic pathogen of global veterinary significance and a challenging candidate for reliable vaccination. Vet Microbiol 122:1–15

Akesson P, Moritz L, Truedsson M, Christensson B, Pawel-Rammingen U (2006) IdeS, a highly specific immunoglobulin G (IgG)-cleaving enzyme from *Streptococcus pyogenes*, is inhibited by specific IgG antibodies generated during infection. Infect Immun 74:497–503

Akineden O, Alber J, Lammler C, Weiss R, Siebert U, Foster G, Tougaard S, Brasseur SM, Reijnders PJ (2007) Relatedness of *Streptococcus equi subsp. zooepidemicus* strains isolated from harbour seals (*Phoca vitulina*) and grey seals (*Halichoerus grypus*) of various origins of the North Sea during 1988–2005. Vet Microbiol 121:158–162

Akineden O, Hassan AA, Alber J, El Sayed A, Estoepangestie AT, Lammler C, Weiss R, Siebert U (2005) Phenotypic and genotypic properties of *Streptococcus equi subsp. zooepidemicus* isolated from harbor seals (*Phoca vitulina*) from the German North Sea during the phocine distemper outbreak in 2002. Vet Microbiol 110:147–152

Allgaier A, Goethe R, Wisselink HJ, Smith HE, Valentin-Weigand P (2001) Relatedness of *Streptococcus suis* isolates of various serotypes and clinical backgrounds as evaluated by macrorestriction analysis and expression of potential virulence traits. J Clin Microbiol 39:445–453

Amass SF, SanMiguel P, Clark LK (1997) Demonstration of vertical transmission of *Streptococcus suis* in swine by genomic fingerprinting. J Clin Microbiol 35:1595–1596

Arends JP, Hartwig N, Rudolphy M, Zanen HC (1984) Carrier rate of *Streptococcus suis* capsular type 2 in palatine tonsils of slaughtered pigs. J Clin Microbiol 20:945–947

Arends JP, Zanen HC (1988) Meningitis caused by *Streptococcus suis* in humans. Rev Infect Dis 10:131–137

Aubry P, Swor TM, Lohr CV, Tibary A, Barrington GM (2000) Septic orchitis in an alpaca. Can Vet J 41:704–706

Baele M, Chiers K, Devriese LA, Smith HE, Wisselink HJ, Vaneechoutte M, Haesebrouck F (2001) The gram-positive tonsillar and nasal flora of piglets before and after weaning. J Appl Microbiol 91:997–1003

Baiano JC, Barnes AC (2009) Towards control of *Streptococcus iniae*. Emerg Infect Dis 15:1891–1896

Baiano JC, Tumbol RA, Umapathy A, Barnes AC (2008) Identification and molecular characterisation of a fibrinogen binding protein from *Streptococcus iniae*. BMC Microbiol 8:67

Balter S, Benin A, Pinto SW, Teixeira LM, Alvim GG, Luna E, Jackson D, LaClaire L, Elliott J, Facklam R, Schuchat A (2000) Epidemic nephritis in Nova Serrana, Brazil. Lancet 355:1776–1780

Bannister MF, Benson CE, Sweeney CR (1985) Rapid species identification of group C streptococci isolated from horses. J Clin Microbiol 21:524–526

Barnham M, Cole G, Efstratiou A, Tagg JR, Skjold SA (1987a) Characterization of *Streptococcus zooepidemicus* (Lancefield group C) from human and selected animal infections. Epidemiol Infect 98:171–182

Barnham M, Ljunggren A, McIntyre M (1987b) Human infection with *Streptococcus zooepidemicus* (Lancefield group C): three case reports. Epidemiol Infect 98:183–190

Barnham M, Thornton TJ, Lange K (1983) Nephritis caused by *Streptococcus zooepidemicus* (Lancefield group C). Lancet 1:945–948

Barquero N, Chanter N, Laxton R, Wood JL, Newton JR (2010) Molecular epidemiology of *Streptococcus zooepidemicus* isolated from the respiratory tracts of thoroughbred racehorses in training. Vet J 183:348–351

Baums CG, Hermeyer K, Ohlemeyer S, Adamek M, Czerny CP, Hoerstgen-Schwark C, Valentin-Weigand P, Baumgärtner W, Steinhagen D (2012) Establishment of a model of Streptococcus iniae meningoencephalitis in Nile tilapia (Oreochromis niloticus). J Comp Pathol (in press)

Baums CG, Kaim U, Fulde M, Ramachandran G, Goethe R, Valentin-Weigand P (2006) Identification of a novel virulence determinant with serum opacification activity in *Streptococcus suis*. Infect Immun 74:6154–6162

Baums CG, Valentin-Weigand P (2009) Surface-associated and secreted factors of *Streptococcus suis* in epidemiology, pathogenesis and vaccine development. Anim Health Res Rev 10:65–83

Baums CG, Verkuhlen GJ, Rehm T, Silva LM, Beyerbach M, Pohlmeyer K, Valentin-Weigand P (2007) Prevalence of *Streptococcus suis* genotypes in wild boars of Northwestern Germany. Appl Environ Microbiol 73:711–717

Benga L, Friedl P, Valentin-Weigand P (2005) Adherence of *Streptococcus suis* to porcine endothelial cells. J Vet Med B Infect Dis Vet Public Health 52:392–395

Benga L, Fulde M, Neis C, Goethe R, Valentin-Weigand P (2008) Polysaccharide capsule and suilysin contribute to extracellular survival of *Streptococcus suis* co-cultivated with primary porcine phagocytes. Vet Microbiol 132:211–219

Benga L, Goethe R, Rohde M, Valentin-Weigand P (2004) Non-encapsulated strains reveal novel insights in invasion and survival of *Streptococcus suis* in epithelial cells. Cell Microbiol 6:867–881

Beres SB, Sesso R, Pinto SW, Hoe NP, Porcella SF, Deleo FR, Musser JM (2008) Genome sequence of a Lancefield group C *Streptococcus zooepidemicus* strain causing epidemic nephritis: new information about an old disease. PLoS One 3:e3026

Bergmann R, Dinkla K, Nitsche-Schmitz DP, Graham RM, Luttge M, Sanderson-Smith ML, Nerlich A, Rohde M, Chhatwal GS (2011) Biological functions of GCS3, a novel plasminogen-binding protein of *Streptococcus dysgalactiae ssp. equisimilis*. Int J Med Microbiol 301:157–164

Bert F, Lambert-Zechovsky N (1997) Septicemia caused by *Streptococcus canis* in a human. J Clin Microbiol 35:777–779

Berthelot-Herault F, Gottschalk M, Labbe A, Cariolet R, Kobisch M (2001) Experimental airborne transmission of *Streptococcus suis* capsular type 2 in pigs. Vet Microbiol 82:69–80

Biberstein EL, Brown C, Smith T (1980) Serogroups and biotypes among beta-hemolytic streptococci of canine origin. J Clin Microbiol 11:558–561

Blank LM, Hugenholtz P, Nielsen LK (2008) Evolution of the hyaluronic acid synthesis (has) operon in *Streptococcus zooepidemicus* and other *pathogenic streptococci*. J Mol Evol 67:13–22

Blum S, Elad D, Zukin N, Lysnyansky I, Weisblith L, Perl S, Netanel O, David D (2010) Outbreak of *Streptococcus equi subsp. zooepidemicus* infections in cats. Vet Microbiol 144:236–239

Blunden AS, Hannant D, Livesay G, Mumford JA (1994) Susceptibility of ponies to infection with *Streptococcus pneumoniae* (capsular type 3). Equine Vet J 26:22–28

Bonifait L, Grignon L, Grenier D (2008) Fibrinogen induces biofilm formation by *Streptococcus suis* and enhances its antibiotic resistance. Appl Environ Microbiol 74:4969–4972

Bordes-Benitez A, Sanchez-Onoro M, Suarez-Bordon P, Garcia-Rojas AJ, Saez-Nieto JA, Gonzalez-Garcia A, Alamo-Antunez I, Sanchez-Maroto A, Bolanos-Rivero M (2006) Outbreak of *Streptococcus equi subsp. zooepidemicus* infections on the island of Gran Canaria associated with the consumption of inadequately pasteurized cheese. Eur J Clin Microbiol Infect Dis 25:242–246

Britton AP, Davies JL (2010) Rhinitis and meningitis in two shelter cats caused by *Streptococcus equi subspecies zooepidemicus*. J Comp Pathol 143:70–74

Bromage ES, Thomas A, Owens L (1999) *Streptococcus iniae*, a bacterial infection in barramundi Lates calcarifer. Dis Aquat Organ 36:177–181

Buchanan JT, Stannard JA, Lauth X, Ostland VE, Powell HC, Westerman ME, Nizet V (2005) *Streptococcus iniae* phosphoglucomutase is a virulence factor and a target for vaccine development. Infect Immun 73:6935–6944

Butaye R (1956) De breed beta haemolytische streptococcen afkomstig van honden. Vlaams Diergeneeskd Tijdschr 15:127–131

Byun JW, Yoon SS, Woo GH, Jung BY, Joo YS (2009) An outbreak of fatal hemorrhagic pneumonia caused by *Streptococcus equi subsp. zooepidemicus* in shelter dogs. J Vet Sci 10:269–271

Chabot-Roy G, Willson P, Segura M, Lacouture S, Gottschalk M (2006) Phagocytosis and killing of *Streptococcus suis* by porcine neutrophils. Microb Pathog 41:21–32

Chaffer M, Friedman S, Saran A, Younis A (2005) An outbreak of *Streptococcus canis* mastitis in a dairy herd in Israel. N Z Vet J 53:261–264

Charland N, Harel J, Kobisch M, Lacasse S, Gottschalk M (1998) *Streptococcus suis* serotype 2 mutants deficient in capsular expression. Microbiology 144(Pt 2):325–332

Charland N, Nizet V, Rubens CE, Kim KS, Lacouture S, Gottschalk M (2000) Streptococcus suis serotype 2 interactions with human brain microvascular endothelial cells. Infect Immun 68:637–643

Chen C, Tang J, Dong W, Wang C, Feng Y, Wang J, Zheng F, Pan X, Liu D, Li M, Song Y, Zhu X, Sun H, Feng T, Guo Z, Ju A, Ge J, Dong Y, Sun W, Jiang Y, Wang J, Yan J, Yang H, Wang X, Gao GF, Yang R, Wang J, Yu J (2007) A glimpse of streptococcal toxic shock syndrome from comparative genomics of *S. suis* 2 Chinese isolates. PLoS One 2: e315

Clark C, Greenwood S, Boison JO, Chirino-Trejo M, Dowling PM (2008) Bacterial isolates from equine infections in western Canada (1998–2003). Can Vet J 49:153–160

Clifton-Hadley FA (1983) *Streptococcus suis* type 2 infections. Br Vet J 139:1–5

Clifton-Hadley FA, Alexander TJ (1980) The carrier site and carrier rate of *Streptococcus suis* type II in pigs. Vet Rec 107:40–41

Corning BF, Murphy JC, Fox JG (1991) Group G streptococcal lymphadenitis in rats. J Clin Microbiol 29:2720–2723

de Greeff A, Buys H, Verhaar R, Dijkstra J, Van Alphen L, Smith HE (2002) Contribution of fibronectin-binding protein to pathogenesis of *Streptococcus suis* serotype 2. Infect Immun 70:1319–1325

Devriese LA, Hommez J, Klipper-Balz R, Schleifer K-H (1986) *Streptococcus canis sp. nov.*: a species of group G Streptococci from Animals. Int J Syst Bacteriol 36:422–425

DeWinter LM, Low DE, Prescott JF (1999) Virulence of *Streptococcus canis* from canine streptococcal toxic shock syndrome and necrotizing fasciitis. Vet Microbiol 70:95–110

DeWinter LM, Prescott JF (1999) Relatedness of *Streptococcus canis* from canine streptococcal toxic shock syndrome and necrotizing fasciitis. Can J Vet Res 63: 90–95

Dogan B, Schukken YH, Santisteban C, Boor KJ (2005) Distribution of serotypes and antimicrobial resistance genes among *Streptococcus agalactiae* isolates from bovine and human hosts. J Clin Microbiol 43:5899–5906

Efstratiou A, Colman G, Hahn G, Timoney JF, Boeufgras JM, Monget D (1994) Biochemical differences among human and animal streptococci of Lancefield group C or group G. J Med Microbiol 41:145–148

Egesten A, Frick IM, Morgelin M, Olin AI, Bjorck L (2011) Binding of albumin promotes bacterial survival at the epithelial surface. J Biol Chem 286:2469–2476

Eldar A, Ghittino C (1999) Lactococcus garvieae and *Streptococcus iniae* infections in rainbow trout Oncorhynchus mykiss: similar, but different diseases. Dis Aquat Organ 36:227–231

Enright MR, Alexander TJ, Clifton-Hadley FA (1987) Role of houseflies (Musca domestica) in the epidemiology of *Streptococcus suis* type 2. Vet Rec 121:132–133

Erol E, Locke SJ, Donahoe JK, Mackin MA, Carter CN (2012) Beta-hemolytic *Streptococcus spp.* from horses: a retrospective study (2000–2010). J Vet Diagn Invest 24:142–147

Esgleas M, Li Y, Hancock MA, Harel J, Dubreuil JD, Gottschalk M (2008) Isolation and characterization of alpha-enolase, a novel fibronectin-binding protein from *Streptococcus suis*. Microbiology 154:2668–2679

Eyngor M, Chilmonczyk S, Zlotkin A, Manuali E, Lahav D, Ghittino C, Shapira R, Hurvitz A, Eldar A (2007) Transcytosis of *Streptococcus iniae* through skin epithelial barriers: an in vitro study. FEMS Microbiol Lett 277:238–248

Eyngor M, Tekoah Y, Shapira R, Hurvitz A, Zlotkin A, Lublin A, Eldar A (2008) Emergence of novel *Streptococcus iniae* exopolysaccharide-producing strains following vaccination with nonproducing strains. Appl Environ Microbiol 74:6892–6897

Facklam R (2002) What happened to the streptococci: overview of taxonomic and nomenclature changes. Clin Microbiol Rev 15:613–630

Facklam R, Elliott J, Shewmaker L, Reingold A (2005) Identification and characterization of sporadic isolates of *Streptococcus iniae* isolated from humans. J Clin Microbiol 43:933–937

Fan H, Wang Y, Tang F, Lu C (2008) Determination of the mimic epitope of the M-like protein adhesin in swine *Streptococcus equi subsp. zooepidemicus*. BMC Microbiol 8: 170

Feng Y, Zhang H, Ma Y, Gao GF (2010) Uncovering newly emerging variants of *Streptococcus suis*, an important zoonotic agent. Trends Microbiol 18:124–131

Finkelstein R, Oren I (2011) Soft tissue infections caused by marine bacterial pathogens: epidemiology, diagnosis, and management. Curr Infect Dis Rep 13:470–477

Fittipaldi N, Segura M, Grenier D, Gottschalk M (2012) Virulence factors involved in the pathogenesis of the infection caused by the swine pathogen and zoonotic agent *Streptococcus suis*. Future Microbiol 7:259–279

Fittipaldi N, Sekizaki T, Takamatsu D, de la Dominguez-Punaro Cruz, Harel J, Bui NK, Vollmer W, Gottschalk M (2008a) Significant contribution of the pgdA gene to the virulence of *Streptococcus suis*. Mol Microbiol 70:1120–1135

Fittipaldi N, Sekizaki T, Takamatsu D, Harel J, Dominguez-Punaro ML, Von Aulock S, Draing C, Marois C, Kobisch M, Gottschalk M (2008b) D-alanylation of lipoteichoic acid contributes to the virulence of *Streptococcus suis*. Infect Immun 76:3587–3594

Flanagan J, Collin N, Timoney J, Mitchell T, Mumford JA, Chanter N (1998) Characterization of the haemolytic activity of *Streptococcus equi*. Microb Pathog 24:211–221

Flock M, Karlstrom A, Lannergard J, Guss B, Flock JI (2006) Protective effect of vaccination with recombinant proteins from *Streptococcus equi* subspecies equi in a strangles model in the mouse. Vaccine 24:4144–4151

Francis AJ, Nimmo GR, Efstratiou A, Galanis V, Nuttall N (1993) Investigation of milk-borne *Streptococcus zooepidemicus* infection associated with glomerulonephritis in Australia. J Infect 27:317–323

Fulde M, Rohde M, Hitzmann A, Preissner KT, Nitsche-Schmitz DP, Nerlich A, Chhatwal GS, Bergmann S (2011a) SCM, a novel M-like protein from *Streptococcus canis*, binds (mini)-plasminogen with high affinity and facilitates bacterial transmigration. Biochem J 434:523–535

Fulde M, Willenborg J, de Greeff A, Benga L, Smith HE, Valentin-Weigand P, Goethe R (2011b) ArgR is an essential local transcriptional regulator of the arcABC operon in *Streptococcus suis* and is crucial for biological fitness in an acidic environment. Microbiology 157:572–582

Fuller JD, Bast DJ, Nizet V, Low DE, de Azavedo JC (2001) *Streptococcus iniae* virulence is associated with a distinct genetic profile. Infect Immun 69:1994–2000

Gaede W, Reckling KF, Schliephake A, Missal D, Hotzel H, Sachse K (2010) Detection of *Chlamydophila caviae* and *Streptococcus equi subsp. zooepidemicus* in horses with signs of rhinitis and conjunctivitis. Vet Microbiol 142:440–444

Galperine T, Cazorla C, Blanchard E, Boineau F, Ragnaud JM, Neau D (2007) *Streptococcus canis* infections in humans: retrospective study of 54 patients. J Infect 55:23–26

Garnett NL, Eydelloth RS, Swindle MM, Vonderfecht SL, Strandberg JD, Luzarraga MB (1982) Hemorrhagic streptococcal pneumonia in newly procured research dogs. J Am Vet Med Assoc 181:1371–1374

Gottschalk M, Petitbois S, Higgins R, Jacques M (1991) Adherence of *Streptococcus suis* capsular type 2 to porcine lung sections. Can J Vet Res 55:302–304

Gottschalk M, Segura M (2000) The pathogenesis of the meningitis caused by *Streptococcus suis*: the unresolved questions. Vet Microbiol 76:259–272

Gottschalk M, Segura M, Xu J (2007) *Streptococcus suis* infections in humans: the Chinese experience and the situation in North America. Anim Health Res Rev 8:29–45

Gottschalk M, Xu J, Calzas C, Segura M (2010) *Streptococcus suis*: a new emerging or an old neglected zoonotic pathogen? Future Microbiol 5:371–391

Gruening P, Fulde M, Valentin-Weigand P, Goethe R (2006) Structure, regulation, and putative function of the arginine deiminase system of *Streptococcus suis*. J Bacteriol 188:361–369

Hamilton CA, Stark DM (1970) Occurrence and characterization of Lancefield group G streptococci in bovine mastitis. Am J Vet Res 31:397–398

Hare T, Frye RM (1938) Preliminary observations of an infection of dogs by beta-haemolytic streptococci. Vet Rec 50:213–218

Hassan AA, Akineden O, Usleber E (2005) Identification of *Streptococcus canis* isolated from milk of dairy cows with subclinical mastitis. J Clin Microbiol 43:1234–1238

Hassan AA, Khan IU, Abdulmawjood A, Lammler C (2003) Development of PCR assays for detection of *Streptococcus canis*. FEMS Microbiol Lett 219:209–214

Hayashi M, Matsuzaki Y, Shimonaka M (2009) Impact of plasminogen on an in vitro wound healing model based on a perfusion cell culture system. Mol Cell Biochem 322:1–13

Hewson J, Cebra CK (2001) Peritonitis in a llama caused by *Streptococcus equi subsp. zooepidemicus*. Can Vet J 42:465–467

Hill JE, Gottschalk M, Brousseau R, Harel J, Hemmingsen SM, Goh SH (2005) Biochemical analysis, cpn60 and 16S rDNA sequence data indicate that *Streptococcus suis* serotypes 32 and 34, isolated from pigs, are *Streptococcus orisratti*. Vet Microbiol 107:63–69

Hitzmann A, Bergmann S, Rohde M, Chhatwal GS, Fulde M (2012) Identification and characterization of the arginine deiminase system of *Streptococcus canis* (in press)

Holden MT, Hauser H, Sanders M, Ngo TH, Cherevach I, Cronin A, Goodhead I, Mungall K, Quail MA, Price C, Rabbinowitsch E, Sharp S, Croucher NJ, Chieu TB, Mai NT, Diep TS, Chinh NT, Kehoe M, Leigh JA, Ward PN, Dowson CG, Whatmore AM, Chanter N, Iversen P, Gottschalk M, Slater JD, Smith HE, Spratt BG, Xu J, Ye C, Bentley S, Barrell BG, Schultsz C, Maskell DJ, Parkhill J (2009) Rapid evolution of virulence and drug resistance in the emerging zoonotic pathogen *Streptococcus suis*. PLoS One 4:e6072

Holden MT, Heather Z, Paillot R, Steward KF, Webb K, Ainslie F, Jourdan T, Bason NC, Holroyd NE, Mungall K, Quail MA, Sanders M, Simmonds M, Willey D, Brooks K, Aanensen DM, Spratt BG, Jolley KA, Maiden MC, Kehoe M, Chanter N, Bentley SD, Robinson C, Maskell DJ, Parkhill J, Waller AS (2009) Genomic evidence for the evolution of *Streptococcus equi*: host restriction, increased virulence, and genetic exchange with human pathogens. PLoS Pathog 5:e1000346

Hong-Jie F, Fu-yu T, Ying M, Cheng-ping L (2009) Virulence and antigenicity of the szp-gene deleted *Streptococcus equi ssp. zooepidemicus* mutant in mice. Vaccine 27:56–61

Hulting G, Flock M, Frykberg L, Lannergard J, Flock JI, Guss B (2009) Two novel IgG endopeptidases of *Streptococcus equi*. FEMS Microbiol Lett 298:44–50

Iglauer F, Kunstyr I, Morstedt R, Farouq H, Wullenweber M, Damsch S (1991) *Streptococcus canis* arthritis in a cat breeding colony. J Exp Anim Sci 34:59–65

Igwe EI, Shewmaker PL, Facklam RR, Farley MM, van Beneden C, Beall B (2003) Identification of superantigen genes speM, ssa, and smeZ in invasive strains of beta-hemolytic group C and G streptococci recovered from humans. FEMS Microbiol Lett 229:259–264

Jacobs AA, Loeffen PL, van den Berg AJ, Storm PK (1994) Identification, purification, and characterization of a thiol-activated hemolysin (suilysin) of *Streptococcus suis*. Infect Immun 62:1742–1748

Jensen A, Kilian M (2012) Delineation of *Streptococcus dysgalactiae*, its subspecies, and its clinical and phylogenetic relationship to *Streptococcus pyogenes*. J Clin Microbiol 50:113–126

Jones M, Miesner M, Grondin T (2009) Outbreak of *Streptococcus equi ssp. zooepidemicus* polyserositis in an alpaca herd. J Vet Intern Med 23:220–223

Jonsson H, Lindmark H, Guss B (1995) A protein G-related cell surface protein in *Streptococcus zooepidemicus*. Infect Immun 63:2968–2975

Kawamura Y, Hou XG, Sultana F, Miura H, Ezaki T (1995) Determination of 16S rRNA sequences of *Streptococcus mitis* and *Streptococcus gordonii* and phylogenetic relationships among members of the genus Streptococcus. Int J Syst Bacteriol 45:406–408

Kim MK, Jee H, Shin SW, Lee BC, Pakhrin B, Yoo HS, Yoon JH, Kim DY (2007a) Outbreak and control of haemorrhagic pneumonia due to *Streptococcus equi* subspecies *zooepidemicus* in dogs. Vet. Rec. 161:528–530

Kim MS, Choi SH, Lee EH, Nam YK, KimS K, KimK H (2007b) a-enolase, a plasmin(ogen) binding protein and cell wall associating protein from a fish pathogenic *Streptococcus iniae* strain. Aquaculture 265:55–60

Kruger EF, Byrne BA, Pesavento P, Hurley KF, Lindsay LL, Sykes JE (2010) Relationship between clinical manifestations and pulsed-field gel profiles of *Streptococcus canis* isolates from dogs and cats. Vet Microbiol 146:167–171

Kuusi M, Lahti E, Virolainen A, Hatakka M, Vuento R, Rantala L, Vuopio-Varkila J, Seuna E, Karppelin M, Hakkinen M, Takkinen J, Gindonis V, Siponen K, Huotari K (2006) An outbreak of *Streptococcus equi* subspecies zooepidemicus associated with consumption of fresh goat cheese. BMC Infect Dis 6: 36

Lalonde M, Segura M, Lacouture S, Gottschalk M (2000) Interactions between *Streptococcus suis* serotype 2 and different epithelial cell lines. Microbiology 146(Pt 8):1913–1921

Lam MM, Clarridge JE III, Young EJ, Mizuki S (2007) The other group G Streptococcus: increased detection of *Streptococcus canis* ulcer infections in dog owners. J Clin Microbiol 45:2327–2329

Lamm CG, Ferguson AC, Lehenbauer TW, Love BC (2010) Streptococcal infection in dogs: a retrospective study of 393 cases. Vet Pathol 47:387–395

Lammler C, Frede C, Gurturk K, Hildebrand A, Blobel H (1988) Binding activity of *Streptococcus canis* for albumin and other plasma proteins. J Gen Microbiol 134:2317–2323

Lannergard J, Guss B (2006) IdeE, an IgG-endopeptidase of *Streptococcus equi* ssp. equi. FEMS Microbiol Lett 262:230–235

Las HA, Vela AI, Fernandez E, Legaz E, Dominguez L, Fernandez-Garayzabal JF (2002) Unusual outbreak of clinical mastitis in dairy sheep caused by *Streptococcus equi* subsp. *zooepidemicus*. J Clin Microbiol 40:1106–1108

Lau SK, Woo PC, Luk WK, Fung AM, Hui WT, Fong AH, Chow CW, Wong SS, Yuen KY (2006) Clinical isolates of *Streptococcus iniae* from Asia are more mucoid and beta-hemolytic than those from North America. Diagn Microbiol Infect Dis 54:177–181

Lefebure T, Richards VP, Lang P, Pavinski-Bitar P, Stanhope MJ (2012) Gene repertoire evolution of *Streptococcus pyogenes* inferred from phylogenomic analysis with *Streptococcus canis* and *Streptococcus dysgalactiae*. PLoS One 7:e37607

Lewis MJ, Meehan M, Owen P, Woof JM (2008) A common theme in interaction of bacterial immunoglobulin-binding proteins with immunoglobulins illustrated in the equine system. J Biol Chem 283:17615–17623

Lindmark H, Guss B (1999) SFS, a novel fibronectin-binding protein from *Streptococcus equi*, inhibits the binding between fibronectin and collagen. Infect Immun 67:2383–2388

Lindmark H, Jacobsson K, Frykberg L, Guss B (1996) Fibronectin-binding protein of *Streptococcus equi subsp. zooepidemicus*. Infect Immun 64:3993–3999

Literak O, Mraz O (1991) *Streptococcus equi subsp. zooepidemicus* as the cause of mass mortality among laboratory mice. Acta Vet Brno 60:51–60

Ljungh A (2000) Helicobacter pylori interactions with plasminogen. Methods 21:151–157

Locke JB, Aziz RK, Vicknair MR, Nizet V, Buchanan JT (2008) *Streptococcus iniae* M-like protein contributes to virulence in fish and is a target for live attenuated vaccine development. PLoS One 3:e2824

Locke JB, Colvin KM, Datta AK, Patel SK, Naidu NN, Neely MN, Nizet V, Buchanan JT (2007a) *Streptococcus iniae* capsule impairs phagocytic clearance and contributes to virulence in fish. J Bacteriol 189:1279–1287

Locke JB, Colvin KM, Varki N, Vicknair MR, Nizet V, Buchanan JT (2007b) Streptococcus iniae beta-hemolysin streptolysin S is a virulence factor in fish infection. Dis Aquat Organ 76:17–26

Lowe BA, Marsh TL, Isaacs-Cosgrove N, Kirkwood RN, Kiupel M, Mulks MH (2011) Microbial communities in the tonsils of healthy pigs. Vet Microbiol 147:346–357

Lowe BA, Miller JD, Neely MN (2007) Analysis of the polysaccharide capsule of the systemic pathogen *Streptococcus iniae* and its implications in virulence. Infect Immun 75:1255–1264

Lun ZR, Wang QP, Chen XG, Li AX, Zhu XQ (2007) *Streptococcus suis*: an emerging zoonotic pathogen. Lancet Infect Dis 7:201–209

Lyskova P, Vydrzalova M, Mazurova J (2007) Identification and antimicrobial susceptibility of bacteria and yeasts isolated from healthy dogs and dogs with otitis externa. J Vet Med A Physiol PatholClin Med 54:559–563

Ma Z, Zhang H, Zheng J, Li Y, Yi L, Fan H, Lu C (2012) Interaction between M-like protein and macrophage thioredoxin facilitates antiphagocytosis for *Streptococcus equi ssp. zooepidemicus*. PLoS One 7:e32099

Mai NT, Hoa NT, Nga TV, Linh lD, Chau TT, Sinh DX, Phu NH, Chuong LV, Diep TS, Campbell J, Nghia HD, Minh TN, Chau NV, de Jong MD, Chinh NT, Hien TT, Farrar J, Schultsz C (2008) *Streptococcus suis* meningitis in adults in Vietnam. Clin Infect Dis 46:659–667

Martinez G, Harel J, Higgins R, Lacouture S, Daignault D, Gottschalk M (2000) Characterization of *Streptococcus agalactiae* isolates of bovine and human origin by randomly amplified polymorphic DNA analysis. J Clin Microbiol 38:71–78

Matz-Rensing K, Winkelmann J, Becker T, Burckhardt I, Van der LM, Kondgen S, Leendertz F, Kaup FJ (2009) Outbreak of *Streptococcus equi* subsp. zooepidemicus infection in a group of rhesus monkeys (Macaca mulatta). J Med Primatol 38:328–334

McDonald TJ, McDonald JS (1976) Streptococci isolated from bovine intramammary infections. Am J Vet Res 37:377–381

Meehan M, Lynagh Y, Woods C, Owen P (2001) The fibrinogen-binding protein (FgBP) of *Streptococcus equi* subsp. equi additionally binds IgG and contributes to virulence in a mouse model. Microbiology 147:3311–3322

Messier S, Lacouture S, Gottschalk M (2008) Distribution of *Streptococcus suis* capsular types from 2001 to 2007. Can Vet J 49:461–462

Miller CW, Prescott JF, Mathews KA, Betschel SD, Yager JA, Guru V, DeWinter L, Low DE (1996) Streptococcal toxic shock syndrome in dogs. J Am Vet Med Assoc 209:1421–1426

Miller JD, Neely MN (2004) Zebrafish as a model host for streptococcal pathogenesis. Acta Trop 91:53–68

Minces LR, Brown PJ, Veldkamp PJ (2011) Human meningitis from *Streptococcus equi* subsp. zooepidemicus acquired as zoonoses. Epidemiol Infect 139:406–410

Mwaniki CG, Robertson ID, Hampson DJ (1994a) The prevalence of *Streptococcus suis* type 2 in Western Australian piggeries. Aust Vet J 71:385–386

Mwaniki CG, Robertson ID, Trott DJ, Atyeo RF, Lee BJ, Hampson DJ (1994b) Clonal analysis and virulence of Australian isolates of *Streptococcus suis* type 2. Epidemiol Infect 113:321–334

Nghia HD, Hoa NT, Linh lD, Campbell J, Diep TS, Chau NV, Mai NT, Hien TT, Spratt B, Farrar J, Schultsz C (2008) Human case of *Streptococcus suis* serotype 16 infection. Emerg Infect Dis 14:155–157

Ngo TH, Tran TB, Tran TT, Nguyen VD, Campbell J, Pham HA, Huynh HT, Nguyen VV, Bryant JE, Tran TH, Farrar J, Schultsz C (2011) Slaughterhouse pigs are a major reservoir of *Streptococcus suis* serotype 2 capable of causing human infection in southern Vietnam. PLoS One 6:e17943

Nicholson ML, Ferdinand L, Sampson JS, Benin A, Balter S, Pinto SW, Dowell SF, Facklam RR, Carlone GM, Beall B (2000) Analysis of immunoreactivity to a *Streptococcus equi* subsp. zooepidemicus M-like protein To confirm an outbreak of poststreptococcal glomerulonephritis, and sequences of M-like proteins from isolates obtained from different host species. J Clin Microbiol 38:4126–4130

Nitsche DP, Johansson HM, Frick IM, Morgelin M (2006) Streptococcal protein FOG, a novel matrix adhesin interacting with collagen I in vivo. J Biol Chem 281:1670–1679

Nitsche-Schmitz DP, Johansson HM, Sastalla I, Reissmann S, Frick IM, Chhatwal GS (2007) Group G streptococcal IgG binding molecules FOG and protein G have different impacts on opsonization by C1q. J Biol Chem 282:17530–17536

O'Sullivan T, Friendship R, Blackwell T, Pearl D, McEwen B, Carman S, Slavic D, Dewey C (2011) Microbiological identification and analysis of swine tonsils collected from carcasses at slaughter. Can.J.Vet.Res. 75: 106-111

Oikawa M, Kamada M, Yoshikawa Y, Yoshikawa T (1994) Pathology of equine pneumonia associated with transport and isolation of *Streptococcus equi* subsp. zooepidemicus. J Comp Pathol 111:205–212

Oikawa M, Takagi S, Anzai R, Yoshikawa H, Yoshikawa T (1995) Pathology of equine respiratory disease occurring in association with transport. J Comp Pathol 113:29–43

Okamoto S, Kawabata S, Terao Y, Fujitaka H, Okuno Y, Hamada S (2004) The *Streptococcus pyogenes* capsule is required for adhesion of bacteria to virus-infected alveolar epithelial cells and lethal bacterial-viral superinfection. Infect Immun 72:6068–6075

Oliveira IC, de Mattos MC, Pinto TA, Ferreira-Carvalho BT, Benchetrit LC, Whiting AA, Bohnsack JF, Figueiredo AM (2006) Genetic relatedness between group B streptococci originating from bovine mastitis and a human group B Streptococcus type V cluster displaying an identical pulsed-field gel electrophoresis pattern. Clin Microbiol Infect 12:887–893

Paillot R, Darby AC, Robinson C, Wright NL, Steward KF, Anderson E, Webb K, Holden MT, Efstratiou A, Broughton K, Jolley KA, Priestnall SL, Marotti Campi MC, Hughes MA, Radford A, Erles K, Waller AS (2010) Identification of three novel superantigen-encoding genes in *Streptococcus equi* subsp. zooepidemicus, szeF, szeN, and szeP. Infect Immun 78:4817–4827

Palmieri C, Varaldo PE, Facinelli B (2011) *Streptococcus suis*, an emerging drug-resistant animal and human pathogen. Front Microbiol 2:235

Panchaud Y, Gerber V, Rossano A, Perreten V (2010) Bacterial infections in horses: a retrospective study at the University Equine Clinic of Bern. Schweiz Arch Tierheilkd 152:176–182

Pancholi V, Chhatwal GS (2003) Housekeeping enzymes as virulence factors for pathogens. Int J Med Microbiol 293:391–401

Pawel-Rammingen U (2012) Streptococcal IdeS and its impact on immune response and inflammation. J Innate Immun 4:132–140

Pawel-Rammingen U, Johansson BP, Bjorck L (2002) IdeS, a novel streptococcal cysteine proteinase with unique specificity for immunoglobulin G. EMBO J 21:1607–1615

Perch B, Pedersen KB, Henrichsen J (1983) Serology of capsulated streptococci pathogenic for pigs: six new serotypes of *Streptococcus suis*. J Clin Microbiol 17:993–996

Pereira UP, Mian GF, Oliveira IC, Benchetrit LC, Costa GM, Figueiredo HC (2010) Genotyping *of Streptococcus agalactiae* strains isolated from fish, human and cattle and their virulence potential in Nile tilapia. Vet Microbiol 140:186–192

Pesavento PA, Hurley KF, Bannasch MJ, Artiushin S, Timoney JF (2008) A clonal outbreak of acute fatal hemorrhagic pneumonia in intensively housed (shelter) dogs caused by *Streptococcus equi* subsp. zooepidemicus. Vet Pathol 45:51–53

Pier GB, Madin SH, Al-Nakeeb S (1978) Isolation and characterization of a second isolate of *Streptococcus iniae*. Int J Syst Bacteriol 28:311–314

Pisoni G, Zadoks RN, Vimercati C, Locatelli C, Zanoni MG, Moroni P (2009) Epidemiological investigation of *Streptococcus equi* subspecies zooepidemicus involved in clinical mastitis in dairy goats. J Dairy Sci 92:943–951

Poyart C, Quesne G, Coulon S, Berche P, Trieu-Cuot P (1998) Identification of streptococci to species level by sequencing the gene encoding the manganese-dependent superoxide dismutase. J Clin Microbiol 36:41–47

Prescott JF, Mathews K, Gyles CL, Matsumiya L, Miller C, Rinkhardt N, Yager JA, Hylands RH, Low DE (1995) Canine streptococcal toxic shock syndrome in Ontario: an emerging disease. Can Vet J 36:486–487

Priestnall S, Erles K (2011) *Streptococcus zooepidemicus*: an emerging canine pathogen. Vet J 188:142–148

Proft T, Fraser JD (2007) Streptococcal superantigens. Chem Immunol Allergy 93:1–23

Rojas M, Labrador I, Concepcion JL, Aldana E, Avilan L (2008) Characteristics of plasminogen binding to Trypanosoma cruzi epimastigotes. Acta Trop 107:54–58

Rosenkranz M, Elsner HA, Sturenburg HJ, Weiller C, Rother J, Sobottka I (2003) *Streptococcus suis* meningitis and septicemia contracted from a wild boar in Germany. J Neurol 250:869–870

Ryu SH, Koo HC, Lee YW, Park YH, Lee CW (2011) Etiologic and epidemiologic analysis of bacterial infectious upper respiratory disease in thoroughbred horses at the Seoul Race Park. J Vet Sci 12:195–197

Schrager HM, Alberti S, Cywes C, Dougherty GJ, Wessels MR (1998) Hyaluronic acid capsule modulates M protein-mediated adherence and acts as a ligand for attachment of group A Streptococcus to CD44 on human keratinocytes. J Clin Invest 101:1708–1716

Schroeder B, Boyle MD, Sheerin BR, Asbury AC, Lottenberg R (1999) Species specificity of plasminogen activation and acquisition of surface-associated proteolytic activity by group C streptococci grown in plasma. Infect Immun 67:6487–6495

Schwarz-Linek U, Hook M, Potts JR (2006) Fibronectin-binding proteins of gram-positive cocci. Microbes Infect 8:2291–2298

Seele J, Singpiel A, Spoerry C, von Pawel-Rammingen U, Valentin-Weigand P, Baums CG Identification of a novel host-specific IgM protease in Streptococcus suis. J Bact (in revision)
Segura M, Gottschalk M (2002) Streptococcus suis interactions with the murine macrophage cell line J774: adhesion and cytotoxicity. Infect Immun 70:4312–4322
Segura M, Stankova J, Gottschalk M (1999) Heat-killed Streptococcus suis capsular type 2 strains stimulate tumor necrosis factor alpha and interleukin-6 production by murine macrophages. Infect Immun 67:4646–4654
Segura M, Vadeboncoeur N, Gottschalk M (2002) CD14-dependent and -independent cytokine and chemokine production by human THP-1 monocytes stimulated by *Streptococcus suis* capsular type 2. Clin Exp Immunol 127:243–254
Segura M, Vanier G, Al Numani D, Lacouture S, Olivier M, Gottschalk M (2006) Proinflammatory cytokine and chemokine modulation by *Streptococcus suis* in a whole-blood culture system. FEMS Immunol Med Microbiol 47:92–106
Sesso R, Pinto SW (2005) Five-year follow-up of patients with epidemic glomerulonephritis due to *Streptococcus zooepidemicus*. Nephrol Dial Transplant 20:1808–1812
Shoemaker CA, Klesius PH, Evans JJ (2001) Prevalence of *Streptococcus iniae* in tilapia, hybrid striped bass, and channel catfish on commercial fish farms in the United States. Am J Vet Res 62:174–177
Silva LM, Baums CG, Rehm T, Wisselink HJ, Goethe R, Valentin-Weigand P (2006) Virulence-associated gene profiling of *Streptococcus suis* isolates by PCR. Vet Microbiol 115:117–127
Smeesters PR, Mardulyn P, Vergison A, Leplae R, Van Melderen L (2008) Genetic diversity of Group A Streptococcus M protein: implications for typing and vaccine development. Vaccine 26:5835–5842
Smith HE, Damman M, van d V, Wagenaar F, Wisselink HJ, Stockhofe-Zurwieden N, Smits MA (1999) Identification and characterization of the cps locus of *Streptococcus suis* serotype 2: the capsule protects against phagocytosis and is an important virulence factor. Infect Immun 67:1750–1756
Smith TC, Capuano AW, Boese B, Myers KP, Gray GC (2008) Exposure to *Streptococcus suis* among US swine workers. Emerg Infect Dis 14:1925–1927
Soedarmanto I, Pasaribu FH, Wibawan IW, Lammler C (1996) Identification and molecular characterization of serological group C streptococci isolated from diseased pigs and monkeys in Indonesia. J Clin Microbiol 34:2201–2204
Staats JJ, Feder I, Okwumabua O, Chengappa MM (1997) *Streptococcus suis*: past and present. Vet Res Commun 21:381–407
Su Y, Yao W, Perez-Gutierrez ON, Smidt H, Zhu WY (2008a) 16S ribosomal RNA-based methods to monitor changes in the hindgut bacterial community of piglets after oral administration of Lactobacillus sobrius S1. Anaerobe 14:78–86
Su Y, Yao W, Perez-Gutierrez ON, Smidt H, Zhu WY (2008b) Changes in abundance of *Lactobacillus spp.* and *Streptococcus suis* in the stomach, jejunum and ileum of piglets after weaning. FEMS Microbiol Ecol 66:546–555
Sukhnanand S, Dogan B, Ayodele MO, Zadoks RN, Craver MP, Dumas NB, Schukken YH, Boor KJ, Wiedmann M (2005) Molecular subtyping and characterization of bovine and human *Streptococcus agalactiae* isolates. J Clin Microbiol 43:1177–1186
Sun H, Ringdahl U, Homeister JW, Fay WP, Engleberg NC, Yang AY, Rozek LS, Wang X, Sjobring U, Ginsburg D (2004) Plasminogen is a critical host pathogenicity factor for group A streptococcal infection. Science 305:1283–1286
Sun JR, Yan JC, Yeh CY, Lee SY, Lu JJ (2007) Invasive infection with *Streptococcus iniae* in Taiwan. J Med Microbiol 56:1246–1249
Sundberg JP, Hill D, Wyand DS, Ryan MJ, Baldwin CH (1981) *Streptococcus zooepidemicus* as the cause of septicemia in racing greyhounds. Vet Med Small Anim Clin 76:839–842
Swindle MM, Narayan O, Luzarraga M, Bobbie DL (1980) Contagious streptococcal lymphadenitis in cats. J Am Vet Med Assoc 177:829–830

Takeda N, Kikuchi K, Asano R, Harada T, Totsuka K, Sumiyoshi T, Uchiyama T, Hosoda S (2001) Recurrent septicemia caused by *Streptococcus canis* after a dog bite. Scand J Infect Dis 33:927–928

Takeuchi D, Kerdsin A, Pienpringam A, Loetthong P, Samerchea S, Luangsuk P, Khamisara K, Wongwan N, Areeratana P, Chiranairadul P, Lertchayanti S, Petcharat S, Yowang A, Chaiwongsaen P, Nakayama T, Akeda Y, Hamada S, Sawanpanyalert P, Dejsirilert S, Oishi K (2012) Population-based study of Streptococcus suis infection in humans in Phayao Province in northern Thailand. PLoS One 7:e31265

Tang J, Wang C, Feng Y, Yang W, Song H, Chen Z, Yu H, Pan X, Zhou X, Wang H, Wu B, Wang H, Zhao H, Lin Y, Yue J, Wu Z, He X, Gao F, Khan AH, Wang J, Zhao GP, Wang Y, Wang X, Chen Z, Gao GF (2006) Streptococcal toxic shock syndrome caused by Streptococcus suis serotype 2. PLoS Med 3:e151

Tapp J, Thollesson M, Herrmann B (2003) Phylogenetic relationships and genotyping of the genus Streptococcus by sequence determination of the RNase P RNA gene, rnpB. Int J Syst Evol Microbiol 53:1861–1871

Tenenbaum T, Adam R, Eggelnpohler I, Matalon D, Seibt A, GE KN, Galla HJ, Schroten H (2005) Strain-dependent disruption of blood-cerebrospinal fluid barrier by Streptoccocus suis in vitro. FEMS Immunol Med Microbiol 44:25–34

Tenenbaum T, Papandreou T, Gellrich D, Friedrichs U, Seibt A, Adam R, Wewer C, Galla HJ, Schwerk C, Schroten H (2009) Polar bacterial invasion and translocation of *Streptococcus suis* across the blood-cerebrospinal fluid barrier in vitro. Cell Microbiol 11:323–336

Tian Y, Aarestrup FM, Lu CP (2004) Characterization of *Streptococcus suis* serotype 7 isolates from diseased pigs in Denmark. Vet Microbiol 103:55–62

Tikofsky LL, Zadoks RN (2005) Cross-infection between cats and cows: origin and control of *Streptococcus canis* mastitis in a dairy herd. J Dairy Sci 88:2707–2713

Tillman PC, Dodson ND, Indiveri M (1982) Group G streptococcal epizootic in a closed cat colony. J Clin Microbiol 16:1057–1060

Timoney JF (2004) The pathogenic equine streptococci. Vet Res 35:397–409

Timoney JF, Artiushin SC, Boschwitz JS (1997) Comparison of the sequences and functions of *Streptococcus equi* M-like proteins SeM and SzPSe. Infect Immun 65:3600–3605

Timoney JF, DeNegri R, Sheoran A, Forster N (2010) Affects of N-terminal variation in the SeM protein of *Streptococcus equi* on antibody and fibrinogen binding. Vaccine 28:1522–1527

Timoney JF, Walker J, Zhou M, Ding J (1995) Cloning and sequence analysis of a protective M-like protein gene from *Streptococcus equi* subsp. zooepidemicus. Infect Immun 63:1440–1445

Ullberg M, Karlsson I, Wiman B, Kronvall G (1992) Two types of receptors for human plasminogen on group G streptococci. APMIS 100:21–28

Vadeboncoeur N, Segura M, Al Numani D, Vanier G, Gottschalk M (2003) Pro-inflammatory cytokine and chemokine release by human brain microvascular endothelial cells stimulated by *Streptococcus suis* serotype 2. FEMS Immunol Med Microbiol 35:49–58

Valentin-Weigand P, Chhatwal GS, Blobel H (1988) Adherence of streptococcal isolates from cattle and horses to their respective host epithelial cells. Am J Vet Res 49:1485–1488

Valentin-Weigand P, Traore MY, Blobel H, Chhatwal GS (1990) Role of alpha 2-macroglobulin in phagocytosis of group A and C streptococci. FEMS Microbiol Lett 58:321–324

Van der LM, Al Lahham A, Nicklas W, Reinert RR (2009) Molecular characterization of pneumococcal isolates from pets and laboratory animals. PLoS One 4:e8286

Vanier G, Segura M, Friedl P, Lacouture S, Gottschalk M (2004) Invasion of porcine brain microvascular endothelial cells by *Streptococcus suis* serotype 2. Infect Immun 72:1441–1449

Vanier G, Segura M, Gottschalk M (2007) Characterization of the invasion of porcine endothelial cells by Streptococcus suis serotype 2. Can J Vet Res 71:81–89

Wada A (2012) Streptococcus pneumoniae infection and drug resistance in Japan. Nihon Rinsho 70:221–225

Walker JA, Timoney JF (1998) Molecular basis of variation in protective SzP proteins of *Streptococcus zooepidemicus*. Am J Vet Res 59:1129–1133

Waller AS, Jolley KA (2007) Getting a grip on strangles: recent progress towards improved diagnostics and vaccines. Vet J 173:492–501

Webb K, Barker C, Harrison T, Heather Z, Steward KF, Robinson C, Newton JR, Waller AS (2012) Detection of *Streptococcus equi* subspecies equi using a triplex qPCR assay (in press)

Webb K, Jolley KA, Mitchell Z, Robinson C, Newton JR, Maiden MC, Waller A (2008) Development of an unambiguous and discriminatory multilocus sequence typing scheme for the *Streptococcus zooepidemicus* group. Microbiology 154:3016–3024

Weinstein MR, Litt M, Kertesz DA, Wyper P, Rose D, Coulter M, McGeer A, Facklam R, Ostach C, Willey BM, Borczyk A, Low DE (1997) Invasive infections due to a fish pathogen, *Streptococcus iniae*. *S. iniae* study group. N Engl J Med 337:589–594

Wertheim HF, Nghia HD, Taylor W, Schultsz C (2009a) *Streptococcus suis*: an emerging human pathogen. Clin Infect Dis 48:617–625

Wertheim HF, Nguyen HN, Taylor W, Lien TT, Ngo HT, Nguyen TQ, Nguyen BN, Nguyen HH, Nguyen HM, Nguyen CT, Dao TT, Nguyen TV, Fox A, Farrar J, Schultsz C, Nguyen HD, Nguyen KV, Horby P (2009) *Streptococcus suis*, an important cause of adult bacterial meningitis in northern Vietnam. PLoS One 4: e5973

Whatmore AM, Engler KH, Gudmundsdottir G, Efstratiou A (2001) Identification of isolates of *Streptococcus canis* infecting humans. J Clin Microbiol 39:4196–4199

Wibawan IW, Pasaribu FH, Utama IH, Abdulmawjood A, Lammler C (1999) The role of hyaluronic acid capsular material of *Streptococcus equi* subsp. zooepidemicus in mediating adherence to HeLa cells and in resisting phagocytosis. Res Vet Sci 67:131–135

Willenborg J, Fulde M, de Greeff A, Rohde M, Smith HE, Valentin-Weigand P, Goethe R (2011) Role of glucose and CcpA in capsule expression and virulence of *Streptococcus suis*. Microbiology 157:1823–1833

Williams AE, Blakemore WF, Alexander TJ (1988) A murine model of *Streptococcus suis* type 2 meningitis in the pig. Res Vet Sci 45:394–399

Winterhoff N, Goethe R, Gruening P, Rohde M, Kalisz H, Smith HE, Valentin-Weigand P (2002) Identification and characterization of two temperature-induced surface-associated proteins of *Streptococcus suis* with high homologies to members of the Arginine Deiminase system of *Streptococcus pyogenes*. J Bacteriol 184:6768–6776

Wisselink HJ, Smith HE, Stockhofe-Zurwieden N, Peperkamp K, Vecht U (2000) Distribution of capsular types and production of muramidase-released protein (MRP) and extracellular factor (EF) of *Streptococcus suis* strains isolated from diseased pigs in seven European countries. Vet Microbiol 74:237–248

Wistedt AC, Ringdahl U, Muller-Esterl W, Sjobring U (1995) Identification of a plasminogen-binding motif in PAM, a bacterial surface protein. Mol Microbiol 18:569–578

Ye C, Zheng H, Zhang J, Jing H, Wang L, Xiong Y, Wang W, Zhou Z, Sun Q, Luo X, Du H, Gottschalk M, Xu J (2009) Clinical, experimental, and genomic differences between intermediately pathogenic, highly pathogenic, and epidemic *Streptococcus suis*. J Infect Dis 199:97–107

Ye C, Zhu X, Jing H, Du H, Segura M, Zheng H, Kan B, Wang L, Bai X, Zhou Y, Cui Z, Zhang S, Jin D, Sun N, Luo X, Zhang J, Gong Z, Wang X, Wang L, Sun H, Li Z, Sun Q, Liu H, Dong B, Ke C, Yuan H, Wang H, Tian K, Wang Y, Gottschalk M, Xu J (2006) *Streptococcus suis* sequence type 7 outbreak, Sichuan China. Emerg Infect Dis 12:1203–1208

Younan M, Estoepangestie AT, Cengiz M, Alber J, El Sayed A, Lammler C (2005) Identification and molecular characterization of *Streptococcus equi* subsp. zooepidemicus isolated from camels (Camelus dromedarius) and camel milk in Kenya and Somalia. J Vet Med B Infect Dis Vet Public Health 52:142–146

Yu H, Jing H, Chen Z, Zheng H, Zhu X, Wang H, Wang S, Liu L, Zu R, Luo L, Xiang N, Liu H, Liu X, Shu Y, Lee SS, Chuang SK, Wang Y, Xu J, Yang W (2006) Human *Streptococcus suis* outbreak, Sichuan China. Emerg Infect Dis 12:914–920

Yuen KY, Seto WH, Choi CH, Ng W, Ho SW, Chau PY (1990) *Streptococcus zooepidemicus* (Lancefield group C) septicaemia in Hong Kong. J Infect 21:241–250

Zhao Y, Liu G, Li S, Wang M, Song J, Wang J, Tang J, Li M, Hu F (2011) Role of a type IV-like secretion system of *Streptococcus suis* 2 in the development of streptococcal toxic shock syndrome. J Infect Dis 204:274–281

Zlotkin A, Chilmonczyk S, Eyngor M, Hurvitz A, Ghittino C, Eldar A (2003) Trojan horse effect: phagocyte-mediated *Streptococcus iniae* infection of fish. Infect Immun 71:2318–2325

Adherence and Invasion of Streptococci to Eukaryotic Cells and Their Role in Disease Pathogenesis

Manfred Rohde and G. Singh Chhatwal

Abstract Streptococcal adhesion, invasion, intracellular trafficking, dissemination, and persistence in eukaryotic cells have a variety of implications in the infection pathogenesis. While cell adhesion establishes the initial host contact, adhering bacteria exploit the host cell for their own benefit. Internalization into the host cell is an essential step for bacterial survival and subsequent dissemination and persistence, thus playing a key role in the course of infection. This chapter summarizes the current knowledge about the diverse mechanisms of streptococcal adhesion to and invasion into different eukaryotic cells and the impact on dissemination and persistence which is reflected by consequences for the pathogenesis of streptococcal infections.

Contents

M. Rohde (✉) · G. S. Chhatwal
Department of Medical Microbiology, Helmholtz Centre for Infection Research, Braunschweig, Germany
e-mail: manfred.rohde@helmholtz-hzi.de

Current Topics in Microbiology and Immunology (2013) 368: 83–110
DOI: 10.1007/82_2012_281

Published Online: 01 December 2012

1 Introduction

Streptococci play an important role as clinically relevant pathogens worldwide. Streptococci cause approximately 700 million human infections each year, with estimated 500,000 deaths (Carapetis 2005). This clinical importance of streptococci was recognized as early as 1879 by Louis Pasteur when studying puerperal fever causing high mortality rates in pregnant women. Thus, streptococci have been recognized as one of the first microorganisms causing contagious disease. Historically, the classification of streptococci was based on Rebecca Lancefield's scheme grouping streptococcal strains according to the different carbohydrate compositions of the cell wall. Nowadays, M- and T- typing is performed or the 5'-end variability of genes for the M protein, *emm* genes, are used for molecular typing. Streptococci consist of a heterogeneous genus that comprises important human pathogens like *Streptococcus pyogenes,* also referred to as Group A Streptococci (GAS), *Streptococcus dysgalactiae* subsp. *equisimilis* as a member of Group C Streptococci (GCS) and Group G Streptococci (GGS), *S. agalactiae* also known as Group B Streptococci (GBS), as well as *S. pneumoniae* (pneumococci). Today, streptococci are also often organized into six groupings (pyogenic, anginosus, mitis, mutans, salivarius, and bovis) based on their 16S rRNA gene sequence. Members of the pyogenic group, *S. pyogenes* and *S. dysgalactiae subsp. equisimilis*, are often associated with a range of diseases including pharyngitis, tonsillitis, impetigo, mastitis, and resulting sequelae like rheumatic fever or glomerulonephritis (Henningham et al. 2012a). Streptococci are also able to cause recurrent infections such as erysipel and tonsillitis. This phenomenon has been described as the carrier stage of streptococcal infection and it is believed that streptococci have a safe ecological niche within the human body, most properly residing intracellularly in eucaryotic cells, which allow them not only to persist but also to resist antibiotic treatment. Moreover, they can lead to severe invasive infections like bacteraemia, cellulites, and necrotizing fasciitis with high mortality due to streptococcal toxic shock syndrome (STSS). This demonstrates that streptococci are able to infect otherwise sterile deep tissue in the human host. For this purpose streptococci express a highly variable and extensive repertoire of virulence determinants that are differentially regulated and expressed in response to signals from the environment within the human host. In the past 15 years more and more evidence has become available which suggests that the long time considered extracellular pathogenic bacteria of the genus *Streptococcus* adhere to host cells, and subsequently invade these cells for their own benefit, namely escaping antibiotic treatment and the host immune system (for reviews see Cunningham 2000; Nitsche-Schmitz et al. 2007; Nobbs et al. 2009; Courtney et al. 2002). This chapter will mainly focus on Group A streptococci highlighting the repertoire of adhesins and invasins of streptococci, and their impact on the colonization and invasion of the human host the persistence of streptococci within the host.

2 Adhesion of GAS

The initial and essential step of streptococcal infectious colonization of the host depends on adhesion of the pathogen to the host tissue. To allow colonization of the site of infection, streptococci express an arsenal of surface bound components, named adhesins, which establish the contact to the host's extracellular matrix (ECM) or directly to host cells. Since streptococci possess multiple adhesins, they are able to colonize various sites in the human body. Adhesion connects the bacteria firmly to the host cells and allows the pathogen to withstand host cleansing mechanisms like mechanical forces by excretion or peristalsis. Nevertheless, in some instances it might be an advantage for the bacteria to detach from a surface in case the growth conditions become unfavorable. Thus, adhesion can sometimes also be considered as being a dynamic process. Most host surfaces are covered by a protective epithelial or endothelial cell layer which is covered with extracellular matrix proteins like collagen, fibrinogen, laminin, vitronectin, or fibronectin (Cremer et al. 1998; Debelle and Tamburro 1999; Schvartz et al. 1999; Pankov and Yamada 2002; Dempfle and Mosesson 2003). Many streptococcal adhesins function by specifically binding to various compounds of the extracellular matrix. Since streptococci express multiple adhesins, it is most likely that different adhesins are expressed under the different environmental conditions encountered throughout an infection. Once adhesion has been established, streptococci can multiply extracellularly, establish colonization of the infection side, internalize into host cells, traffic intracellularly in the host cell and multiply within the cell, or escape from the host cell, or use the host cell as a Trojan horse for systemic spread of the infection.

2.1 Cell Wall-Anchored Adhesins

Adhesins can be attached in four different ways to the streptococcal surface (i) covalently linked through the C-terminus to the cell wall peptidoglycan through an LPxTz motif, (ii) tethered to the cell membrane through N-terminal modifications with lipid, (iii) retained on the surface by a yet unknown mechanism, or (iv) bound back to the cell surface through noncovalent interactions with cell surface compounds like polysaccharides or other proteins. Most adhesins belong to the family of cell wall-anchored proteins. A prerequisite for those proteins is a functioning membrane-associated transpeptidase, sortase A (Maraffini et al. 2006). Many streptococcal adhesins function by specifically recognizing and binding to the various components found in the ECM of the host. The ECM is the major support of cells and tissue and it is responsible for maintaining strength and elasticity of the body. Thus, it is ubiquitously present and frequently exposed under conditions such as trauma and injury, making the then exposed components an ideal prime target for streptococcal adhesion. In addition to the adhesins, other surface structures of streptococci like fimbriae, pili, lipoteichoic acid, and polysaccharides may also play an important role during the first steps of adhesion to the host cell.

2.1.1 Hyaluronic Acid Capsule and Lipoteichoic Acid

Many of the β-hemolytic streptococci are producing a polysaccharide capsule, consisting of hyaluronic acid (HA), a glycosaminoglycan that is a linear polymer of alternating monosaccharide units of *N*-acetylglucosamine and glucuronic acid. Streptococci show differently expressed HA capsules. While some strains are barely encapsulated when grown ex vivo on agar plates, others produce high amounts of the HA capsule that confers a mucoid appearance to the bacterial colonies that were grown on solid media (Wilson 1959). When compared to the host HA, the streptococcal HA is chemically very similar. The host HA is a widely distributed and abundant component of the host's ECM. The chemical similarity might be the reason for the low immunogenicity of the streptococcal HA capsule in the infected patients. It is assumed that the HA capsule masks the bacteria and therefore provides protection against the host immune system. An epidemiological investigation demonstrated that encapsulated strains can often be isolated from severe invasive infections, whereas isolates from uncomplicated pharyngitis were only rarely of mucoid character. A total of 42 % of isolates from patients with acute rheumatic fever were found to be of the mucoid type (Johnson et al. 1992). In addition, the HA capsule can act as a nonprotein adhesin by binding to the hyaluronic acid receptor CD44 on keratinocytes (Schrager et al. 1998), and the HA capsule serves as a ligand for collagen IV (Dinkla et al. 2003a). These studies highlight that the HA capsule is not only a nonprotein adhesin, but can also be described as a virulence factor associated with invasive infections and their severe sequelae like acute rheumatic heart disease. Studies in the US concluded that outbreaks of acute rheumatic fever were associated with the spread of mainly M-type 18 highly encapsulated strains (Veasy et al. 2004; Kaplan et al. 1989). In addition, the contribution of the HA capsule for pharyngeal colonization was also demonstrated (Wessels et al. 1991). Most noteworthy, mucoid strains seem to help to breach epithelial cell layers for entering underlying sterile tissue in the process of dissemination in the host. Binding of the HA capsule to CD44 leads to cytoskeletal rearrangements in human epithelial cells (Schrager et al. 1998; Cywes et al. 2000) that cause disruption of intracellular junctions and allow the passage into deeper tissue. Streptococci therefore travel paracellularly through an epithelial cell barrier, such as keratinocytes, instead of an intracellular passage after internalization into the epithelial host cells (Cywes and Wessels 2001).

Evidence has been acquired demonstrating that the HA capsule is not always representing a nonprotein adhesin, since it was demonstrated that the HA capsule can also counteract internalization into human epithelial cells and keratinocytes (Jadoun et al. 2002; Darmstadt et al. 2000; Schrager et al. 1996). The observed decreased internalization is accompanied by an increase in tissue damage as observed in a mouse model (Schrager et al. 1996). The increased virulence was also attributed to an antiphagocytic effect of the HA capsule (Wessels et al. 1991, 1994; Wessels and Bronze 1994). The HA capsule was reported to decrease the association with PMNs, thereby counteracting phagocytosis (Dale et al. 1996).

Lipoteichoic acid (LTA) is composed of chain-like glycerol phosphate polymers that are covalently coupled to glycolipids of the plasma membrane and represent a component of the cell wall. LTA is thought to allow streptococci to come into first contact with the ECM or directly with the host cell. A two-step adhesion mechanism has been postulated. LTA has the role of a first-step adhesive component with low cell type selectivity (Beachey and Ofek 1976; Leon and Panos 1990; Courtney and Hasty 1991; Courtney et al. 1992), whereas the highly abundant extracellular ECM protein fibronectin functions as a ligand or "bridging-molecule" for LTA mediating the initial streptococcal cell adhesion (Simpson and Beachey 1983). For establishing a firm adhesion to host cells, a second step with high avidity and cell specificity is required (Courtney et al. 1992; Hasty et al. 1992). This second step has a crucial influence on the observed tissue tropism of streptococcal infection and the pathway of dissemination in the host. More importantly, interactions of the proposed second step were shown to promote internalization of the bacteria into eukaryotic cells (LaPenta et al. 1994, Greco et al. 1995). The involvement of LTA in adhesion and invasion during streptococcal pathogenesis was recently demonstrated. In a cell culture model consisting of human brain microvascular endothelial cells, Group B streptococci (GBS), the leading pathogen in neonatal meningitis, with an impaired anchoring of LTA in the cell wall due to an inactivated glycosyltransferase gene showed a decreased invasion capability. The mutant strain exhibited less virulence in the meningitis mouse model of infection (Doran et al. 2005). Although LTA promotes adhesion in a nonselective way, LTA adhesion might contribute to the onset of the subsequent infection pathway.

2.1.2 Fimbrious Structures Like Pili

Examination of negatively stained streptococci reveals in some groups filamentous structures like fibrils and pili or fimbriae. Fibrils have been detected especially in oral streptococci with a peritrich or specifically localized structure on the cell wall. These structures have been related to growth and survival of oral streptococci since all fresh isolates express fibrils (Handley et al. 1984, 1987).

For many years, bacteriologists had evidence that Gram-negative bacteria adhere to host cells and pili, also sometimes referred to as fimbriae, serving as the primary colonization factor in the adhesion process. Pili are long hair-like extensions of the cell surface and can best be made visible by negative-staining of the bacteria and by electron microscopic observation. Although pili in Gram-positive bacteria were described first in 1968 in Corynebacteria and in the 1990s in streptococci (Yanagawa et al. 1968; Yanagawa and Honda 1976; Wu and Fives-Taylor 1999), they were more or less neglected in research in the years to follow. Just recently, pili in all three pathogenic streptococcal species causing invasive disease, *S. pyogenes, S. agalactiae,* and *S. pneumonia*, have been described (Mora et al. 2005; Lauer et al. 2005; Rosini et al. 2006; Barocchi et al. 2006). Since then, pili have come into the research focus and the newly detected pili in streptococci have been considered as most promising candidates for vaccine development against streptococci (Gianfoldoni et al. 2007).

In Gram-negative pathogens, pili consist of multiple subunits which are non-covalently attached to each other and the adhesin molecule is located exclusively at the tip of the pili. There are remarkable differences in the pili assembly in Gram-negative and Gram-positive bacteria. Whereas in Gram-negative bacteria the Sec-dependent secretion and chaperones are involved in the formation of the pili subunits (backbone) and the tip adhesin subunit (the most common assembly way is the chaperone/usher pathway) (Saulino et al. 2000), in Gram-positive bacteria pili subunits are covalently linked to each other and pili are covalently attached to the peptidoglycan. As in Gram-negative bacteria, the Sec-dependent machinery is also involved, but the polymerization of the subunit is controlled by an enzyme with homologies to sortases, called sortase A, which sorts and attaches proteins covalently to the peptidoglycan. For covalent linkage to the peptidoglycan, proteins exhibit a specific motif, the cell wall sorting signal (CWSS), in the C-terminus of the protein. The main characteristic is the LPxTG motif, a sequence that is conserved in all surface proteins anchored by sortase A. The LPxTG motive is followed by a hydrophobic membrane-spanning domain and a positively charged tail which is needed for the sortase A catalyzed anchoring with the peptidoglycan. For anchoring a surface protein to the cell wall, sortase A cleaves the LPxTG motif between threonine (T) and glycine (G), and the threonine residue of the cleaved peptide is covalently linked to the amino group of the cross-bridge within the peptidoglycan (Schneewind et al. 1993; Ton-That et al. 2004; Telford et al. 2006; Scott and Zähner 2006; Mandlik et al. 2007a). During the polymerization process, self-generated intramolecular isopeptide bonds are formed which give additional stabilization to the thin pilus structure and allow to withstand the tensile forces during the first steps in adhesion to the host cells. Remarkably, similar isopeptide bonds have been found in other surface displayed adhesins (Kang et al. 2007). In addition, genes encoding transpeptidases of the sortase C subfamily have been found functioning in polymerization of the subunits (Proft and Baker 2009).

For all streptococci it was found that the genes encoding for pilus proteins are located on pathogenicity islands (PI) and clustered at the same genetic locus in an operon. Sortase genes are in close proximity to the pili genes, implying that they might also belong to the operon. Up to now, nine PIs have been identified in GAS. Remarkably, the genes in GAS occur in the fibronectin binding, collagen binding, T-antigen region of the chromosome (FCT-region). Interestingly, this locus actually encodes the Lancefield T-antigen which has been used for serotype typing (Falugi et al. 2008). Mora et al. (2005) first identified that the T-antigen of serotyping actually represents the shaft of the pilus in a set of four GAS strains of different serotypes. Immune negative-staining with an antibody against the T-antigen revealed staining alongside the entire shaft of the pilus, but no labeling was detected at the pilus tip. The tip region was only stained by an antibody against the presumed adhesin. Thus, the pilus shaft represents the T-antigen used by R. Lancefield for serotyping. Additionally, a third component was identified in Gram-positive pili which is added at intervals into the shaft region of the pilus, the AP1 (ancillary protein 1) component, representing a collagen binding protein moiety (Podbielski et al. 1999). This observation confirms that pili are involved in the adhesion process especially with components of the ECM. Since pili were detected in streptococci, more evidence has been accumulating that pili are involved in the adhesion process, and in fact pili might actually mediate the first contact with the host cell. Subsequently, it was demonstrated that pili are involved in the adhesion process to a broad range of host epithelial cells including cells from the nasopharynx, tonsils, lung, cervix, and intestine (Abbot et al. 2007; Crotty et al. 2010). Remarkably, for GBS it was found that despite being involved in the adhesion process, pili in GBS are responsible for a paracellular passage through an epithelial cell barrier as well as for promoting invasion into brain microvascular endothelial cells (Maisey et al. 2007; Pezzicoli et al. 2008). Recent pili research has highlighted that pilus protein components of invasive streptococci have amino acid sequences similar to those of the members of the microbial surface components recognizing adhesive matrix molecules (MSCRAMMs) which have been shown to interact with the ECM proteins like fibronectin, fibrinogen, collagen, laminin, and vitronectin (Patti and Höök 1994). Pili have also been shown to be responsible for tissue tropism during infections. Besides the long stretched extended pili, the so-called minor pilins have been identified in *Corynebacterium* by mutation analyses of pili. For these minor pilins, it has been demonstrated that they are displayed on the cell surface. Recombinant protein, when coated onto inert latex beads, clearly exhibited binding only to pharyngeal cells, but no binding resulted to cells of lung or laryngeal origin when tested in a cell culturing model (Mandlik et al. 2007b). In addition, the role of minor pilins in streptococcal tissue tropism has also been elucidated. These minor pilins mediate adhesion of GAS to human tonsil epithelium and primary keratinocytes, the prime colonization targets of GAS (Abbott et al. 2007).

From these observations, a putative model of cell adhesion through pili has been deduced (Telford et al. 2006). The first contact with the ECM of the host or directly with the host cell surface is mediated by the extended pili with the presumed adhesin (AC2-protein) on the tips of the pili. This interaction might be reversible to allow streptococci to find the specific tissue (cell tropism). Once the first contact has been established, the Ap1-protein can interact with collagen to attach the bacterium more firmly to the host cell. Due to this interaction, the bacterium comes closer to the host cell surface (in Gram-negative bacteria it has been shown that after the first attachment pili are retracted to get the bacteria closer to the cell surface) and cell surface-anchored adhesins can establish the intimate contact with the host cell and subsequently colonize the cell or tissue. The intimate contact favors three different functions which are needed to progress with the infection: (i) the proximity of the bacterial cell wall with the host cell membrane allows further ligand–receptor interactions, (ii) the close contact favors the efficient delivery of virulence factors as for example for the type three secretion system, TTSS, in Gram-negative bacteria, and (iii) the intimate adherence between streptococci and host cell membrane is the prerequisite for an successful invasion of the host.

2.1.3 M Proteins

M proteins were the first reported adhesins of GAS (Ellen and Gibbons 1972, 1974) and represent multifunctional virulence factors and adhesins of the streptococcal surface. Although structurally closely related, M proteins represent a heterogeneous group of adhesins with respect to their ligand molecules or target cells (Berkower et al. 1999). Some M proteins exhibit very distinct binding properties, whereas the interaction with glycosaminoglycanes represents a more common adherence mechanism displayed by several M protein serotypes. These interactions are mediated via the conserved C-terminal part of M protein (Frick et al. 2003). M proteins are hypervariable in the N-terminal region of the protein and today more than 200 emm-serotypes can be distinguished which differ in their adhesion capabilities (Beall et al. 1996). M proteins consist of four repeat regions with the A repeats region being hypervariable and the B repeat regions being semivariable. Structural-based studies revealed that M proteins are entirely α-helical coiled-coil dimers and can be detected on the bacterial surface as hair-like projections in ultra-thin sections. The N-terminal region is further stabilized by antiparallel interactions, which is thought to result in bacterial aggregation (Frick et al. 2000; Oemcke et al. 2010). The hypervariable regions of M proteins bind to a wide range of different human proteins including plasminogen, IgA, IgG, factor H, and C4b-binding protein (C4BP) which inhibit complement activation (Andre et al. 2006, McArthur and Walker 2006). The B repeats can bind to fibrinogen, human serum albumin, and IgG. For the ability to bind fibrinogen, it was shown that structural irregularities and instabilities in the coiled-coil structure of M1 protein facilitate binding to fibrinogen (McNamara et al. 2008). In addition, serotype M6 protein can bind to the membrane-bound cofactor protein CD46 on keratinocytes. The C-terminal region of M6 protein as well as the short consensus domains 3 and 4 of CD46 was shown to be crucial for M6/CD46-mediated keratinocyte attachment (Okada et al. 1995; Giannakis et al. 2002). The direct involvement of M proteins in cellular adhesion has been demonstrated for several M protein serotypes. For example, for M1 protein, it was shown that the ECM protein fibronectin is necessary for efficient adhesion to epithelial HeLa cells. As the cellular receptor for fibronectin $\alpha_5\beta_1$ integrins have been identified. An M1 protein mutant showed significant reduced adhesion which could be blocked by specific M1 protein antibodies (Cue et al. 2000, 2001). Similar results were observed for the M24 serotype protein in which the serotype M24 protein mediated adhesion to HEp-2 cells, whereas the M24 protein deficient mutant was unable to bind to HEp-2 cells but was able to adhere to buccal cells demonstrating the cell type specificity (cell tropism) of the M24 protein. The mutant studies also revealed that another adhesin or adhesins other than M24 protein must be displayed on the bacterial surface (Courtney et al. 1994). Using the cell culture approach, other M proteins have been tested for their involvement in adhesion to epithelial cells. For example, for the M6 protein it was shown that M6 protein does not contribute to the adhesion to buccal and tonsillar epithelial cells. In these studies, again the involvement of fibronectin in the adhesion process was demonstrated (Caparon et al. 1991). The M3 protein serves as an important adhesin for soluble type I and type IV collagen as well as to native collagen matrices in the host. The N-terminal variable region, though specific for the M3 protein, was identified as the collagen binding region. This explains why all other M proteins lack the collagen binding properties of the M3 protein. Besides direct collagen binding, M3 protein is responsible for aggregation of soluble collagen, resulting in large aggregates of streptococci that are involved in the colonization process. Due to this aggregate formation streptococci can better resist phagocytosis and antibiotic treatment (Dinkla et al. 2003b). *S. pyogenes* expresses only one collagen binding protein, Cpa, which was identified in a serotype M49 isolate and was shown to mediate attachment to immobilized type I collagen (Podbielski et al. 1999). Nevertheless, the potential role of collagen binding by serotype M3 protein in the adhesion to host tissue has to be demonstrated. That collagen can play a role in cell adhesion can be deduced from investigations on GGS. In strains of this group, the M-like protein FOG is expressed which has a high affinity for collagen I and allows GGS to bind to human foreskin fibroblasts (Nitsche et al. 2006). Since it is known that two distinct collagen-like proteins, termed SclA/Scl1 and SclB/Scl2, can bind to pharyngeal and fibroblast cells reacting with the $\alpha 2$-domain of $\alpha_2\beta_1$-integrin, and thereby triggering cellular signaling in lung fibroblasts, the conclusion could be draw that GAS employs collagen binding integrins to adhere to host cells (Humtsoe et al. 2005).

Taken together, the role of M protein-mediated adhesion varies obviously with the M-serotype and represents a concerted action of the different ligand binding properties and sites within the M protein. We are just beginning to understand parts of the complex cell adhesion interactions mediated by the M proteins with respect to ligand–receptor binding. Once this process is fully understood, one might gain an insight into the so far mostly unknown interactions in streptococcal cell tropism and the onset of the cell tropism on the development of the disease.

2.1.4 Fibronectin-Binding Proteins

Fibronectin is a large glycoprotein which exists both as a soluble protein in plasma and as a fibrillar polymer in the ECM. Fibronectin is composed as a dimer of two 250 kDa subunits which are C-terminal linked via disulfide bonds. Each subunit exhibits three distinct modules, the type I, II, and III modules. Fibronectin interacts with host cell surface exposed integrins of which the $\alpha_5\beta_1$ integrin is the classical fibronectin-binding integrin. Integrin binding is mediated through the RGD sequence within the fibronectin subunits (Pankov and Yamada 2002). Fibronectin can be seen as one of the prime targets in streptococcal adhesion and subsequent internalization when exploring the ECM during first steps of adhesion. Fibronectin acts as a bridging molecule, connecting bacterial adhesins with integrins on the surface of eukaryotic cells. All streptococci express fibronectin-binding proteins, but the proteins show differences in the fibronectin-binding capacities. Some serotype strains bind soluble fibronectin with high affinities (in the nanomolar range), whereas other strains can only adhere to immobilized fibronectin for a successful adherence. *S. pyogenes* possesses at least 11 distinct fibronectin-binding adhesins including SfbI/F1, protein F2, serum opacity factor (SOF), FbaA, and several M proteins. Some of these are present in a large number of serotypes, such as SfbI protein or FBP54, whereas others like the M1 or M3 proteins are exclusively expressed by M1 or M3 serotype streptococci (Talay et al. 1992, 1994; Hanski and Caparon 1992; Sela et al. 1993; Natanson et al. 1995; Ozeri et al. 1996; Cue et al. 2000; Kreikemeyer et al. 2004a). The expression of these fibronectin-binding proteins is highly regulated in response to environmental signals. For example, SfbI expression is regulated by a superoxide signal. At high partial pressures of O_2, expression of SfbI is increased while expression of M protein is upregulated in a CO_2-rich environment, which suggests that SfbI protein is expressed at the prime target sites of *S. pyogenes* adhesion, namely in the O_2-rich environment of the respiratory tract and skin, where it is required for host adhesion and colonization. Whereas in deeper infections protection against the immune system becomes more important, the increased partial pressure of CO_2 upregulates M protein (Gibson et al. 1995; Gibson and Caparon 1996; Kreikemeyer et al. 2003, 2004b).

Among all fibronectin-binding adhesins of *S. pyogenes*, SfbI protein and its allelic variant F1 are studied most extensively. Identified in 1992, SfbI/F1 was shown to act as an adhesin on epithelial cells (Talay et al. 1992; Hanski and Caparon 1992). SfbI protein exhibits a modular architecture with an aromatic domain, rich in aromatic amino acids, ARO, at the N-terminus, a proline-rich repeat region (PRR) in the middle of the molecule and a fibronectin-binding repeat region (FnBR) at the C-terminus as major modules (Talay et al. 1994). SfbI protein is very variable in the number of repeats in the PRR and FnBR regions. One study revealed that 34 distinct alleles of SfbI were found in 54 *S. pyogenes* strains as a result of horizontal gene transfer (Towers et al. 2003). The ARO region showed a high degree of sequence variability, whereas deletion or duplication of repeat units in PRR and FnBR resulted in variable numbers of PRR (1–11 repeats) and FnBR (1–5 repeats). Thus, SfbI protein exhibits a high antigenic variation, which might also reflect possible variable functional capabilities.

Binding to fibronectin is mediated by two distinct domains (Sela et al. 1993; Ozeri et al. 1996) and proceeds in the following way: the C-terminal fibronectin-binding repeat region and the adjacent nonrepetitive domain termed spacer 2 or UR synergistically bind to two distinct regions on the fibronectin molecule: the N-terminal fibrin-binding fragment harboring fibronectin F1 modules 1–5, and the gelatine/collagen-binding fragment harboring F1 modules 6–9 and the two F2 modules. Due to this cooperative binding, the RGD region in fibronectin gets exposed and can bind to integrins (Talay et al. 2000).

The molecular interaction between SfbI and fibronectin is based on three-dimensional structural data. Three-dimensional structure data are available describing a bacterial fibronectin-binding fragment of *S. dysgalactiae* bound to its target (Schwarz-Linek et al. 2003, 2004a, b). Since SfbI shows high similarity in respect to structural composition, conclusion can be drawn on the binding mechanism for SfbI and fibronectin. That this can indeed be observed was recently shown by Marjenberg et al. (2011) by applying microcalorimetry to reveal cooperative binding of fibronectin fragments to arrays of binding in SfbI (Marjenberg et al. 2011). The so far existing structural model for fibronectin-binding proteins is comprehensively reviewed by Schwarz-Linek et al. (2004a, b). Briefly, SfbI and fibronectin bind to each other in an antiparallel fashion. The C-terminal FnBR in SfbI recognizes the N-terminal domain of fibronectin with high specificity and high affinity (in the nanomolar range) by forming a novel protein–protein interaction mechanism, termed tandem ß-zipper. According to the tandem ß-zipper model, FnBRs can bind multiple copies of fibronectin. For SfbI from a *S. pyogenes* strain, it was demonstrated that a single SfbI molecule can bind up to five fibronectin molecules. The observed high affinity is considerably important since high-affinity binding is a prerequisite for firm bacterial attachment, because adherent streptococci have to withstand shear forces occurring on the mucosal surfaces or during the internalization process itself.

The C-terminal fibronectin-binding repeat region of SfbI was demonstrated to be sufficient to mediate adherence to epithelial cells (Talay et al. 2000). Several studies revealed that SfbI mediates attachment to epithelial cells of the oral mucosa and the lung, but also to endothelial cells (Rohde et al. 2003). Besides its potential to bind to cell exposed integrins, SfbI has the ability to recruit collagen via prebound fibronectin. This allows *S. pyogenes* to form aggregates and renders the organism capable of colonizing collagen matrices within the body (Dinkla et al. 2003a) as depicted in Fig. 1.

Protein F2 or PFBP are homologous but distinct fibronectin-binding proteins, found in most isolates of *S. pyogenes* lacking the *sfbI/prtF1* gene (Jaffe et al. 1996; Rocha and Fischetti 1999; Kreikemeyer et al. 2004a). Protein F2 possesses two binding domains that interact with fibronectin. The most abundant fibronectin-binding protein found in all *S. pyogenes* isolates is FBP54. Although lacking a classical membrane anchor motif of Gram-positive surface proteins, it seems to be localized on the streptococcal surface by a distinct mechanism, thereby acting as an adhesin for buccal epithelial cells but not HEp2 cells (Courtney et al. 1996; Chhatwal 2002). In summary, this data suggests that distinct fibronectin-binding proteins target different cell types, and therefore might contribute to streptococcal cell tropism.

Streptococci can express two other fibronectin-binding proteins, Fba and FbaB. The *fba* gene was found only in five serotypes of *S. pyogenes* including serotype M1 and M49. An Fba mutant showed reduced adhesion to HEp2 cells, suggesting that this protein has adhesive properties. Interestingly, FbaB protein was only found in serotype M3/M18 *S. pyogenes* isolates and appears to be genetically most closely related to protein F2 (Terao et al. 2001, 2002). FbaB protein from serotype M3 *S. pyogenes* exhibits cell tropism since it mediates adhesion only to endothelial and not to epithelial cells (Amelung et al. 2011).

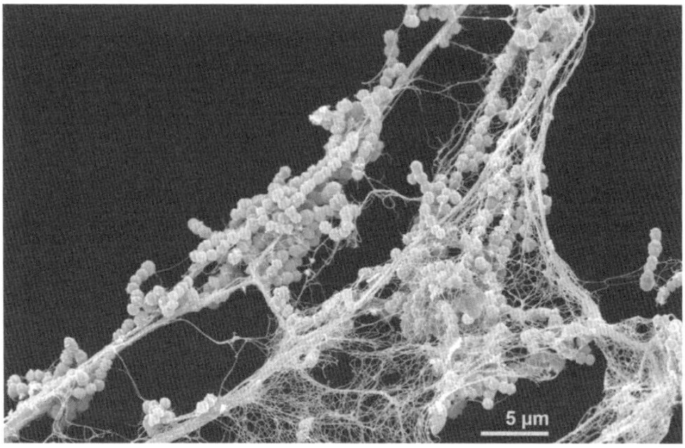

Fig. 1 Field emission scanning electron microscopic (FESEM) image of direct binding of serotype M3-type *Streptococcus pyogenes* to mouse collagen I

Another fibronectin-binding protein exposed on the bacterial surface is Protein H, a member of the M protein family. Protein H binds to the type III modules of fibronectin, unlike the so far described proteins that mainly interact with the type I or type II module containing domains of fibronectin. In addition, Protein H was shown to mediate streptococcal aggregation through a so called AHP sequence that also promoted adhesion to epithelial cells (Frick et al. 1995, 2000). M1 protein, another member of the M protein family, also binds to fibronectin and as in the case with SfbI, $\alpha_5\beta_1$ integrins are the terminal receptor proteins on the cellular surface. M1 specific antibodies efficiently blocked adherence to epithelial HeLa cells. Moreover, an M1-deficient mutant showed reduced adherence indicating that M1 protein acts as an adhesin in serotype M1 *S. pyogenes* strains (Cue et al. 1998, 2000; Dombek et al. 1999).

2.2 Anchorless Adhesins

Most streptococcal surface proteins are attached through the C-termini via an LPxTz motif to the peptidoglycan of the cell wall. In recent years, a few proteins have been identified that bind to the cell surface without the LPxTz motif- the so called anchorless proteins. Noteworthy, these proteins also lack the N-terminal signal sequence. How these proteins are exported from the cytoplasm to the cell surface is not understood and remains an enigma as well as the attachment mechanism to the cell wall. Since the anchorless adhesins do not contain choline-binding repeats and can be removed from the bacterial surface with chaotropic agents, the interaction seems to be of hydrophobic nature, van der Waals forces, or of less-defined charges (Derbise et al. 2004). Anchorless adhesins are structurally and functionally very diverse and bind to different ligands. Most important is the fact that many anchorless adhesins display an enzymatic function mostly in the cytoplasm of the bacterium, especially in the glycolytic cycle. Triosephosphate isomerase, glyceraldehyde 3-phosphate dehydrogenase (GAPDH), 3-phosphoglycerate kinase, 3-phosphoglycerate mutase, and α-enolase are the last five enzymes in the glycolytic cycle and all have been found to be associated with the cell wall surface. Up to date, from these GAPDH and α-enolase are the best studied examples (Pancholi and Fischetti 1992, 1998; Kinnby et al. 2008). GAPDH can be detected in several species of streptococcus and targets multiple matrix proteins like fibronectin, fibrinogen, plasmin, and plasminogen. It is discussed that GAPDH contributes significantly to the colonization capabilities of GAS. Surprisingly, GAPDH seems also to be involved in interactions with the cytoskeletal proteins actin and myosin and with urokinase plasminogen activator receptor (Pancholi and Fischetti, 1992; Seifert et al. 2003; Jin et al. 2005). The above described anchorless adhesins have another feature in common when they function as a complex in the glycolytic cycle namely generation of ATP. Thus, the anchorless adhesins allow streptococci to produce extracellular ATP on the surface. It can be speculated that streptococci might use this ATP for host modulatory effects, since it is known that ATP is able to bind to $P2X_7$ receptors on immune cells and can inhibit apoptosis. Streptococci seem to have evolved a system to manipulate the host immune cells for the benefit of a better progression of the infection after the first steps of adhesion to the host tissue (Yilmaz et al. 2008).

α-enolase serves as the major plasmin and plasminogen-binding protein of streptococci. After binding plasminogen, it can be converted into plasmin by plasminogen activators. Plasmin represents a very potent serine protease which degrades ECM matrix proteins to allow streptococci to gain access into deeper tissue. It is also discussed if plasmin is involved in facilitating bacterial dissemination through epithelial or endothelial barriers. Other members of anchorless adhesins are GtfG, SpeB, HtrA, Sib35, and FBP54. GtfG, SpeB, and Sib35 are secreted proteins which bind back to the cell wall surface (Hytonen et al. 2001; Kawabat et al. 2002). Other anchorless proteins like ADI, TF, and FBA were found by a post-proteomic approach. For ADI and TF, it was shown that both proteins elicited a protective immune response against intraperitoneal challenge with a serotype M1 isolate (Henningham et al. 2012b).

In summary, anchorless adhesins allow streptococci more flexibility compared to cell wall-linked adhesins since the anchorless proteins can be released and then bound back to the cell wall surface for better exploiting the environment by interacting with host molecules in a certain distance away from the bacterium. The question whether the glycolytic enzymes produce extracellular ATP or not and the function of extracellular ATP in the infection process has to be answered in further studies.

3 Invasion of Streptococci into Eukaryotic Cells

As outlined above, receptor recognition via ECM proteins as bridging molecules for adhesins is essential for a successful adherence to host tissue or cells. During the last two decades, it has become clear that after adhesion a subsequent process is triggered by the firmly attached streptococci, namely internalization into host cells. As mentioned, integrins play an important role in adhesion to the ECM proteins bound by streptococci and the integrin signaling pathways are then exploited by the bacteria to promote their own uptake by an outside–inside signal.

In 1994, LaPenta and colleagues first demonstrated in cell culture infection models that GAS can invade into nonphagocytic human epithelial cells at frequencies equal or greater than the classical intracellular pathogens such as *Listeria* or *Salmonella* (LaPenta et al. 1994). These experiments revealed that the long standing view of streptococci as being only extracellular human pathogenic bacteria has to be changed. These cell culture model studies were followed by the observation that excised tonsils from patients with recurrent tonsillitis contained viable streptococci (Oesterlund and Engstrand 1997; Oesterlund et al. 1997). Furthermore, it was demonstrated that besides GAS, other streptococci also invade nonphagocytic epithelial host cells such as Group B, Group C, and Group G streptococci, *S. suis* as well as *S. pneumoniae* and oral streptococci of the viridans group (Rubens et al. 1992; Talbot et al. 1996; Norton et al. 1999; Molinari and Chhatwal 1999; Haidan et al. 2000; Stinson et al. 2003).

Scanning electron microscopy (SEM) imaging revealed that streptococci use multiple morphological ways of invasion. This became evident when different serotypes with most probably different adhesins were tested for invasion. All invasion mechanism known so far of the classical intracellular bacteria could be detected. Some isolates exhibited the classical membrane-ruffling pattern (triggering mechanism) as described for *Salmonella*, whereas other isolates showed a well-defined zipper-like mechanism found in invading Listeria. Remarkably, a high number of streptococcal isolates expressing SfbI protein induced a third, so far unknown invasion pathway, with an easily detectable morphological feature, namely the formation of large invaginations during the internalization process, sometimes looking like a "hole" in the host cell membrane (Molinari et al. 2000). Interestingly, these different invasion patterns were also detected for Group C and Group G streptococci (Haidan et al. 2000) as well as for nonencapsulated strains of *S. suis* (Benga et al. 2004) and for *S. aureus* which also invades via $\alpha 5\beta 1$ integrins (Agerer et al. 2005).

These studies revealed that streptococci also possess invasins, like the classical intracellular pathogens *Shigella*, *Listeria,* or *Salmonella.* As for the classical intracellular bacteria, streptococcal invasins aid bacteria in their internalization process into nonphagocytic host cells like fibroblasts, keratinocytes, endothelial, and epithelial cells. Invasins are surface exposed and/or diffusible proteins that can promote (i) actin rearrangement of the host cytoskeleton that stimulates engulfment of the bacteria by membrane ruffles (Dombek et al. 1999), or (ii) interact with specific host cell receptors triggering signaling events which result in uptake of streptococci (Rezcallah et al. 2005; Ozeri et al. 1998), or (iii) co-opt host cell endocytotic pathways, caveolae, for their own internalization (Rohde et al. 2003).

The two streptococcal invasins are the streptococcal fibronectin-binding protein SfbI protein or F1 and the M proteins which are widely distributed in clinical isolates. Despite being only adhesins, these two proteins act also as invasins. SfbI/F1 protein, for example, is detectable in nearly 70–85 % of all clinical isolates and the M protein provides the basis of the Lancefield's method of *emm* gene serotyping. As mentioned before, both proteins bind to fibronectin which acts as a bridging molecule for mediating adhesion to the host cell receptor. Another member of the invasin group of proteins is the surface exposed streptococcal dehydrogenase (SDH) which is involved in the invasion of M6 serotype streptococci. SDH triggers signaling events resulting in phosphorylation of several eukaryotic proteins (Pancholi and Fischetti 1997).

3.1 SfbI/F1 Invasion of Epithelial Cells

The work of Talay et al. (2000) explained the molecular mechanism of the binding properties of the distinct domains of SfbI protein with regard to fibronectin and dissected the adherence from the invasion process. The cooperative interaction, further explained by the tandem ß-zipper mechanism (Schwarz-Linek et al. 2003), subsequently resulted in exposure of the RGD region of the fibronectin molecule allowing to bind to $\alpha_5\beta_1$ integrins on the host cell surface. The interaction between the RGD region of fibronectin and integrins could be blocked by antibodies against the ß-subunit or with RGD peptides (Jadoun et al. 1998; Ozeri et al. 1998; Molinari et al. 2000). Ozeri et al. also demonstrated that the amount of bound fibronectin on the bacterial surface is responsible for the uptake efficiency into the host cell. Most properly, a certain threshold of bound fibronectin to integrins is needed to start the integrin signaling. According to the tandem ß-zipper mechanism, a single SfbI molecule is able to bind up to five fibronectin molecules. This means that the surface of SfbI-expressing streptococci is most properly covered by a high density of fibronectin. This allows streptococci to bind many integrins on the cell. A clustering of integrins underneath attached streptococci could be demonstrated by staining the ß-subunits of integrins or by a microscopic approach applying high-resolution field emission scanning electron microscopy (FESEM) to visualize integrin clustering directly on the host cell surface. Recombinant SfbI protein was coated onto 15 nm colloidal gold particles and the fate of the SfbI-gold particles on endothelial cells was observed. After 1 h, aggregates of SfbI-gold particles could be detected, meaning integrin clustering, and after 2 h SfbI-gold particle aggregates were taken up by invaginations into the cell comparable to the internalization of the SfbI-expressing *S. pyogenes isolate* (Rohde et al. 2003). This observation is in accordance with findings for other pathogens which also enter cells via integrins like *Neisseria gonorrhoeae* (van Putten et al. 1998), *S. aureus*

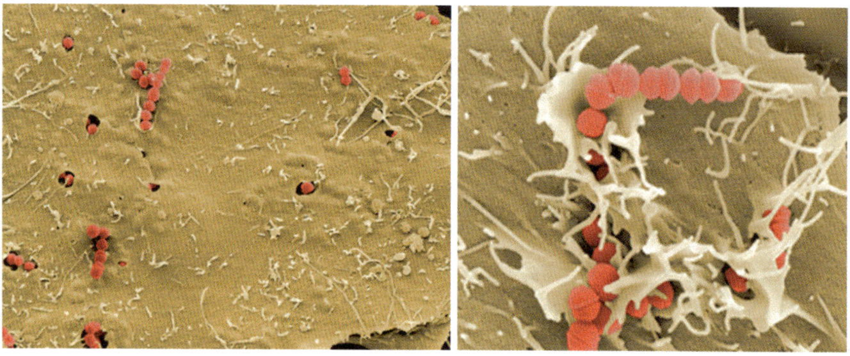

Fig. 2 FESEM image of invading Group A streptococci via SfbI-mediated fibronectin binding, integrin clustering, and formation of invaginations in the host epithelial cell (*left* image). In a mutant strain lacking SfbI protein, streptococci invade via induction of cytoskeletal rearrangements, membrane ruffles (*right* image) giving evidence that a single strain has the potential to invade via two morphological distinct pathways depending on the protein repertoire expressed on the surface

(Fowler et al. 2000), and *Yersiniae* species (Isberg and Barnes 2001). Engagement and clustering of integrin receptors is the prerequisite for sufficient uptake and subsequent integrin signaling (Isberg and Leong 1990; Isberg 1991) (Fig. 2).

In further studies, it was shown that formation of large invaginations is triggered by the SfbI-fibronectin-integrin signaling complex which is responsible for the aggregation of caveolae around adherent bacteria. Caveolae fuse with each other to form the invagination. Once inside the host cell, SfbI-expressing streptococci traffic in a new compartment called "caveosome". The intriguing aspect of this type of streptococcal invasion is the fact that caveosomes do not fuse with lysosomes. By co-opting this caveolae-mediated cellular pathway, SfbI-carrying streptococci bypass the degradation machinery of the host cell, since no fusion with lysosomes is detectable (Rohde et al. 2003).

GGS exhibit GfbA, another adhesin and invasin, on their surface which binds equal amounts of fibronectin as SfbI and interacts with $\alpha_5\beta_1$ integrins. But GfbA-mediated invasion leads to the formation of membrane ruffles, i.e. rearrangement of the host cell cytoskeleton. Immunofluorescence studies demonstrated that GfbA-expressing streptococci follow the classical endocytic pathway with subsequent fusion with lysosomes to form phagolysosomes. By heterologous surface expression of GfbA in the nonpathogenic *S. gordonii,* it was demonstrated that GfbA alone is responsible for the morphologically distinct invasion mechanism compared to the SfbI-mediated caveolae dependent invasion mechanism. Sequencing of the GfbA gene demonstrated that the FnBR repeat region in the C-terminal part is very similar to SfbI. Only the N-terminal part, including the aromatic domain of GfbA (ARO) and SfbI, shows significant differences. Therefore, it was assumed that the ARO region might be responsible for the observed morphological differences in the invasion mechanism. To test this hypothesis, a

GfbA protein was constructed without the aromatic domain, and the aromatic domain in SfbI was replaced by the aromatic domain of GfbA. FESEM studies revealed that GfbA with a deleted ARO region invades like SfbI-expressing strains, whereas the SfbI strain containing the ARO region of GfbA now induced membrane ruffles. By repeating the FESEM studies for integrin clustering, it was shown that only GfbA without the aromatic domain induced integrin clustering and signaling, whereas SfbI with the ARO region of GfbA was impaired in integrin clustering and signaling. Thus, for the first time a biological function for the ARO region in a fibronectin-binding protein was demonstrated (Rohde et al. 2011).

3.2 M Protein Invasion of Epithelial Cells

For M protein-mediated invasion, fibronectin, laminin, or plaminogen/plasmin (Colognato and Yurchenco 2000; Siemens et al. 2011) binding on the bacterial surface are the prerequisites. Serotypes M1, M3, M5, M6, M12, M18, and M49 (Berkower et al. 1999; Dombeck et al. 1999; Molinari et al. 2000; Nerlich et al. 2009) have been studied with respect to HEp-2 cell invasion and all serotypes were found to be invasive. Only the highly capsulated M18 serotype was found to be less invasive because of interference of the capsule with the initial adherence process to the cells (Hagman et al. 1999).

After binding to fibronectin and $\alpha_5\beta_1$-integrins, the serotype M1 promotes invasion of human lung epithelial cells. A M1 negative mutant had a significant reduced invasion capability. M1 protein can also bind to laminin with subsequent internalization in host cells (Cue et al. 1998). The mechanism of entrance was accompanied by cytoskeletal rearrangements and represented a zipper-like mechanism in HeLa cells as shown by scanning EM (Dombek et al. 1999). This observation is in contrast to the caveolae-mediated internalization of SfbI/F1 carrying strains, despite the fact that also M1 serotype strains bind fibronectin and subsequently $\alpha_5\beta_1$ integrins.

The zipper-like invasion was mediated by an early intimate contact of streptococci with host cell microvilli and invading streptococci were associated with polymerized actin at their site of entry. Identical observations of actin polymerization underneath the entry port were observed for a M5 serotype/A8 strain (Molinari et al. 2000) which does not express SfbI. M1 protein-mediated invasion subsequently leads to the fusion with lysosomes to form a phagolysosome.

4 Dissemination and Persistence

After a first episode of infection with streptococci, a small percentage of patients will experience recurrent tonsillitis or erysipelas episodes. These recurrent infections may be attributed to a bacterial subpopulation surviving beyond the first

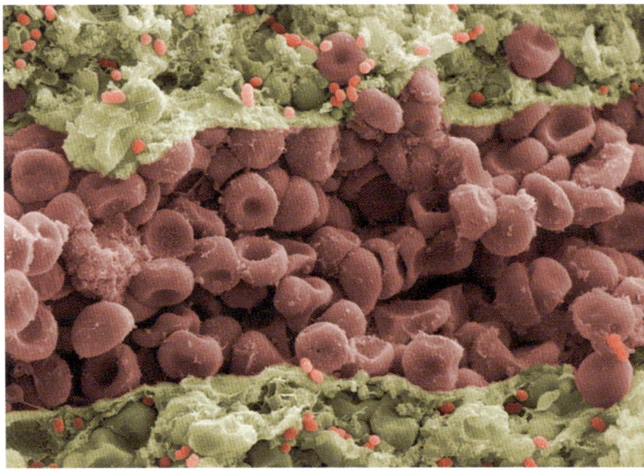

Fig. 3 FESEM image of dissemination of Group A streptococci. Streptococci were administered intravenously. From the blood stream, streptococci passed through the endothelial layer of the blood vessels into deeper tissue (streptococci in *pink*, *red blood* cells in *red*)

episode. There are two plausible reasons why the streptococci might persist: (i) by hiding in the matrix proteins of the human mucosa, especially serotype M3 and M18 by binding to collagen, (ii) or intracellularly by residing in an intracellular compartment or in the cytoplasm (Fig. 3).

For dissemination in the host, two possible ways are conceivable. Firstly, streptococci might invade certain cells and travel unrecognized in these cells through the body until reaching a safe niche in the body—Trojan horse theory. Secondly, after bacteraemia streptococci might be able to passage through the dense endothelial cell layer of the blood vessels to gain access into deeper tissue. For both ways, first experimental evidence has been obtained during recent years. Nevertheless, we are far away from understanding the entire multiple strategies of streptococci during dissemination in the body. For example, it was shown that GAS, which are resistant to killing in PMNs, may exploit the ability of the neutrophils to transmigrate through endothelial cell barrier of the vascular system for spread via the blood stream (Medina et al. 2003a, b). Surprisingly, streptococci were also found residing in macrophages when biopsies of patients with soft tissue infections were examined, supporting the Trojan horse theory (Thulin et al. 2006). These studies conclusively showed that host phagocytic cells, especially macrophages, were the reservoir for intracellular GAS during infection. Remarkably, the authors found evidence that localization of GAS varied and depended on the severity of tissue infection. Intracellular streptococci were predominantly found in noninflamed tissue with a low bacterial load, whereas in inflamed tissue, even after prolonged intravenous treatment with antibiotics, high loads of GAS were detected. It has been suggested that internalization could indeed promote the spread of GAS within the tissue. Macrophages might then play the role of the Trojan horse

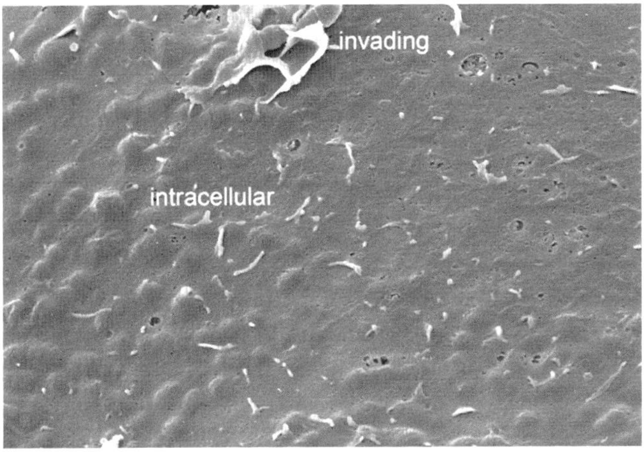

Fig. 4 FESEM image of a highly invasive Group G isolate, *Streptococcus dysgalactiae* subspec. *equisimilis*, from an aneurysm of a patient. Despite antibiotic treatment, this strain was persistent for several years

in the infected tissue. On the other hand, work by Cywes et al. (2000) described another type of streptococcal invasion. They demonstrated that binding to CD44 on a confluent layer of human keratinocytes via HA (capsule) of encapsulated strains induced cytoskeletal changes, membrane ruffling, and a disruption of cellular junctions after binding to the host cell lamellipodia. In contrast, nonencapsulated strains attached to host membrane regions devoid of lamellipodia. This type of invasion opens up the way for streptococci to gain access into deeper tissue layers. For Group B streptococci (GBS), an identical paracellular translocation through the epithelial barrier during onset of infection has been reported. Soriani et al. (2008) applied differentiated epithelial cells grown in cell culture transwell inserts for their studies. Translocation of bacteria occurred without measurable decrease in the transepithelial resistance and analysis of ultra-thin sections revealed an intimate association of GBS with the intercellular junctions (Soriani et al. 2008). Recently, a third mechanism has been found demonstrating that streptococci mediate an exocytosis process in which streptococci residing in a phagolysosome can trigger their own exocytosis process, with Rab27 being involved, to transmigrate through an endothelial barrier (Talay et al. personal communication).

For persistence in the host, only scant data are available. For example, *S. dysgalactiae* subsp. *equisimilis* (Group C or Group D) are relatively common in invasive infections, especially in older patients (Sylvetsky et al. 2002). This species (type stG485.0) was involved in several recurrent bacteremic sepsis episodes despite prolonged antibiotic treatment over years and using opsonizing antibodies against the isolate. The isolate was suspected to reside in an aortic aneurysm of the patient. With a cell culturing model, it was shown that the isolate possesses a very high invasiveness in endothelial HUVEC cells (Rohde et al. 2012) (Fig. 4).

That streptococci actually can survive for a long period of time was shown in batch cultures using sugar-limited Todd-Hewitt broth. Bacteria were still viable and culturable after 1 year (Wood et al. 2005). Other experiments extended the survival time to over 45 months. An overnight culture with an inoculation with 5 μl of the sedimented bacteria, grown in TBS, was sufficient to obtain fully viable streptococci. Even more surprising was the fact that FESEM imaging of the bacteria revealed no visible difference of 45-month-old streptococci to the overnight regrown bacteria; only the interior morphology of the bacteria was remarkably different (Rohde, unpublished). It is speculated that during persistence *S. pyogenes* might enter a kind of a quiescent state, due to conditions which do not support rapid bacterial growth, such as nutrition limitations. If a trauma from outside destroys areas of intracellular streptococci containing tissue bacteria, allowing access to new nutrition due to infiltration of immune cells etc., to the damaged tissue side, then bacteria can respond with a fulminant growth thereby causing a recurrent infection.

5 Conclusions

Adhesion of streptococci through the multiple adhesins and internalization into human epithelial and endothelial cells by invasins creates a safe sanctuary for streptococci protecting them against the host immune system and antibiotic treatment. Together, with the capability to penetrate cellular barrier for dissemination and hijacking immune cells like PMNs for their own spread in the body makes streptococci a very versatile and not easy to eradicate human pathogen. Since the work of Österlund and colleagues (Oesterlund and Engstrand 1997; Oesterlund et al. 1997), who provided first evidence for the existence of persisting intracellular streptococci, evidence has accumulated that streptococcal invasion into host cells is a severe therapeutic problem (Thulin et al. 2006; Rohde et al. 2012). The routinely used β-lactam antibiotics, penicillins, are not efficiently permeating into eukaryotic cells and fail to completely eradicate intracellular GAS. Recently, macrolides like erythromycin or derivatives thereof or an azalide, azithromycin were found to be more efficient in eradicating intercellular streptococci in vitro and may present an attractive therapeutic alternative for eradication of streptococci (Kaplan et al. 2006). Nevertheless, the use of macrolides faces another drawback since streptococci have developed a strong resistance against macrolides in some countries. The mostly in vitro obtained observations of streptococcal adhesion, internalization, trafficking inside the cell, dissemination, and persistence, taken together with in vivo studies on persistence of streptococci, should allow for a more effective approach for clinical management of Group A *S. pyogenes* carriers with the aim to eradicate bacteria at the onset of the infection.

References

Abbot EL, Smith WD, Siou GP, Chriboga C, Smith RJ, Wilson JA, Hirts BH, Kehoe MA (2007) Pili mediate specific adhesion of *Streptococcus pyogenes* to human tonsils and skin. Cell Microbiol 9:1822–1833

Agerer F, Lux S, Michel A, Rohde M, Ohlsen K, Hauck CR (2005) Cellular invasion by *Staphylococcus aureus* reveals a functional link between focal adhesion kinase and cortactin in integrin-mediated internalisation. J Cell Sci 118:2189–2200

Amelung S, Nerlich A, Rohde M, Spellerberg B, Cole JN, Nizet V, Chhatwal GS, Talay SR. (2011) The FbaB-type fibronectin-binding protein of *S. pyogenes* promote specific invasion into endothelial cells. Cell Microbiol 13:1200–1211

Andre I, Persson J, Blom AM, Nilsson H, Drakenberg, T, Lindahl G, Linse S (2006) Streptococcal M protein: structural studies of the hypervariable regions, free and bound to human C4BP. Biochemistry 45:4559–4568

Barocchi MA, Ries J, Zogaj X, Hemsley C, Albiger B, Kanth A, Dahlberg S, Fernebro J, Moschioni M, Masignani V, Hultenby K, Taddei AR, Beiter K, Wartha F, von Euler A, Covacci A, Holden DW, Normark S, Rappuoli R, Henriques-Normark B (2006) A pneumococcal pilus influences virulence and host inflammatory responses. Proc Natl Acad Sci USA 103: 2857–2862

Beachey EH, Ofek I (1976) Epithelial cell binding of group A streptococci by lipoteichoic acid on fimbriae denuded of M protein. J Exp Med 143:759–771

Beall B, Facklam R, Thompson T (1996) Sequencing emm-specific PCR products for routine and accurate typing of group A streptococci. J Clin Microbiol 34:9533–958

Benga L, Goethe R, Rohde M, Valentin-Weigand P (2004) Non-encapsulated strains reveal novel insights in invasion and survival of *Streptococcus suis* in epithelial cells. Cell Microbiol 6:867–881

Berkower C, Ravins M, Moses AE, Hanski E (1999) Expression of different group A streptococcal M proteins in an isogenic background demonstrates diversity in adherence to and invasion of eukaryotic cells. Mol Microbiol 31:1463–1475

Caparon MG, Stephens S, Olsen A, Scott JR (1991) Role of M protein in adherence of group A streptococci. Infect Immun 59:1811–1817

Carapetis (2005) Acute rheumatic fever. Lancet 366:155-168

Chhatwal GS (2002) Anchorless adhesins and invasion of Gram-positive bacteria: a new class of virulence factors. Trends in Microbiol 10:205–208

Colognato H, Yurchenco PD (2000) Form and function: the laminin family of heterotrimers. Dev Dyn 218:213–234

Courtney HS, Hasty DL (1991) Aggregation of group A streptococci by human saliva and effect of saliva on streptococcal adherence to host cells. Infect Immun 59:1661–1666

Courtney HS, von Hunolstein C, Dale JB, Bronze MS, Beachey EH, Hasty DL (1992) Lipoteichoic acid and M protein: dual adhesins of group A streptococci. Microb Pathog 12:199–208

Courtney HS, Bronze M S, Dale JB, Hasty DL (1994) Analysis of the role of M24 protein in group A streptococcal adhesion and colonization by use of omega-interposon mutagenesis. Infect Immun 62:4868–4873

Courtney HS, Dale JB, Hasty DI (1996) Differential effects of the streptococcal fibronectin-binding protein, FBP54, on adhesion of group A streptococci to human buccal cells and HEp-2 tissue culture cells. Infect Immun 64:2415–2419

Courtney HS, Hasty DL, Dale JB (2002) Molecular mechanisms of adhesion, colonization, and invasion of group A streptococci. Ann Med 34:77–87

Cremer MA, Rosloniec EF, Kang AH (1998) The cartilage collagens: a review of their structure, organization, and role in the pathogenesis of experimental arthritis in animals and in human rheumatic disease. J Mol Med 76:275–288

Crotty Alexander LE, Maisey HC, Timmer AM, Rooijakkers SHM, Gallo RL, von Köckritz-Blickwede M, Nizet V. (2010) M1T1 group A streptococci pili promote epithelial colonization but diminish systemic virulence through neutrophil extracellular entrapment. J Mol Med 88:371–381

Cue D, Dombek PE, Lam H, Cleary PP (1998) *Streptococcus pyogenes* serotype M1 encodes multiple pathways for entry into human epithelial cells. Infect Immun 66:4593–4601

Cue D, Southern SO, Southern PJ, Prabhakar J, Lorelli W, Smallheer JM, Mousa SA, Cleary PP (2000) A nonpeptide integrin antagonist can inhibit epithelial cell ingestion of *Streptococcus pyogenes* by blocking formation of integrin alpha 5beta 1-fibronectin-M1 protein complexes. Proc Natl Acad Sci USA 97:2858–2863

Cue D, Lam H, Cleary PP (2001) Genetic dissection of the Streptococcus pyogenes M1 protein: regions involved in fibronectin-binding and intracellular invasion. Microb Pathog 31:231–242

Cunningham MW (2000) Pathogenesis of group A streptococcal infections. Clin Microbiol Rev 13:470–511

Cywes C, Wessels MR (2001) Group A Streptococcus tissue invasion by CD44-mediated cell signalling. Nature 414:648–652

Cywes C, Stamenkovic I, Wessels MR (2000) CD44 as a receptor for colonization of the pharynx by group A Streptococcus. J Clin Invest 106:995–1002

Dale JB, Washburn RG, Marques MB, Wessels MR (1996) Hyaluronate capsule and surface M protein in resistance to opsonization of group A streptococci. Infect Immun 64:1495–1501

Darmstadt GL, Mentele L, Podbielski A, Rubens CE (2000) Role of group A streptococcal virulence factors in adherence to keratinocytes. Infect Immun 68:1215–1221

Debelle L, Tamburro AM (1999) Elastin: molecular description and function. Int J Biochem Cell Biol 31:261–272

Dempfle CE, Mosesson MW (2003) Theme issue: Fibrinogen and fibrin - structure, function, interactions and clinical applications. Thromb Haemost 89:599–600

Derbise A, Song YP, Parikh S, Fischetti VA, Pancholi V (2004) Role of the C-terminal lysine residues of streptococcal surface enolase in Glu- and Lys-plasminogen-binding activities of group A streptococci. Infect Immun 72:94–105

Dinkla K, Rohde M, Jansen WTM, Kaplan EL, Chhatwal GS, Talay SR (2003a) Rheumatic fever-associated *Streptococcus pyogenes* isolates aggregate collagen. J Clin Invest 111:1905–1912

Dinkla K, Rohde M, Jansen WTM, Carapetis JR, Chhatwal GS, Talay SR (2003b) *Streptococcus pyogenes* recruits collagen via surface-bound fibronectin: a novel colonization and immune evasion mechanism. Mol Microbiol 47:861–869

Dombek PE, Cue D, Sedgewick J, Lam H, Ruschkowski S, Finlay BB, Cleary PP (1999) High-frequency intracellular invasion of epithelial cells by serotype M1 group A streptococci: M1 protein-mediated invasion and cytoskeletal rearrangements. Mol Microbiol 31:859–870

Doran KS, Engelson EJ, Khosravi A, Maisey HC, Fedtke I, Equils O, Michelsen KS, Arditi, M, Peschel A, Nizet V (2005) Blood-brain barrier invasion by group B Streptococcus depends upon proper cell-surface anchoring of lipoteichoic acid. J Clin Invest 115:2499–2507

Ellen RP, Gibbons RJ (1972) M protein-associated adherence of *Streptococcus pyogenes* to epithelial surfaces: prerequisite for virulence. Infect Immun 5:826–830

Ellen RP, Gibbons RJ (1974) Parameters affecting the adherence and tissue tropisms of *Streptococcus pyogenes*. Infect Immun 9:85–91

Falugi F, Zingaretti C, Pinto V, Mariani M, Amodeo A, Manetti G, Capo S, Musser JM, Orefici G, Margarit I, Telford L, Grandi G, Mora M (2008) Sequence variation in group A Streptococcus pili and associatition of pilus backbone types with Lancefield T serotypes. J Infect Dis 198:1834–1841

Fowler T, Wann ER, Joh D, Johansson S, Foster TJ, Höök M (2000) Cellular invasion by Staphylococcus aureus involves a fibronectin bridge between the bacterial fibronectin-binding MSCRAMMs and host cell beta1 integrins. Eur J Cell Biol 79:672–679

Frick IM, Crossin KL, Edelman GM, Björck L (1995) Protein H-a bacterial surface protein with affinity for both immunoglobulin and fibronectin type III domains. EMBO J 14:1674–1679

Frick IM, Mörgelin M, Björck L (2000) Virulent aggregates of *Streptococus pyogenes* are generated by hemophilic protein–protein interactions. Mol Microbiol 37:1232–1247

Frick IM, Schmidtchen A, Sjobring U (2003) Interactions between M proteins of *Streptococcus pyogenes* and glycosaminoglycans promote bacterial adhesion to host cells. Eur J Biochem 270:2303–2311

Gianfoldoni C, Censini S, Hilleringmann M, Moschioni M, Facciotti C, Pansegrau W, Masigani V, Covacci A, Rappuoli R, Barocchi MA, Rugiero P (2007) *Streptococcus pneumoniae* pilus subunits protect mice against lethal challange. Infect Immun 75:1059–1062

Giannakis E, Jokiranta TS, Ormsby RJ, Duthy TG, Male DA, Christiansen D, Fischetti VA, Bagley C, Loveland BE, Gordon DL (2002) Identification of the streptococcal M protein binding site on membrane cofactor protein (CD46). J Immunol 168:4585–4592

Gibson CM, Caparon MG (1996) Insertional inactivation of *Streptococcus pyogenes* sod suggests that prtF is regulated in response to a superoxide signal. J Bacteriol 178:4688–4695

Gibson CM, Fogg G, Okada N, Geist RT, Hanski E, Caparon MG (1995) Regulation of host cell recognition in *Streptococcus pyogenes*. Dev Biol Stand 85:137–144

Greco R, De Martino L, Donnarumma G, Conte MP, Seganti L, Valenti P (1995) Invasion of cultured human cells by *Streptococcus pyogenes*. Res Microbiol 146:551–560

Hagman MM, Dale JB, Stevens DL (1999) Comparison of adherence to and penetration of a human laryngeal epithelial cell line by group A streptococci of various M protein types. FEMS Immunol Med Microbiol 23:195–204

Haidan A, Talay SR, Rohde M, Sriprakash KS, Currie BJ, Chhatwal GS (2000) Pharyngeal carriage of group C and group G streptococci and acute rheumatic fever in an Aboriginal population. Lancet 356:1167–1169

Handley PS, Carter PL, Fielding J (1984) *Streptococcus salivarius* strains carry either fibrils or fimbriae on the cell surface. J Bacteriol 157:64–72

Handley PS, Harty DW, Wyatt JE, Brown CR, Doran JP, Gibbs AC (1987) A comparison of the adhesion, coaggregation and cell-surface hydrophobicity properties of fibrillar and fimbriae strains of *Streptococcus salivarius*. J Gen Microbiol 133:3207–3217

Hanski E, Caparon M (1992) Protein F, a fibronectin-binding protein, is an adhesin of the group A streptococcus. Proc Natl Acad Sci USA 89:6172–6176

Hasty DL, Ofek I, Courtney HS, Doyle RJ (1992) Multiple adhesins of streptococci. Infect Immun 60:2147–2152

Henningham A, Barnett TC, Maamary PG, Walker MJ (2012a) Pathogenesis of group A infections. Discov Med 13:329–342

Henningham A, Chiarot E, Gillen CM, Cole JN, Rohde M, Fulde M, Ramachandran V, Cork AJ, Hartas J, Magor G, Djordjevic SP, Cordwell SJ, Kobe B, Sriprakash KS, Nizet V, Chhatwal GS, Margarit IY, Batzloff MR, Walker MJ (2012b) Conserved anchorless surface proteins as group A vaccine candidates. J Mol Med (Epub ahead)

Humtsoe JO, Kim JK, Xu,Y, Keene DR, Höök M, Lukomski S, Wary KK (2005) A streptococcal collagen-like protein interacts with the alpha2beta1 integrin and induces intracellular signaling. J Biol Chem 280:13848–13857

Hytonen J, Haataja S, Gerlach D, Podbielski A, Finne J (2001) The SpeB virulence factor of streptococcus pyogens, a multifuntional secreted and cell surface molecule with strepadhesin, laminin-binding and cysteine protease activity. Mol Microbiol 39:512–519

Isberg RR (1991) Discrimination between intracellular uptake and surface adhesion of bacterial pathogens. Science 252:934–938

Isberg RR, Barnes P (2001) Subversion of integrins by enteropathogenic Yersinia. J Cell Sci 114:21–28

Isberg RR, Leong JM (1990). Multiple beta 1 chain integrins are receptors for invasin, a protein that promotes bacterial penetration into mammalian cells. Cell 60:861–871

Jadoun J, Ozeri V, Burstein E, Skutelsky E, Hanski E, Sela S (1998) Protein F1 is required for efficient entry of *Streptococcus pyogenes* into epithelial cells. J Infect Dis 178:147–158

Jadoun J, Eyal O, Sela S (2002) Role of CsrR, hyaluronic acid, and SpeB in the internalization of *Streptococcus pyogenes* M type 3 strain by epithelial cells. Infect Immun 70:462–469

Jaffe J, Natanson-Yaron S, Caparon MG, Hanski E (1996) Protein F2, a novel fibronectin-binding protein from *Streptococcus pyogenes*, possesses two binding domains. Mol Microbiol 21:373–384

Jin H, Song YP, Boel G, Kochar J, Pancholi V (2005) Group A streptococcal surface GAPDH, SDH, recognizes uPAR/CD87 as its receptor on the human pharyngeal cell and mediates adherence to host cells J Biol Chem 350:27–41

Johnson DR, Stevens DL, Kaplan EL (1992) Epidemiologic analysis of group A streptococcal serotypes associated with severe systemic infections, rheumatic fever, or uncomplicated pharyngitis. J Infect Dis 166: 374–382

Kang HJ, Coulibaly F, Clow F, Proft T, Baker EN (2007) Stabilizing isopeptide bonds revealed in gram-positive bacterial pilus structure. Science 318:1625–1628

Kaplan EL, Johnson DR, Cleary PP (1989) Group A streptococcal serotypes isolated from patients and sibling contacts during the resurgence of rheumatic fever in the United States in the mid-1980s. J Infect Dis 159:101–103

Kaplan EL, Chhatwal GS, Rohde M (2006) Reduced ability of penicillin to eradicate ingested group A streptococci from epithelial cells: clinical and pathogenetic implication. Clin Infect Dis 43:1398–1406

Kawabat S, Tamura Y, Murakami J, Teroa Y, Nakagawa I, Hamada S (2002) A novel, anchorless streptococcal surface protein that binds to human immunoglobulins. Biochem Biophys Res Comm 296:1329–1333

Kinnby B, Booth NA, Svensater G (2008) Plasminogen binding by oral streptococci from dental plaque and inflammatory lesions. Microbiology 154:924–931

Kreikemeyer B, McIver KS, Podbielski A (2003) Virulence factor regulation and regulatory networks in *Stretococcus pyogenes* and their impact on pathogen-host interactions. Trends Microbiol 11:224–232

Kreikemeyer B, Oemcke S, Nakata M, Hoffrogge R, Podbielski A (2004a) *Streptococcus pyogenes* fibronectin-bindung protein F2: expression profile, binding characteristics, and impact on eukaryotic cell interactions. J Biol Chem 279:15850–15859

Kreikemeyer B, Klenk M, Podbielski A (2004b) The intracellular status of *Streptococcus pyogenes*; role of extracellular matrix-binding proteins and their regulation. Int J Med Micriobiol 294:177–188

LaPenta D, Rubens C, Chi E, Cleary PP (1994) Group A streptococci efficiently invade human respiratory epithelial cells. Proc Natl Acad Sci USA 91:12115–12119

Lauer P, Rinaudo CD, Soriani M, Margarit I, Maione D, Rosini R, Taddei AR, Mora M, Rappuoli R, Grandi G, Telford JL (2005) Genome analysis reveals pili in Group B *Streptococcus*. Science 309:105

Leon O, Panos C (1990) *Streptococcus pyogenes* clinical isolates and lipoteichoic acid. Infect Immun 58:3779–3787

Maisey HC, Hensler M, Nizet V, Doran KS (2007) Group B streptococcal pilus proteins contribute to adherence to and invasion of brain microvascular endothelial cells. J Bacteriol 189:1464–1467

Mandlik A, Swierczynski A, Das A., Ton-That H. (2007a) Pili in Gram-positive bacteria: assembly, involvement in colonization and biofilm development. Trends in Microbiol 16:33–39

Mandlik A, Swierczynski A, Das A, Ton-That H (2007b) *Corynebacterium diphtheria* employs specific minor pilins to target human pharyngeal epithelial cells. Mol Microbiol 64:11–124

Maraffini LA, Dedent AC, Schneewind O (2006) Sortases and the art of anchoring proteins to the envelope of gram-positive bacteria. Microbiol Mol Biol Rev 70:192–221

Marjenberg ZR, Ellis IR, Hagan RM, Prabhakaran S, Höök M, Talay SR, Potts JR, Staunton D, Schwarz-Linek U (2011) Cooperative binding and activation of fibronectin by a bacterial surface protein. J Biol Chem 286:1884–1894

McArthur JD, Walker MJ (2006) Domains of group A streptococcal M protein that confer resistance to phagocytosis, opsonozation and protection: implications for vaccine development. Mol Microbiol 59:1–4

McNamara C, Zinkernagel AS, Machebeouf M, Cunningham M, Nizet V. (2008) Coiled-coil irregularities and instabilities in group A streptococcus M1 are required for virulence. Science 319: 1405–1408
Medina E, Rohde M, Chhatwal GS (2003a) Intracellular survival of *Streptococcus pyogenes* in polymorphonuclear cells results in increased bacterial virulence. Infect Immun 71:5376–5380
Medina E, Goldmann O, Toppel AW, Chhatwal GS (2003b) Survival of Streptococcus pyogenes within host phagocytic cells: a pathogenic mechanism for persistence and systemic invasion. J Infect Dis 187:597–603
Molinari G, Chhatwal GS (1999) Streptococcal invasion. Curr Opin Microbiol 2:56–61
Molinari G, Rohde M, Guzman CA, Chhatwal GS (2000) Two distinct pathways for the invasion of *Streptococcus pyogenes* in non-phagocytic cells. Cell Microbiol 2:145–154
Mora M, Bensi G, Capo S, Falugi F, Zingaretti C, Manetti AG, Maggi T, Taddei AR, Grandi G, Telford JL (2005) Group A streptococcus produce pilus-like structures containing protective antigens and Lancefield T antigens. Proc Natl Acad Sci USA 102:15641–15646
Natanson S, Sela S, Moses AE, Musser JM, Caparon MG, Hanski, E (1995) Distribution of fibronectin-binding proteins among group A streptococci of different M types. J Infect Dis 171:871–878
Nerlich A, Rohde M, Talay, SR, Genth H, Just I, Chhatwal GS (2009) Invasion of endothelial cells by tisuue-invasive M3 type Group A streptococci requires Src kinase and activation of Rac1 by a phosphatidylinositol 3-kinase-independent mechanism. J Biol Chem 284:20319–20328
Nitsche DP, Johansson HM, Frick IM, Mörgelin, M (2006) Streptococcal protein FOG, a novel matrix adhesin interacting with collagen I in vivo. J Biol Chem 281:1670–1679
Nitsche-Schmitz DP, Rohde M, Chhatwal GS (2007) Invasion mechanisms of gram-positive pathogenic cocci. Thromb Haemost 98:488–496
Nobbs AH, Lamont RJ, Jenkinson HF (2009) Streptococcus adherence and colonization. Microbiol Mol Biol Rev 73:407–450
Norton PM, Rolph C, Ward PN, Bentley RW, Leigh JA (1999) Epithelial invasion and cell lysis by virulent strains of *Streptococcus suis* is enhanced by the presence of suilysin. FEMS Immunol Med Microbiol 26:25–35
Oemcke S, Shannon O, Mörgelin M, Herwald H. (2010) Streptococcal M proteins and their role as virulence determninants. Clinica Chemica Acta 411:1172–1180
Oesterlund A, Engstrand L (1997a) An intracellular sanctuary for *Streptococcus pyogenes* in human tonsillar epithelium-studies of asymptomatic carriers and in vitro cultured biopsies. Acta Otolaryngol 117:883–888
Oesterlund A, Popa R, Nikkila T, Scheynius, A, Engstrand L (1997b) Intracellular reservoir of *Streptococcus pyogenes* in vivo: a possible explanation for recurrent pharyngotonsillitis. Laryngoscope 107:640–647
Okada N, Liszewski MK, Atkinson JP, Caparon M (1995) Membrane cofactor protein (CD46) is a keratinocyte receptor for the M protein of the group A Streptococcus. Biochem Biophys Res Commun 377:1128–1134
Ozeri V, Tovi A, Burstein I, Natanson-Yaron S, Caparon MG, Yamada KM, Akiyama SK, Vlodavsky I, Hanski E (1996) A two-domain mechanism for group A streptococcal adherence through protein F to the extracellular matrix. EMBO J 15:989–998
Ozeri V, Rosenshine I, Mosher DF, Fässler R, Hanski E (1998) Roles of integrins and fibronectin in the entry of *Streptococcus pyogenes* into cells via protein F1. Mol Microbiol 30:625–637
Pancholi V, Fischetti VA (1992) A major surface protein on group A streptococci is a glyceraldehydes-3-phosphate-dehydrogenase with multiple binding activity. J Exp Med 176:415–426
Pancholi V, Fischetti VA (1997) Regulation of the phosphorylation of human pharyngeal cell proteins by group A streptococcal surface dehydrogenase: signal transduction between streptococci and pharyngeal cells. J Exp Med 186:1633–1643
Pancholi V, Fischetti VA (1998) alpha-enolase, a novel strong plasminogen-(ogen) binding protein on the surface of pathogenic streptococci. J Biol Chem 273:14503–14515

Pankov R, Yamada KM (2002) Fibronectin at a glance. J Cell Sci 115:3861–3863

Patti JM, Höök M (1994) Microbial adhesions recognizing extracellular matrix molecules. Curr Opin Cell Biol 6:752–758

Pezzicoli A, Santi I, Lauer P, Rosini R, Rinaudo D, Grandi, G, Telford JL, Soriani M (2008) Pilus backbone contributes to group B Streptococcus paracellular translocation through epithelial cells. Infect Immun 198:890–898

Podbielski A, Woischnik M, Leonard BA, Schmidt KH (1999) Characterization of nra, a global negative regulator gene in group A streptococci Mol Microbiol 31:1051–1064

Proft T, Baker EN (2009) Pili in gram-negative and gram-positive bacteria-structure, assembly and role in disease. Cell Mol Life Sci 66:613–635

Rezcallah M S, Hodges K, Gill DB, Atkinson JP, Wang B, Cleary PP (2005) Engagement of CD46 and alpha5beta1 integrin by group A streptococci is required for efficient invasion of epithelial cells. Cell Microbiol 7:645–653

Rocha CL, Fischetti VA (1999) Identification and characterization of a novel fibronectin-binding protein on the surface of group A streptococci. Infect Immun 67:2720–2728

Rohde M, Müller E, Chhatwal GS, Talay SR (2003) Host cell caveolae act as an entry-port for Group A streptococci. Cell Microbiol 5:323–342

Rohde M, Graham RM, Branitzki-Heinemann K, Borchers P, Preuss C, Schleicher I, Zähner D, Talay SR, Fulde M, Dinkla K, Chhatwal GS (2011) Differences in the aromatic domain of homologous streptococcal fibronectin-binding proteins trigger different cell invasion mechanisms and survival rates. Cell Microbiol 13:450–468

Rohde M, Talay SR, Rasmussen M (2012) Molecular mechanisms of *Streptococcus dysgalactiae* subsp. *Equisimilis* enabling intravascular persistence. Microbes Infect 14:329–334

Rosini R, Rinaudo CD, Soriani M, Lauer P, Mora M, Maione D, Taddei A, Santi I, Ghezzo C, Brettoni C, Buccato S, Margarit I, Grandi G, Telford JL. (2006) Identification of novel genomic islands coding for antigenic pilus-like structures in *Streptococcus agalacticae*. Mol Microbiol 61:126–141

Rubens, CE, Smith S, Hulse M, Chi EY, van Belle G (1992) Respiratory epithelial cell invasion by group B streptococci. Infect Immun 60:5157–5163

Saulino ET, Bullitt E, Hultgren SJ (2000) Snapshots of usher mediated protein secretion and ordered pilus assembly. Proc Natl Acad Sci USA 97:9240–9245

Schneewind O, Mihaylova-Petkov D, Model P (1993) Cell wall sorting signals in surface proteins of gram-positive bacteria EMBO J 12:4803–4811

Schrager HM, Rheinwald JG, Wessels MR (1996) Hyaluronic acid capsule and the role of streptococcal entry into keratinocytes in invasive skin infection. J Clin Invest 98:1954–1958

Schrager HM, Alberti S, Cywes C, Dougherty GJ, Wessels MR (1998) Hyaluronic acid capsule modulates M protein-mediated adherence and acts as a ligand for attachment of group A Streptococcus to CD44 on human keratinocytes. J Clin Invest 101:1708–1716

Schvartz I, Seger D, Shaltiel S (1999) Vitronectin. Int J Biochem Cell Biol 31:539–544

Schwarz-Linek U, Werner JM, Pickford AR, Gurusiddappa S, Kim JH, Pilka ES, Briggs JA, Gough TS, Höök M, Campbell ID, Potts JR (2003) Pathogenic bacteria attach to human fibronectin through a tandem beta-zipper. Nature 423:177–181

Schwarz-Linek U, Höök M, Potts JR (2004a) The molecular basis of fibronectin-mediated bacterial adherence to host cells Mol Microbiol 52:631–641

Schwarz-Linek U, Pilka ES, Pickford AR, Kim JH, Höök M, Campbell ID, Potts J R (2004b) High affinity streptococcal binding to human fibronectin requires specific recognition of sequential F1 modules. J Biol Chem 279:39017–39025

Scott JR, Zähner D (2006) Pili with strong attachments: Gram-positive bacteria do it differently Mol Microbiol 62:320–330

Seifert KN, McArthur WP, Bleiweis AS, Brady LJ (2003) Characterization of group B streptococcal glyceraldehydes-3-phosphate dehydrogenase: surface localization, enzymatic activity, and protein–protein interactions. Can J Microbiol 49:350–356

Sela S, Aviv A, Tovi A, Burstein I, Caparon MG, Hanski E (1993) Protein F: an adhesin of *Streptococcus pyogenes* binds fibronectin via two distinct domains. Mol Microbiol 10:1049–1055

Siemens N, Patenge N, Otto J, Fiedler T, Kreikemeyer B (2011) *Streptococcus pyogenes* M49 plasminogen/plasmin binding facilitates keratinocyte invasion via integrin–integrin-linked kinase (ILK) pathways and protects from macrophage killing. J Biol Chem 286:21612–21622

Simpson WA, Beachey EH (1983) Adherence of group A streptococci to fibronectin on oral epithelial cells. Infect Immun 39:275–279

Soriani M, Santi I, Taddei A, Rappuoli R, Grandi G, Telford JL (2008) Group B streptococci crosses human epithelial cells by a paracellular route. J Infect Dis 193:241–250

Stinson M W, Alder S, Kumar S (2003) Invasion and killing of human endothelial cells by viridans group streptococci. Infect Immun 71:2365–2372

Sylvestky N, Raveh D, Schlesinger Y, Rudensky B, Yinnon AM (2002) Bacteremia due to beta-hemolytic Streptococcus group G: increasing incidence and clinical characteristics of patients. Am J Med 112:622–626

Talay SR, Valentin-Weigand P, Jerlstrom PG, Timmis KN, Chhatwal GS (1992) Fibronectin-binding protein of *Streptococcus pyogenes*: sequence of the binding domain involved in adherence of streptococci to epithelial cells. Infect Immun 60: 3837–3844

Talay SR, Valentin-Weigand P, Timmis KN, Chhatwal GS (1994) Domain structure and conserved epitopes of Sfb protein, the fibronectin-binding adhesin of *Streptococcus pyogenes*. Mol Microbiol 13:531–539

Talay SR, Zock A, Rohde M, Molinari G, Oggioni M, Pozzi G, Guzman CA, Chhatwal GS (2000) Co-operative binding of human fibronectin to SfbI protein triggers streptococcal invasion into respiratory epithelial cells. Cell Microbiol 2:521–535

Talbot UM, Paton AW, and Paton JC (1996). Uptake of *Streptococcus pneumoniae* by respiratory epithelial cells. Infect Immun 64:3772–3777

Telford JT, Barocchi MA, Margarit I, Rappuoli R, Grandi G (2006) Pili in Gram-positive pathogens. Nat Rev Microbiol 4:509–519

Terao Y, Kawabata S, Kunitomo E, Murakami J, Nakagawa I, Hamada S (2001) Fba, a novel fibronectin-binding protein from *Streptococcus pyogenes*, promotes bacterial entry into epithelial cells, and the *fba* gene is positively transcribed under the Mga regulator. Mol Microbiol 42:75–86

Terao Y, Kawabata S, Nakata M, Nakagawa I, Hamada S (2002) Molecular characterization of a novel fibronectin-binding protein of *Streptococcus pyogenes* strains isolated from toxic shock-like syndrome patients. J Biol Chem 277:47428–47435

Thulin P, Johansson L, Low DE, Gan BS, Kotb M, McGeer A, Norrby-Teglund A (2006) Viable Group A streptococci in macrophage during acut soft tissue infection. PLos Med 3:e53

Ton-That H, Marraffini LA, Schneewind O (2004) Protein sorting to the cell wall envelope of Gram-positive bacteria. Biochim Biophys Acta 1694:269–278

Towers RJ, Fagan PK, Talay SR, Currie BJ, Sriprakash KS, Walker MJ, Chhatwal GS (2003) Evolution of sfbI encoding streptococcal fibronectin-binding protein I: horizontal genetic transfer and gene mosaic structure. J Clin Microbiol 41:5398–5406

van Putten JP, Duensing, TD, and Cole RL (1998) Entry of OpaA+gonococci into HEp-2 cells requires concerted action of glycosaminoglycans, fibronectin and integrin receptors. Mol Microbiol 29:369–379

Veasy LG, Tani LY, Daly JA, Korgenski K, Miner L, Bale J, Kaplan EL, Musser JM, Hill HR (2004) Temporal association of the appearance of mucoid strains of *Streptococcus pyogenes* with a continuing high incidence of rheumatic fever in Utah. Pediatrics 113, e168–172

Wessels MR, Bronze MS (1994) Critical role of the group A streptococcal capsule in pharyngeal colonization and infection in mice. Proc Natl Acad Sci USA 91:12238–12242

Wessels MR, Moses AE, Goldberg JB, DiCesare TJ (1991) Hyaluronic acid capsule is a virulence factor for mucoid group A streptococci. Proc Natl Acad Sci USA 88:8317–8321

Wessels MR, Goldberg JB, Moses AE, DiCesare TJ (1994) Effects on virulence of mutations in a locus essential for hyaluronic acid capsule expression in group A streptococci. Infect Immun 62:433–441

Wilson AT (1959) The relative importance of the capsule and the M-antigen in determining colony form of group A streptococci. J Exp Med 109:257–270

Wood DN, Chaussee MA, Chaussee MS, Buttaro BA (2005) Persistence of *Streptococcus pyogenes* in stationary-phase cultures. J Bacteriol 187:3319–3328

Wu H, Fives-Taylor PM (1999) Identification of dipeptide repeats and cell wall sorting signal in the fimbriae-associated adhesion, Fap1, of *Streptococcus parasanguis*. Mol Microbiol 34:1070–1081

Yanagawa R, Honda E (1976) Presence of pili in species of human and animal parasites and pathogens of the genus corynebacterium. Infect Immun 13:1293–1295

Yanagawa R, Otsuki K Tokui T (1968) Electron microscopy of fine structure of *Corynebacterium renale* with special reference to pili. Jpn J Vet Res 16:31–37

Yilmaz O, Yao L, Maeda K, Rose TM, Lewis EL, Duman M, Lamont RJ, Ojcius DM (2008) ATP scavenging by the intracellular pathogen *Porphyromonas gingivalis* inhibits $P2X_7$-mediated host-cell apoptosis. Cell Microbiol 10:863–875

Common Regulators of Virulence in Streptococci

Nadja Patenge, Tomas Fiedler and Bernd Kreikemeyer

Abstract Streptococcal species are a diverse group of bacteria which can be found in animals and humans. Their interactions with host organisms can vary from commensal to pathogenic. Many of the pathogenic species are causative agents of severe, invasive infections in their hosts, accounting for a high burden of morbidity and mortality, associated with high economic costs in industry and health care. Among them, *Streptococcus pyogenes*, *Streptococcus agalactiae*, *Streptococcus pneumoniae*, and *Streptococcus suis* are discussed here. An environmentally stimulated and tightly controlled expression of their virulence factors is of utmost importance for their pathogenic potential. Thus, the most universal and widespread regulators from the classes of stand-alone transcriptional regulators, two-component signal transduction systems (TCS), eukaryotic-like serine/threonine kinases, and small noncoding RNAs are the topic of this chapter. The regulatory levels are reviewed with respect to function, activity, and their role in pathogenesis. Understanding of and interfering with transcriptional regulation mechanisms and networks is a promising basis for the development of novel anti-infective therapies.

Contents

N. Patenge · T. Fiedler · B. Kreikemeyer (✉)
Institute of Medical Microbiology, Virology and Hygiene, University Medicine Rostock, Schillingallee 70, 18057 Rostock, Germany
e-mail: bernd.kreikemeyer@med.uni-rostock.de

Current Topics in Microbiology and Immunology (2013) 368: 111–153
DOI: 10.1007/82_2012_295

Published Online: 16 December 2012

1 Introduction

Streptococcal species can colonize and live in animals and humans and are present in almost all body compartments. They mostly live in benign and commensal relationship with their hosts. However, several pathogenic species are known, which act as zoonosis pathogens or represent exclusive human pathogens. Infections are multifaceted and occur in all age groups ranging from newborns to adults and the elderly. The pathogenic streptococcal species reviewed in this chapter can cause severe and live-threatening invasive diseases in humans. They are responsible for substantial mortality and morbidity. In the era of spreading antibiotic resistance, these species are studied intensively to understand their pathogenesis mechanisms, including the regulatory machinery in control of virulence factor expression. This knowledge will help to discover novel targets for innovative therapies.

Streptococcus pyogenes is an exclusive human pathogen, which is transmitted from person to person and causes diseases ranging from mild superficial and self-healing to severe invasive diseases. Sepsis, toxic shock like syndrome and necrotizing fasciitis are the most frightening disease characteristics (Cunningham 2000; Carapetis et al. 2005). Antibiotic treatment is still the therapy of choice and is mandatory to prevent post-streptococcal autoimmune sequelae.

Streptococcus pneumoniae, commonly referred to as the pneumococcus, is known to cause a high burden of human disease and death. *S. pneumoniae* can switch from commensal and asymptomatic stage to infectious stage, thereby infecting the lung, blood, and brain. Otitis media and pneumonia are the most frequent infections caused by these bacteria (Mitchell and Mitchell 2010; Gamez and Hammerschmidt 2012).

Like the other streptococci mentioned above, *Streptococcus agalactiae* is also a common asymptomatic colonizer of healthy adults. As an opportunistic pathogen, it is able to defend itself against all host immune system functions, and to cause severe invasive diseases and tissue damage in unborn and newborn children as well as adults (Rajagopal 2009).

For a long time, *Streptococcus suis* was considered a classical swine pathogen responsible for high economic losses in the swine industry worldwide. *S. suis* associated diseases include meningitis, septicemia and endocarditis (Fittipaldi et al. 2012). Late in the 1960s first cases of human infections were reported and since then *S. suis* gained increasing importance as zoonosis pathogen also affecting humans.

Common to all these severe and invasive disease causing pathogens is environmentally driven, coordinated, and fine-tuned regulation of expression of their armament of virulence factor genes. The activity of the most prevalent stand-alone transcriptional regulators, two-component regulatory systems (TCS), eukaryotic-like serine/threonine kinases and, a rather just recently appreciated level, small noncoding RNAs (sRNAs) is the topic of this review.

Numerous so-called stand-alone transcriptional regulators influence the coordinated expression of virulence factors in all streptococci discussed here. Some of them have been extensively studied, especially in *S. pyogenes* (Kreikemeyer et al. 2003; McIver 2009; Fiedler et al. 2010a). In this review, we will focus on those stand-alone regulators that can be found and are involved in virulence regulation in more than one of the pathogenic streptococcal species, such as the members of the RofA-like protein (RALP) family of regulators, Mga and orthologous proteins, as well as several regulators primarily responsible for controlling metabolism.

A common way for bacteria to adapt to environmental conditions is mediated by two-component signal TCS (Hoch 2000; Stock et al. 2000), which are encoded in varying numbers on the chromosomes of these species. In *S. pyogenes*, the genome sequencing era allowed identification of 11 up to 13 such systems (Kreikemeyer et al. 2003; Fiedler et al. 2010a), in *S. pneumoniae* 13 systems are present (Paterson et al. 2006), in *S. agalactiae* up to 20 TCS have been reported, depending on the strain investigated (Glaser et al. 2002; Tettelin et al. 2002), and in *S. suis* 15 TCS were predicted from comparative genomics analyses (Chen et al. 2007). Common to almost all TCS is the basic composition of two proteins. A membrane-associated sensor histidine kinase (HK) is responsible for receiving external signals and autophosphorylates a conserved histidine residue. The phosphate group is subsequently transferred by the HK to a conserved aspartate residue in the cognate response regulator (RR). The RR undergoes a conformational change enabling DNA binding and regulation of gene expression (Stock et al. 2000; Hoch 2000). There are variations of this common activation theme. In several studies, response regulators were found which are independent of their cognate histidine kinases for phosphorylation. Eukaryotic-like serine/threonine kinase/phosphatase systems have been discovered and studied, which either independently regulate pathogen physiology and virulence, or which are networking with TCS systems (Burnside and Rajagopal 2011). For all stand-alone

Table 1 Distribution and function of stand-alone regulators, TCS, and STK/STP systems in different pathogenic streptococci

Regulator system	*S. pyogenes*	*S. agalactiae*	*S. pneumoniae*	*S. suis*
Mga	Mga locus: *emm* and *emm*-like genes	Presence/function unknown	Designated MgrA, Mga_{Spn}	Presence/function unknown
			Pilus (*rlrA* islet)	
	Allelic variants associated with preferred site of infection (*mga*-1: throat; *mga*-2: skin, generalists)		Crucial for development of pneumoniae in mouse model	
	Central in regulatory network			
LuxS	Hemolysis (SLS), SpeB secretion, capsule biosynthesis	Presence reported, function unknown	Biofilm formation, pneumolysin/ autolysin production	Biofilm formation, capsule biosynthesis,
	fasX, *emm*		Competence	Hemolysis, host cell adherence
CcpA	Binds to P_{mga}, transcriptional activation of *mga*		Also designated RegM	Capsule biosynthesis,
	sag operon, *speB*		Capsule biosynthesis, surface enolase,	*eno*, *ofs*, *cpsA2*
			β-galactosidase, *nanA*/*B*, *sodA*, *pcpA*	
CodY	Activates expression of *mga*, *fasX*, *rofA*, and *rivR*	Presence/function unknown	Deletion in wild types fatal	Presence/function unknown
	Represses expression of *covRS* and *sptRS*		*pcpA* (adherence)	
RALP	Important regulators of pilus protein expression, encoded on pathogenicity like-islands			Presence/function unknown
	Act in species-specific and within species even strain-specific regulatory circuits			
	Frameshift mutations occur bistable phenotypes bistable phenotypes			

(continued)

Table 1 (continued)

Regulator system	*S. pyogenes*	*S. agalactiae*	*S. pneumoniae*	*S. suis*
MsmR	Adversely controls pilus gene transcription together with RALP	Positive control of pilus gene expression on pilus island 1	MgrA (Mga family) regulator adjacent to pilus island found Presence of MsmR-like regulator unknown	Presence/function unknown
CovRS	Networking with other TCS and sRNAS Heat, acid, salt stress Most important virulence-associated TCS Spontaneous mutants allowing switch from pharyngeal to invasive lifestyle Unidirectional activity	Crucial TCS for virulence (rat and mouse infection models) General stress response Bidirectional activity	Presence/function unknown	Orphan response regulator CovR present Acts as global negative regulator and repressor of virulence
CiaHR	In control of virulence factors like hemolysis, MSCRAMMs, DNAses	Promotion of intracellular survival Resistance to host innate immune functions Important for survival in blood and brain	Also designated TCS05 Involved in natural competence Stress response Antibiotic resistance Spontaneous mutants in CiaH found in clinical isolates	Important for host cell adherence Crucial for survival in macrophages Important for mortality in mouse and pig infection models

(continued)

Table 1 (continued)

Regulator system	*S. pyogenes*	*S. agalactiae*	*S. pneumoniae*	*S. suis*
Ihk/Irr	Important for growth and survival in blood, saliva, phagocytes, and macrophages Fine tuning together with CovRS	Presence/function unknown	Presence/function unknown	Important for virulence in mice Host cell adherence Macrophage killing Oxidative stress survival In control of cellular metabolism
VicRK	Crucial for murein biosynthesis, cell division, lipid integrity, exopolysaccharide synthesis, biofilm phenotype, virulence factor expression			
STK/STP	Eukaryotic-like serine/threonine kinase/phosphatase systems with regulatory connections to TCS Play important roles in pathogen physiology and virulence			Presence/function unknown

regulators and TCS discussed in this review, designations and known or putative functions are summarized in Table 1.

In addition to the known protein based transcriptional regulation, the role of sRNAs in the control of bacterial gene expression became increasingly evident. The high number of regulatory RNAs identified in different bacterial species was unexpected (Brantl 2009; Narberhaus and Vogel 2009; Waters and Storz 2009) and the variability in length, structure, and mode of action of the different RNAs is very striking (Gottesman and Storz 2011). Bacterial regulatory RNAs influence the expression of genes involved in processes as diverse as stress response, sugar metabolism, and surface composition (Vanderpool and Gottesman 2005; Gottesman et al. 2006; Heidrich et al. 2006; Gorke and Vogel 2008; Gogol et al. 2011; Sharma et al. 2011). Thus, it is not surprising that pathogens employ regulatory RNAs for the tightly controlled expression of virulence factor genes and in some cases for the fine tuning of conventional, protein-mediated gene regulation (Livny et al. 2006; Papenfort and Vogel 2010). Although the regulatory character is shared by this class of RNAs, the basic properties between subtypes vary immensely. There is an intriguing diversity of regulatory mechanisms. Some *cis*-regulatory RNAs are located on untranslated regions (UTRs) of a coding transcript and act as RNA-thermometers or riboswitches (Klinkert and Narberhaus 2009; Bastet et al. 2011). Another group consists of small sRNAs, which are transcribed independently and function via *cis*- and *trans*-antisense base pairing. Of those, *cis*-acting sRNAs are encoded on the opposite strand of their respective target gene. Consequently, they show a high sequence complementarity to their target RNA, which guarantees very specific binding. Recently, differential sequencing analysis revealed a high antisense transcriptional activity in several model organisms, including *Helicobacter pylori* and *S. pyogenes*, which points to the importance of *cis*-regulatory elements in bacteria in general with implications for specific involvement in bacterial virulence regulation (Sharma et al. 2010; Deltcheva et al. 2011). Compared to their *cis*-acting counterparts, *trans*-acting sRNAs exhibit only a short and imperfect complementarity to their target RNAs, allowing them to control several different target genes. These sRNAs are usually regulated in response to environmental stimuli. Some *trans*-acting sRNA molecules act as repressors of translation and destabilize mRNA transcripts but others activate and stabilize the target mRNAs (Thomason and Storz 2010; Storz et al. 2011). Although the conventional point of view regarded *trans*-acting sRNAs as inhibitory antisense regulators, to date a significant number of sRNAs activating bacterial gene expression is known (Frohlich and Vogel 2009). Furthermore, regulatory mechanisms include the stabilization as well as the destabilization of target transcripts (Podkaminski and Vogel 2010). Later in this chapter, we will focus on the structure and function of regulatory RNAs in streptococci. We will give an overview of already characterized sRNAs known to mediate virulence factor regulation in streptococcal diseases, but we will also summarize recent whole-genome screens for sRNAs in pathogenic streptococci.

2 Streptococcal Stand-Alone Transcriptional Regulators

2.1 *Multiple Gene Regulator of Group A Streptococci—Mga and Orthologous Regulators*

One of the best characterized and probably most important stand-alone virulence regulators in *S. pyogenes* is Mga. Initially, Mga was identified as positive regulator of the expression of the M-protein encoding *emm* gene (Caparon and Scott 1987; Okada et al. 1993; Podbielski et al. 1995). Nowadays, it is known that Mga is a global transcriptional activator in the exponential growth phase (Kreikemeyer et al. 2003). The Mga regulon of *S. pyogenes* comprises multiple genes encoding for virulence factors involved in host cell adhesion (e.g., M- and M-like proteins, fibronectin- and collagen-binding proteins) and immune evasion (e.g., C5a peptidase and other complement inhibitors) (Hondorp and McIver 2007; Fiedler et al. 2010a). Genomic analysis revealed the serotype-dependent presence of two allelic variants of the *mga* gene in *S. pyogenes*, *mga*-1 and *mga*-2 (Haanes et al. 1992; Hollingshead et al. 1993). The genomic presence of either version is correlated to the *S. pyogenes* strain's M-protein class, serum opacity factor production, and especially to the structure of the adjacent *emm* gene region, i.e., the absence or presence of *emm*-related genes upstream and downstream of the respective *emm* gene (Bessen and Hollingshead 1994; Hollingshead and Bessen 1995; Bessen and Lizano 2010). It has been proposed that the allelic *mga* variants are indicative of tissue tropism of *S. pyogenes* strains, with *mga*-1 primarily associated with throat infecting strains and *mga*-2 associated with skin infecting or "generalist" strains (Bessen et al. 2005; Bessen and Lizano 2010).

Genes directly activated by Mga are referred to as the Mga core regulon and comprise the virulence genes *emm*, *scpA* (C5a peptidase), *sclA* (collagen-like protein), *sic* (secreted inhibitor of complement), *fba* (fibronectin binding protein), and *sof* (serum opacity factor) as well as the *mga* gene itself (Ribardo and McIver 2006). Furthermore, there are several virulence genes that can be regulated by Mga indirectly, such as the capsule biosynthesis (*has*-) operon or the cysteine protease gene *speB* (Ribardo and McIver 2006; Hondorp and McIver 2007). It was also shown that Mga acts as a transcriptional repressor of genes related to sugar metabolism such as the mannose/fructose phosphotransferase system component IIA (Ribardo and McIver 2006; Hondorp and McIver 2007).

For activation of the core regulon genes, a direct binding of Mga in the promoter region of the respective genes is necessary (McIver et al. 1995, 1999; Podbielski et al. 1995; Almengor and McIver 2004). The mechanism of transcriptional activation by Mga binding seems to vary for different genes, depending on the position of the binding sites in relation to the transcriptional start site. Usually, the Mga binding site overlaps at least in part with the -35 region and is thus located in close proximity to the transcriptional start site of the respective genes. Mga bound to these proximal sites probably stabilizes the binding of the RNA polymerase by direct interaction. For some genes (*sof-sfbX*, *sclA*), the Mga binding site is located further upstream, indicating that binding of Mga to these distal sites rather activates transcription via DNA binding than via direct interaction of Mga with the RNA polymerase. Finally, upstream of the *mga* gene there is both, a distal and a proximal Mga binding site combining both activation mechanisms (McIver et al. 1995, 1999; Almengor and McIver 2004; Almengor et al. 2006). Mga activity displays growth phase association with a maximum in the exponential phase (McIver and Scott 1997; Ribardo and McIver 2003). The transcription of the *mga* gene is not entirely depending on autoregulation but is influenced and fine-tuned by numerous other transcriptional regulators of *S. pyogenes*, such as the regulators of the RALP-family, the sugar metabolism regulator CcpA, the CovR repressed response regulator TrxR and the sugar metabolism regulator MsmR (Podbielski et al. 1999; Beckert et al. 2001; Almengor et al. 2007; Kreikemeyer et al. 2007; Kratovac et al. 2007; Leday et al. 2008; Fiedler et al. 2010a).

Since Mga is involved in the activation of many of the major *S. pyogenes* virulence factors, the deletion of the *mga* gene leads to a dramatic loss of virulence of *S. pyogenes* in vitro and in animal models. Mga-defective mutants are more sensitive to phagocytosis, exhibit a decreased adherence to human skin tissue sections, an attenuated virulence in intraperitoneal and skin infection mouse models, and show a reduced ability to bind human serum and matrix proteins (Perez-Casal et al. 1993; Kihlberg et al. 1995; Luo et al. 2008; Fiedler et al. 2010b).

Mga orthologous regulators were found in *S. pneumoniae* (Hemsley et al. 2003; Solano-Collado et al. 2012) and other streptococci, such as *S. dysgalactiae*, *S. equi*, *S. gordonii*, and *S. mitis* (Vahling and McIver 2006). No Mga-like regulators have been described in *S. agalactiae* and *S. suis*. In *S. pneumoniae* the Mga-like regulator is designated MgrA (TIGR4 strain) or Mga_{Spn} (R6 strain) (Hemsley et al. 2003; Solano-Collado et al. 2012). It has been shown that MgrA is involved in virulence, i.e., by repression of the genes of the pilus-encoding *rlrA* pathogenicity islet (Hava and Camilli 2002; Hemsley et al. 2003). Since MgrA/Mga_{Spn} can be found in all currently sequenced pneumococcal strains while the *rlrA* pathogenicity islet is only present in some of them, it is improbable that the *rlrA* pathogenicity islet genes are the major target of MgrA/Mga_{Spn} regulation (Hoskins et al. 2001; Tettelin et al. 2001; Lanie et al. 2007). Transcriptional activation of the operon downstream of the mga_{Spn} gene in *S. pneumoniae* R6 by Mga_{Spn} binding to two binding sites in the promoter region of this operon has been shown. The function of the respective operon is not known to date (Solano-Collado et al. 2012). Although the regulatory mechanisms and the role of the *S. pneumoniae* Mga-like regulators in pathogenesis are not very well understood, it is apparent that they can function as transcriptional activator and repressor as it has also been shown for Mga of *S. pyogenes*. It is likely that the Mga orthologous regulator(s) play a crucial role in development of full virulence in *S. pneumoniae*, as it has been shown that the presence of MgrA is required for development of pneumonia in a mouse model (Hava and Camilli 2002; Hemsley et al. 2003).

2.2 LuxS and AI-2 Dependent Quorum Sensing

LuxS is an enzyme of the activated methyl cycle (AMC) and catalyzes the reaction from S-ribosylhomocysteine to homocysteine and 4,5-dihydroxy-2,3-pentanedione which can spontaneously convert into an active furanosyl borate diester designated autoinducer 2 (AI-2) (Schauder et al. 2001; Zhu et al. 2004). AI-2 has originally been described to be involved in cell density dependent gene regulation in *Vibrio harveyi* (Surette et al. 1999). Nowadays, it is known that AI-2 is produced by numerous gram-negative and gram-positive species and serves the intra- and interspecies quorum sensing (Schauder et al. 2001; Federle and Bassler 2003; Federle 2009). In many pathogenic bacteria, LuxS/AI-2 has been associated with regulatory processes in virulence (Vendeville et al. 2005; Antunes et al. 2010). In *S. pyogenes*, *S. pneumoniae*, and *S. suis* the impact of LuxS/AI-2 on virulence regulation has been investigated to some extent. Also in *S. agalactiae* the presence of *luxS*/AI-2 has been described but a detailed analysis is still lacking (Ou et al. 2005; Ouyang et al. 2006).

In *S. pyogenes*, there is evidence that LuxS/AI-2 is influencing virulence factor expression in a growth phase dependent manner (Lyon et al. 2001). *S. pyogenes luxS* deletion mutants exhibit, in addition to growth deficiencies, an increased transcription of the streptolysin S precursor gene *sagA* accompanied by enhanced hemolytic activity (Lyon et al. 2001). Furthermore, a decreased secretion of the immunoglobulin degrading cysteine protease SpeB can be observed. Transcription of *speB* or secretion of SpeB might depend on LuxS and seems to be *S. pyogenes* strain specific (Lyon et al. 2001; Marouni and Sela 2003). It has been shown that capsule biosynthesis as well as the transcription of the sRNA gene *fasX*, the *emm3* gene and the immunoglobulin-binding protein-encoding *sib* gene can be influenced by LuxS in *S. pyogenes* in a serotype or strain dependent manner (Marouni and Sela 2003; Siller et al. 2008). LuxS deficient mutants are more efficiently internalized into epithelial cells and show better growth in acidic environments or in the presence of human serum (Siller et al. 2008).

LuxS has been proposed to be the key regulator for early biofilm formation in *S. pneumoniae* (Joyce et al. 2004; Romao et al. 2006; Vidal et al. 2011; Trappetti et al. 2011). Biofilm formation in LuxS deficient mutants is drastically hampered and autolysin and pneumolysin production is decreased. While LuxS activates the transcription of the pneumolysin (*ply*) and autolysin (*lytA*) genes (Joyce et al. 2004; Vidal et al. 2011), the *com*-operon genes involved in the regulation of the genetic competence of *S. pneumoniae* are repressed by LuxS (Romao et al. 2006; Trappetti et al. 2011). Consequently, LuxS deficient mutants proved to be less virulent in nasopharyngeal mouse models and were outcompeted by the wild-type strain when co-administered in an intraperitoneal mouse model (Stroeher et al. 2003; Joyce et al. 2004).

In *S. suis*, the deletion of *luxS* was shown to lead to decreased biofilm formation, capsule biosynthesis, adherence to epithelial cells, hemolytic activity and hydrogen peroxide tolerance (Cao et al. 2011; Wang et al. 2011). In a *luxS* deficient mutant of a *S. suis* serotype 2 strain, the transcription of several virulence-associated genes (e.g. genes for fibronectin/fibrinogen-binding protein FbpS, muraminidase released protein Mrp, or extracellular factor EF) was decreased compared to the parental strain (Wang et al. 2011). In zebra fish or piglet models, *luxS* deficient *S. suis* mutants were described to be severely attenuated in virulence (Wang et al. 2011; Cao et al. 2011).

In *S. pyogenes*, *S. suis*, and *S. pneumoniae luxS* is transcribed as a monocistronic mRNA. In batch cultures, the highest *luxS* expression level can be observed during early exponential growth while maximum AI-2 secretion is reached in the transition phase. Hence, the *luxS* gene expression is apparently not depending on (or correlating with) AI-2 levels (Siller et al. 2008; Han and Lu 2009; Vidal et al. 2011). Although it has been shown that *luxS* transcription is induced by iron in *S. pneumoniae* or repressed by the CovR (CsrR) response regulator in *S. pyogenes*, the regulatory mechanisms controlling *luxS* expression and AI-2 production in streptococci are not well understood (Marouni and Sela 2003; Trappetti et al. 2011). Anyway, phenotypes caused by the deletion of *luxS* might not necessarily be associated with the lack of AI-2 production but could be caused by the accumulation of toxic intermediates of the AMC such as S-adenosylhomocystein, as recently described for *Streptococcus sanguinis* (Redanz et al. 2012). Here, transcriptome analysis revealed that in a *luxS* deficient mutant 216 genes were differentially expressed in comparison to the wild-type strain. When this strain was complemented with an alternative route of the AMC, preventing the accumulation of S-adenosylhomocystein, only nine genes showed altered transcription in comparison to the wild type (Redanz et al. 2012). This experimental approach dissects the AI-2 effects and the AMC effects and clearly demonstrates that biofilm formation of *S. sanguinis* is not depending on AI-2 quorum sensing but on an intact AMC. This demonstrates the importance of distinguishing between the quorum sensing and metabolic or toxic effects of *luxS* deletions. Especially in terms of biofilm formation, an impact of AI-2/LuxS has been described not only for *S. pyogenes*, *S. pneumoniae*, and *S. suis* but also for several streptococci residing in the human oral cavity such as *S. mutans*, *S. anginosus*, *S. gordonii*, or *S. intermedius* (Blehert et al. 2003; Yoshida et al. 2005; Petersen et al. 2006; Ahmed et al. 2008). Doubtlessly, LuxS is crucial for biofilm formation in streptococci, but it is necessary to reconsider the role of AI-2 dependent quorum sensing in the above-mentioned context.

2.3 Regulators in Control of Metabolism

Streptococcaceae have a relatively small genome but need to be able to quickly adapt to changing nutritional conditions in the course of infection. Consequently, complex regulatory mechanisms are applied to allow efficient use of the nutrients available at the respective site of infection. Transcriptional changes as a consequence of varying nutritional conditions not only affect metabolism but also virulence-related genes. One global regulator responsible for regulation of metabolism and virulence in streptococci CcpA is the central sugar metabolism regulator. CcpA is primarily responsible for C-catabolite repression (CCR), which means repression of genes involved in catabolism of sugars less favorable than glucose. This is achieved by binding of CcpA to C-responsive elements (*cre*) in the promoter of the respective genes (Price et al. 2011). In *S. pyogenes*, apart from sugar utilization associated genes also numerous virulence-related genes are repressed by CcpA either directly or indirectly (Kinkel and McIver 2008; Shelburne et al. 2008; Kietzman and Caparon 2011). The impact of CcpA on virulence gene regulation is more pronounced under nutrient limitation (Shelburne et al. 2008). One of the major interfaces between CcpA and virulence in *S. pyogenes* is the control of *mga* expression by direct binding of CcpA to at least one *cre* element in the promoter of the *mga* gene (P*mga*). Binding of CcpA to this *cre* element increases *mga* transcription and therefore indirectly induces the expression of Mga-regulated genes (Almengor et al. 2007). Furthermore, CcpA apparently represses the streptolysin S (*sag* operon) genes (Shelburne et al. 2008; Kinkel and McIver 2008; Kietzman and Caparon 2010). Whether this is due to direct binding to the promoter upstream of the *sagA* gene or to an indirect regulatory effect is controversially discussed (Shelburne et al. 2008; Kinkel and McIver 2008; Kietzman and Caparon 2010). Furthermore, direct regulation of *speB* expression by CcpA has been demonstrated (Kietzman and Caparon 2010). Interestingly, data on the contribution of CcpA to virulence in vivo are contradictory. Two studies have been published on the effect of a *ccpA* deletion in the *S. pyogenes* M1 strain MGAS5005 on virulence in a CD-1 mouse model. While one of the studies found the mutant to be less virulent (Shelburne et al. 2008) the other group observed a hypervirulent phenotype (Kinkel and McIver 2008).

Much less is known about the contribution of CcpA on virulence gene regulation in other streptococci. In *S. suis* and *S. pneumoniae* as well as in *S. pyogenes* CcpA was shown to be involved in regulation of the capsule biosynthesis (Giammarinaro and Paton 2002; Shelburne et al. 2008; Willenborg et al. 2011). Furthermore, CcpA regulates the transcription of the plasminogen-binding surface enolase (*eno*) gene and the *β*-galactosidase gene which is crucial for the colonization of the nasopharynx by *S. pneumoniae* (Iyer et al. 2005; Kaufman and Yother 2007; Carvalho et al. 2011). A comprehensive transcriptome study on *S. pneumoniae* wild-type and CcpA deletion strains under several nutritional conditions additionally revealed a CcpA mediated repression of the expression of neuraminidase encoding genes *nanA/B*, the superoxide dismutase gene *sodA* and the choline-binding protein gene *pcpA* (Carvalho et al. 2011). In *S. suis*, a recent publication described CcpA to be involved in capsule biosynthesis, expression of surface enolase and transcriptional activation of virulence factor encoding genes *ofs* and *cpsA2* (Willenborg et al. 2011). Generally, the role of CcpA in virulence regulation is extremely complex, since CcpA activity is strongly influenced by the availability of sugars in the environment of the bacteria. Hence, the impact of a *ccpA* deletion on virulence gene expression might be different in the presence of glucose than in the presence of less-favorable sugars in the growth medium.

Another metabolic regulator commonly involved in virulence gene regulation is the branched chain amino acid (BCAA) activated CodY, which is induced under nutritional deprivation conditions and can exclusively be found among the Firmicutes (Sonenshein 2005; Stenz et al. 2011). With a helix-turn-helix motif the dimeric CodY with one BCAA bound to each monomer can bind to conserved palindromic DNA sequences called CodY boxes. Common targets of CodY regulation in streptococci and other bacteria are i.e., oligopeptide permeases (*opp*) (Malke et al. 2006; Malke and Ferretti 2007; Hendriksen et al. 2008; Stenz et al. 2011) and the *codY* gene itself (Guedon et al. 2005). Additionally, several virulence factors are directly or indirectly regulated by CodY in streptococci. In *S. pyogenes*, CodY has been shown to activate the expression of *mga, fasX, rofA,* and *rivR* and to repress the expression of *covRS* and *sptRS* (Malke et al. 2006; Malke and Ferretti 2007; Kreth et al. 2011). Since all these genes are encoding for important *S. pyogenes* virulence regulators, the effect of CodY on virulence gene expression, although indirect, is very comprehensive. A deletion of *codY* in *S. pyogenes* consequently leads to decreased expression of the genes encoding for M-protein, streptokinase, streptolysin O, NAD-glycohydrolase, C5a peptidase, SpeH and others while an increased transcription of e.g., capsule biosynthesis genes can be observed (Malke et al. 2006; Malke and Ferretti 2007; Kreth et al. 2011). Although CodY binds in the promoter region of its own gene, neither in the promoter regions of the regulator nor of the virulence factor genes, CodY binding motifs have been detected (Malke and Ferretti 2007; Kreth et al. 2011). Hence, the mechanism by which CodY influences the expression of the regulator genes is not known to date. It has been speculated that at least the regulation of CovRS (see Sect. 3.1) could depend on the intracellular potassium level, which is influenced by the impact of CodY on the expression of potassium uptake system genes (Malke and Ferretti 2007). Especially, the interplay between CodY and CovRS has been investigated in more detail by analysis of *codY*, *covR*, and *codY-covR* deletion mutants (Kreth et al. 2011). It has been postulated that the expression of the abovementioned virulence genes in *S. pyogenes* are well balanced by the actions of CcpA (integrating information on sugar availability via PTS), CodY (integrating amino acid nutrition information), and CovRS directly acting at the promoter of many virulence factors to adapt to either invasive or noninvasive phenotypes depending on the nutritional conditions. CodY is postulated to counteract the activity of CovRS by stimulating expression of those genes repressed by CovRS (Kreth et al. 2011), leading to a rather invasive phenotype when nutritional conditions are unfavorable.

In *S. pneumoniae*, the investigation of CodY regulatory effects is complicated by the fact that a deletion of *codY* is fatal unless an ectopic copy of the gene is present (Caymaris et al. 2010). Originally, it had been published that CodY is activating the expression of *pcpA*, a choline-binding protein necessary for successful adherence of the bacteria to nasopharyngeal cells (Hendriksen et al. 2008). The *S. pneumoniae* D39 *codY* deletion strain used in the study of Hendriksen et al. (2008) has later been sequenced and it has been postulated that mutations in genes for the ferric uptake iron permease (*fatC*) and an oligopeptide permease (*amiC*) allowed for tolerance of the *codY* deletion in that strain (Caymaris et al. 2010). The data indicate that in *S. pneumoniae* CodY action to certain extent controls the ability of the bacteria to adhere to host cells.

To the best of our knowledge, the role of CodY in *S. suis* and *S. agalactiae* has not been investigated yet. For an overview on the involvement of CodY in virulence gene regulation in *S. mutans* and other gram-positive bacteria, we refer to a recent review published by Stenz et al. (2011).

Next to CcpA and CodY, there are many other metabolic regulators involved in virulence gene regulation in streptococci, such as the regulatory tagatose-1,6-bisphosphate aldolase LacD.1 in *S. pyogenes* (Loughman and Caparon 2006, 2007), the methionine transport regulator MtaR in *S. agalactiae* (Shelver et al. 2003; Bryan et al. 2008) or the stringent response amino acid metabolism regulator RelA in *S. pyogenes* and *S. pneumoniae* (Malke et al. 2006; Kazmierczak et al. 2009), to name only a few. Apart from the direct or indirect effects of metabolic regulators on virulence gene expression, tight regulation of metabolic processes in response to the changing nutritional conditions encountered by streptococci in the course of infection is crucial for the general fitness of the bacteria and therefore essential for successful infection of the host.

2.4 Transcriptional Regulators of Streptococcal Pilus Genome Regions

Two classes of transcriptional regulators, which are encoded on the chromosomes of several streptococcal species, gained special attention during the last couple of years. The RALP-family of RofA/Nra regulators is present and active in *S. pyogenes*, *S. agalactiae*, *S. pneumoniae* whereas no information is available in *S. suis*. The second family is the MsmR-like stand-alone transcriptional regulator family, which belongs to the large group of AraC/XylS-type regulators. This family is present and active in *S. pyogenes* and *S. agalactiae*. No orthologous genes have been reported in *S. pneumoniae* and *S. suis* yet.

As the most intriguing and unique feature of these regulators, presence of the genes of both regulator families in so-called pathogenicity island-like chromosomal regions can be noted. Such regions encode all necessary proteins enabling streptococci to form long pilus-appendages on their surface. Pili are now recognized to play important and pivotal roles in infections caused by streptococci. The genetic makeup of these discrete genomic regions, their size, their composition, aspects of expression, assembly, and function in virulence have recently been reviewed on a comparative basis (Kreikemeyer et al. 2011). Briefly, in *S. pyogenes* nine different pilus region variants can be found among clinical isolates (FCT1–FCT9), *S. pneumoniae* isolates encode two pilus variant regions (PI-1 and PI-2), in *S. agalactiae*, the pilus proteins are encoded in three related genomic regions (island-1, island-2a, and island-2b), and recently *S. suis* was shown to encode four distinct genomic regions encoding pilus proteins (*srtBCD*, *srtE*, *srtF*, and *srtG*) (Telford et al. 2006; Takamatsu et al. 2009; Kreikemeyer et al. 2011). However, these important virulence factors are not limited to *S. pyogenes*, *S. agalactiae*, *S. pneumoniae*, and *S. suis*. There is a growing list of streptococcal species for which pilus genome islands and gene clusters have been identified and surface-localized pilus expression was proven. Recently, *Streptococcus gallolyticus*, a causative agent of infective endocarditis also associated with colon cancer, was shown to encode three separate pilus loci, named *pil1-pil3* (Danne et al. 2011). Also *S. oralis*, *S. mitis*, and *S. sanguinis*, all members of the Mitis group of streptococci, harbor a pilus-encoding genomic region and express variable pili on their surface (Zahner et al. 2011), resembling the genetic organization of the PI-2 islet of *S. pneumoniae*.

From the virulence point of view, all pili play a similar role in the respective species during the infection process. In case of *S. pyogenes*, numerous studies have shown their importance for attachment to pharyngeal cells, human tonsillar epithelium, and keratinocytes and suggested a role in autoaggregation and biofilm formation. In *S. agalactiae*, pili are important for adherence to human brain microvascular endothelial cells and cervical epithelial cells, and have important functions in biofilm formation and resistance against cationic antimicrobial peptides as well as for survival in phagocytes. Mutational analyses revealed a function for *S. pneumoniae* pili in many pathogenic processes, including adhesion to respiratory cells. Interestingly, mutants in the *S. suis srtF* pili cluster were not attenuated in host cell adherence and a murine model of *S. suis* sepsis, suggesting that pili structures are dispensable for critical steps of *S. suis* pathogenesis and infection (Fittipaldi et al. 2010). The role of pili in the Mitis group of streptococci needs to be elucidated, whereas in *S. gallolyticus* pili are critical for collagen binding, biofilm formation, and virulence in experimental endocarditis (Zahner et al. 2011; Danne et al. 2011).

Since pilus proteins of many streptococcal species are apparently immunogenic—they might even represent excellent candidates for vaccine development—it is critical for streptococcal pathogens to tightly control expression of the pilus-encoding genes. Within the pilus gene clusters, this is achieved by aforementioned stand-alone transcriptional regulators of the RALP- and XylS/AraC-families. However, also stand-alone regulators and two-component signal TCS encoded outside of the core pilus genomic regions are implicated in pilus gene regulation, but are discussed in other sections of this review.

In *S. pyogenes* pilus islands, the RALP-family regulator RofA/Nra and the XylS/AraC-family regulator MsmR, both encoded within the FCT regions, adversely control pilus gene transcription (Nakata et al. 2005; Kreikemeyer et al. 2007). Of note, *S. pyogenes* serotype-dependent regulatory circuits have been reported (Luo et al. 2008), further complicating the attribution of common functions of these regulators. External signals to which pilus expression is responsive include temperature, anaerobic atmosphere, and pH (Granok et al. 2000; Nakata et al. 2009; Manetti et al. 2010). In *S. agalactiae*, three RALP-family regulators have been described and investigated: (I) RogB, which is present 284 bp upstream of island-2a but is absent from island-2b, has been described to positively control transcription of *fbsA*, encoding a *S. agalactiae* fibrinogen-binding protein, negatively regulates capsular polysaccharide gene cluster expression (Gutekunst et al. 2003), and acts as a positive transcriptional regulator of the PI-2a encoded genes (Dramsi et al. 2006), (II) Rga which is located outside of the pilus region and is involved in transcriptional control of the *secA2* locus (Mistou et al. 2009) and was recently identified as the major transcriptional activator of the PI-2a island in *S. agalactiae* (Samen et al. 2011; Dramsi et al. 2012), (III) Gbs1426, which is more distantly related to RogB and Rga. In *S. agalactiae*, the *sag0644* gene, present 380 bp upstream of the island 1, is the respective pilus region AraC/XylS-type regulator with 72 % identity toward the *S. pyogenes* MsmR regulators present in *S. pyogenes* FCT regions 3, 4, 7, and 8 (Nakata et al. 2005). Deletion of this gene in *S. agalactiae* decreased production of pili, suggesting a positive role in pilus expression. Of note, *S. agalactiae* pilus region PI-2b does not encode any RALP- or AraC/XylS-type regulator.

In *S. pneumoniae*, almost nothing is known on regulation of genes encoded in PI-1, and PI-2 regulation is only partially studied. Both pilus genomic islands are flanked by regulator-encoding genes (Scott and Zahner 2006; Telford et al. 2006). One gene encodes the transcriptional regulator RlrA, which belongs to the RALP-family, and which was shown to be required for colonization, lung infection, and systemic infection. RlrA is positively autoregulated, and controls expression of six genes within the PI-2 island, acting on four different promoters (Hava et al. 2003; Kreikemeyer et al. 2011). Downstream of both *S. pneumoniae* pilus islands MgrA is encoded. This transcriptional regulator belongs to the Mga family, best described in *S. pyogenes*. MgrA activity is also important for development of pneumonia and this regulator acts as transcriptional repressor of the same four promoters controlled by RlrA (Hemsley et al. 2003). MgrA is discussed into more detail elsewhere in this review. It should be noted that *S. pneumoniae* is the only species, in which MgrA is encoded so close to the pilus genomic regions. In *S. pyogenes*, the Mga gene is more distantly located to the FCT regions, although there is a clear regulatory circuit connection of Mga and RALP regulators in *S. pyogenes*, which is absent in *S. pneumoniae*. Moreover, also metal-dependent regulators were implicated in pilus gene control. For example, the PsaR (Mn^{2+}-dependent) and MerR (metal-dependent) regulator activities converge in the pilus island control (Rosch et al. 2008).

In the *S. suis* pilus genome islands no transcriptional regulators have been characterized into detail. Only a MerR metal-responsive regulator with homology to *S. agalactiae* MerR-type regulators has been annotated in the SrtE-*S. suis* pilus island (Takamatsu et al. 2009).

Two common aspects regulating pilus gene expression need to be mentioned. First, during investigation of clinical isolates many point mutations in pilus genes were identified leading to either altered pilus protein composition or nonpiliated variants. In *S. agalactiae*, variability in the expression of pili 1 and 2a among isolates are due to frameshift mutations in the aforementioned regulatory genes. One-base-pair deletions inactivating the *rogB* gene, or in-frame-stop codon mutations inactivating the *sag0644* gene were reported. In a *S. suis* strain carrying a *srtF* pilus cluster, a nonsense mutation at the 5′ end of the gene encoding the minor pilin subunit was detected (Fittipaldi et al. 2010).

Second, in *S. pyogenes* and *S. pneumoniae*, a rather bistable expression mode was found. Nakata and colleagues reported higher expression levels of the *S. pyogenes* pilus backbone protein FctA at 30 °C compared to 37 °C (Nakata et al. 2009). Moreover, at 37 °C only 20 % of all *S. pyogenes* cells expressed pili on their surface. This number increases to 47 % at 30 °C, which is a clear indication of a bistable expression mode. In *S. pneumoniae*, the bistable expression of type I pili is dependent on the native *rlrA* promoter, as a nonexpressing population reverts to the previous bimodal distribution, whereas the expressing population retains the same high level expression (Basset et al. 2011). Other studies also reported the positive feedback loop being under control of the transcriptional regulator RlrA expression (De Angelis et al. 2011; Basset et al. 2012).

Together, there are many common themes among transcriptional regulators, their action on pilus gene transcription and their regulatory circuits among different streptococcal species.

3 Streptococcal Two-Component Signal Transduction Systems

3.1 The CovRS/CsrRSS system

The CovRS/CsrRS TCS is the best characterized in *S. pyogenes* and *S. agalactiae*. In *S. pyogenes*, it was first discovered by several groups as a major regulator of capsule gene expression, streptolysin S production, and cysteine protease SpeB expression (Levin and Wessels 1998; Heath et al. 1999; Bernish and van de Rijn 1999). This system is functionally connected to *S. pyogenes* general stress response, including growth in heat, acid environments, and under salt stress. CovRS is also important for growth under iron starvation, in the presence of antimicrobial peptides, and plays a role in the serotype-dependent *S. pyogenes* biofilm formation and keratinocyte adherence (Dalton et al. 2006; Froehlich et al. 2009; Sugareva et al. 2010). Moreover, it is apparently the major sensor of *S. pyogenes* for Mg^{2+} (Gryllos et al. 2007).

Apart from the above-mentioned *S. pyogenes* virulence factors and functions, many more are directly or indirectly under control of this important TCS. A total of 15 % of all genes encoded on the *S. pyogenes* chromosome are repressed by CovR, including its own gene. The activity of CovR on most promoters is direct and involves the consensus binding sequence ATTARA, mutation of which relieved CovR repression. Promoter occlusion is the primary mechanism of repression by CovR (Gusa and Scott 2005). Phosphorylation of CovR is critical to its functional activity and leads to dimerization (Gao et al. 2005; Gusa et al. 2006). However, a subcutaneous mouse infection model revealed that CovR is independent of its cognate sensor kinase CovS to get phosphorylated and exert its control functions, suggesting other sources for phosphotransfer (Dalton et al. 2006). Eukaryotic-type serine/threonine kinases, discussed in the next section, are donors for phosphotransfer. Of note, the HK CovS itself was found to inactivate the RR CovR, thereby allowing *S. pyogenes* to grow under the above-mentioned stress conditions. Thus, the CovRS system has a Janus-like behavior. CovS acts as a kinase to activate the CovR response regulator, which subsequently acts as gene repressor. However, CovS, upon environmental stimulation, can act as a phosphatase inactivating CovR to permit gene transcription from respective promoters. The importance, functional activity and the Janus-like behavior of this system in *S. pyogenes* have recently been reviewed (Churchward 2007).

Although CovRS is among the most important TCS in *S. pyogenes*, it is not a master regulator per se, but is rather integrated into existing regulatory networks and is connected and linked to other *S. pyogenes* stand-alone regulators and sRNAs (Kreikemeyer et al. 2003; Fiedler et al. 2010a). Roberts and Scott investigated the link between CovRS and the Mga regulon (described elsewhere in this review) and found CovRS to repress RivR, a RALP-family regulator (Ralp4). Adjacent to *rivR* these authors identified *rivX*, encoding a small regulatory RNA. RivR enhanced transcription by Mga in vivo and in vitro and the *rivRX* locus was involved in pathogenesis in a mouse soft tissue infection model (Roberts and Scott 2007). Another *S. pyogenes* TCS, termed TrxSR, which is homologous to the *S. pneumoniae* HK07/RR07 system, was found as another CovR repressed system (Leday et al. 2008). TrxR activates transcription of Mga-regulated virulence genes, and thus, TrxR defines a new pathway by which CovR affected virulence via control over Mga. A study conducted by Shelburne and colleagues established partially overlapping transcriptomes of Cov and CcpA mutants, proved that both regulator proteins bound to promoters of co-regulated genes, and identified attenuated phenotypes of the mutants in a myositis model (Shelburne et al. 2010). Moreover, CovRS activity was critical for *S. pyogenes* growth in human blood, during pharyngitis in cynomolgus macaques and mouse soft tissue infection. Pioneering work of Sumby and colleagues linked CovRS function, observed in vitro and from animal studies, to the real situation in clinical isolates (Sumby et al. 2006). Two fundamentally different transcriptome profiles could be detected during comparison of nine clinical isolates. The pharyngeal transcriptome differed by 10 % from the invasive transcriptome. Complete genome sequencing of an invasive transcriptome isolate uncovered a 7-bp frameshift mutation in the CovS encoding gene. Based on these initial observations, further studies revealed that different in vivo-induced *covR* mutations led to different transcriptomes and that the CovS-mediated regulation of CovR activity is critical for *S. pyogenes* cycling between pharyngeal and invasive phenotypes (Trevino et al. 2009). Although also wild-type bacteria undergo extensive transcriptional reprogramming under in vitro and in vivo infection conditions, this rapid host adaptation is not sufficient to survive in the host. Rather the hypervirulent *cov* mutant strains are able to outcompete the wild type, and thus it can be concluded that mutations and rapid reprogramming are critical steps allowing *S. pyogenes* to switch infectious phenotypes and to move between niches. This puts sociomicrobiological aspects into the focus of infection research (Aziz et al. 2010).

What are the functions of the CovRS system in *S. agalactiae*? The genes for the CovRS orthologous system in *S. agalactiae* are part of a seven-gene operon and mutation led to dramatic phenotypic changes (Lamy et al. 2004). Although the mutant adhered much better to epithelial cells, the survival in human serum was attenuated. Moreover, as also reported for *S. pyogenes*, the *S. agalactiae* CovRS system is critical for virulence, as proven by neonate rat sepsis and mouse infection models (Lamy et al. 2004; Jiang et al. 2005). A total of 76 genes are repressed whereas 63 were positively regulated. Transcriptome comparison of several strains uncovered a conserved 39 gene core regulon (Jiang et al. 2008). Moreover, there was a significant overlap of the acid stress transcriptome in *S. agalactiae* with the CovRS regulon (Santi et al. 2009), and CovRS as well as pH-regulated *S. agalactiae* adherence to host cells of vaginal, cervical and respiratory origin (Park et al. 2012).

In contrast to CovRS in *S. pyogenes*, the *S. agalactiae* system can act bidirectional and is not an exclusive repressor system (Jiang et al. 2008). Direct binding of phosphorylated CovR to many of the target gene promoters has been reported (Lamy et al. 2004; Jiang et al. 2008). However, the binding site differs from the consensus sequence reported for *S. pyogenes* CovR, although both share a high AT content. It can be concluded that the general classes of genes under control of both systems are identical in *S. pyogenes* and *S. agalactiae*. It is currently unknown, whether *S. agalactiae* CovR is dependent on CovS activity, but there is clear evidence linking CovRS and serine/threonine kinase systems also in *S. agalactiae* (discussed in Sect. 4).

It is obvious that the CovRS system is central to the virulence of both streptococcal species. Many similar functions have been elucidated, but depending on the pathogens primary niche in the host, both systems also underwent a pathogen-specific evolution and adaptation. It remains to be established if point mutations in the CovRS system in *S. agalactiae* also occur in a disease specific manner in clinical isolates and if such mutants would be as hypervirulent as their *S. pyogenes* counterparts.

No CovRS orthologs were found in *S. pneumoniae*. In *S. suis* exclusively CovR was identified as an orphan response regulator, lacking the adjacent CovS histidine kinase. Virulence studies characterized the *S. suis* CovR as a globally acting negative modulator of virulence, as in a mutant most phenotypes changed in the direction of higher virulence potential (Pan et al. 2009).

3.2 The CiaHR System

The CiaHR TCS is widespread among streptococci. Genes encoding this TCS are found on the chromosomes of *S. pyogenes*, *S. agalactiae*, *S. pneumoniae*, and also *S. suis*. The amino acid sequence identity of CiaH and CiaR from different species ranges between 48–71 and 77–85 %, respectively (Riani et al. 2007). Most data on Cia function have been compiled from studies in *S. pneumoniae*. An increased cefotaxime resistance and impaired natural competence was noted in a *S. pneumoniae* CiaH mutant (Guenzi et al. 1994). The influence on competence involves sensing of Ca^{2+} and oxidative stress (Zahner et al. 2002). A key connection between the *S. pneumoniae* CiaHR regulon, which has been defined in at least three different strains, and competence is the CiaHR control of HtrA expression. This surface expressed serine protease is down-regulated in CiaHR mutants (Paterson et al. 2006). A restoration of *htrA* expression in the Cia mutant background alone restored natural competence (Sebert et al. 2005). *S. pneumoniae* Cia regulons comprise roughly 20 genes and apart from competence genes also stress response genes and operons were identified (Paterson et al. 2006), suggesting natural competence as *S. pneumoniae* stress factor, which is balanced by Cia activity. Of note, many other virulence genes are found in the Cia regulon, which makes it difficult to clearly define its role in virulence. Further studies have revealed that the CiaH acts as kinase/phosphatase, that CiaR needs to be phosphorylated, that CiaR can obtain phosphates independent from CiaH, and that CiaR acts directly on 15 promoters identified by transcriptional mapping (Halfmann et al. 2007, 2011). Those promoters belong to genes involved in teichoic acid biosynthesis, sugar metabolism, chromosome segregation, and protease maturation. Five of those promoters were identified upstream of small noncoding RNAs (Marx et al. 2010). Matching the observation made in the CovRS TCS in *S. pyogenes*, clinical *S. pneumoniae* isolates, particularly those with spontaneous β-lactame resistance, apparently acquired mutations in the CiaH encoding gene. This suggested that *ciaH* alleles, which overstimulate the CiaR regulon expression, are present in clinical isolates (Muller et al. 2011). Although not studied into such a detail as in *S. pneumoniae*, it is apparent that the *S. pyogenes* Cia system has different functions. As *S. pyogenes* is not naturally competent and lacks large parts of the *com* operon genes, regulation of competence, and as a consequence, also major regulation of stress response is not attributed to *S. pyogenes* CiaHR. The CiaH sensor mutants in two *S. pyogenes* serotype M49 strains allowed identification of up to 120 genes as Cia-controlled, with equal numbers up- and down-regulated (Riani et al. 2007). Among those, genes encoding proteins for divalent cation transport, PTS systems, hemolysins, hyaluronidase, matrix-protein-binding factors, and DNAses were identified. Divergent to observations in *S. pneumoniae*, metal ions did not affect Cia expression significantly, antibiotic resistance was not affected, and only a small number of six stress response genes were differentially transcribed in the *S. pyogenes ciaH* mutant (Riani et al. 2007). In *S. suis*, *ciaHR* mutants revealed a decreased adherence toward HEp-2 and PIEC cells, a higher susceptibility toward killing by RAW2647 macrophages, and were attenuated in mice and pig animal models of infection (Li et al. 2011). Not unexpected, CiaR was found to promote intracellular survival and resistance to innate immune defences in *S. agalactiae*. A *ciaR* mutant had a reduced survival in neutrophils, human macrophages and brain microvascular endothelial cells (BMCE) (Quach et al. 2009). In mouse infection models, the mutant was attenuated and showed decreased survival in blood and brain compared to the wild type (Quach et al. 2009). In summary, particularly in *S. pneumoniae* CiaHR is crucial for natural competence and general virulence, whereas in the other species virulence control is the major function of Cia.

3.3 The IhK/Irr System

The Ihk/Irr TCS of *S. pyogenes* is highly important for the pathogenesis of these bacteria and is thus reported to be active during *S. pyogenes* growth in many body fluids and host niches. As human saliva is one of the first liquids encountered after *S. pyogenes* droplet-mediated entry into the host oral cavity, growth, and persistence in saliva was tested. The most important TCS in hierarchy in saliva is the SptRS TCS. Among many others, also the genes encoding the Ihk/Irr TCS were up-regulated during *S. pyogenes* growth in saliva (Shelburne, III et al. 2005). Moreover, a role for Ihk/Irr could be established during *S. pyogenes* growth in human blood, conditions under which up to 75 % of the genes were differentially transcribed, and during phagocytosis by neutrophils, under which varying sets of genes are time-dependently and differentially transcribed (Voyich et al. 2003; Graham et al. 2005). A more recent study discovered 145 differentially transcribed genes during a 2-h *S. pyogenes* persistence in macrophages, of which many belong to metabolic and energy dependent processes. Ihk/Irr was again established as a key regulator important for the early processes in *S. pyogenes*-macrophage interaction (Hertzen et al. 2012). A major shift of response regulator activity from time point 2–6 h during persistence was noted, in which Ihk/Irr was down-regulated and the CovRS system was up-regulated. This hinted toward a TCS driven and fine-tuned temporal expression shift in *S. pyogenes* intracellular life cycle (Hertzen et al. 2012). In *S. suis* serotype 2, an Ihk/Irr ortholog was recently characterized (Han et al. 2012). Deletion of this system notably attenuated virulence in mice. Mutants showed reduced adherence to host cells, an increased susceptibility to macrophages killing, and a decreased survival under oxidative stress conditions. Of note, the important *S. suis* virulence factors suilysin, autolysin, and muraminidase-released protein encoding genes were not under Ihk/Irr control. Rather cell metabolism and superoxide dismutase were repressed in the mutant (Han et al. 2012). It is conceivable, that in the other streptococcal species discussed here, other TCS have similar functions although they cannot be classified as Ihk/Irr orthologs based on sequence homology.

3.4 The VicRK System

The VicRK system is noteworthy as it is present and conserved in many Firmicutes and for a long time was considered essential for growth in all organisms after many failed mutation attempts. In the literature, there are a couple of synonymous names, like YycFG, MicAB, and WalRK. An unconditional *vicR* insertional mutant generated in *S. pyogenes* allowed insight into functional aspects of this system (Liu et al. 2006). The mutant grew well in rich lab media, but was severely attenuated in growth in nonimmune human blood and serum, had attenuated virulence, and was unstable in mice. However, as phagocytosis and killing was normal, a general influence on evasion of host defences was excluded. Among Vic-controlled genes, those involved in cell-wall hydrolysis (*pcsB*), phosphotransferase systems, osmoprotectant transporters, and nutrient uptake were identified by transcriptomics (Liu et al. 2006). A view across the species barriers linked the essentiality in some organisms to VicRK exerted control over functions like murein biosynthesis, cell division, lipid integrity, exopolysaccharide synthesis, biofilm formation, and also virulence factor expression. However, the signals stimulating the Vic-system are largely unknown (Winkler and Hoch 2008). In *S. pneumoniae*, the essentiality of the Vic-system could only be bypassed by a constitutive PcsB cell-wall hydrolase expression, suggesting that control over cell-wall biosynthesis and osmosis is the critical function (Winkler and Hoch 2008). Recently, also in *S. suis* in vivo swine infection the VikRK system was found up-regulated together with its target gene *pcsB* (Li et al. 2010). Since not many virulence factor encoding genes were found to be directly controlled by the VicRK TCS, it cannot be designated a general virulence regulator. However, considering that central cell-wall turnover and metabolic functions are also directly linked to fitness of the pathogens in vivo there is a clear link of VicRK to virulence. Apparently, homologs and orthologs were also found in Actinobacteria like *Streptomyces*, *Mycobacterium*, and *Corynebacterium* species, which places the Vic-system in the focus as excellent target for novel antimicrobial therapy strategy developments (Winkler and Hoch 2008). First structure-based virtual screenings of potential inhibiting compounds from the SPECS library have been done in *S. pneumoniae*, and interesting target compounds with activity against *S. pneumoniae* were identified in vitro and also in a mouse sepsis infection model. These compounds decreased mortality in mice and had no general cytotoxic effect (Li et al. 2009).

4 Eukaryotic-Type Serine/Threonine Kinases in Streptococci

Results discussed in the paragraphs above have already indicated that some response regulators can be phosphorylated independently of their cognate kinases, suggesting a potential link to other TCS within the same organism. Just recently, a whole new level of signal recognition and signal processing has been discovered in prokaryotes thanks to advances in genetic strategies and genome sequencing approaches. This includes eukaryotic-like serine/threonine kinases (STKs) which were found to be linked to adjacent phosphatases (STPs). An in-depth review focussing on discovery and functional analysis of these systems in various species has been published (Burnside and Rajagopal 2011). What places these signaling and response systems into the research focus is the fact that their eukaryotic counterparts are excellent targets for therapy. Many STK inhibitors are approved by the United States Food and Drug Administration (USFDA) or at various stages in clinical trials. Thus, the prospect that such inhibitors also work for the bacterial STK/STP systems is promising. The STKs in *S. pyogenes*, *S. pneumoniae*, and *S. agalactiae* have been identified at nearly the same time (Rajagopal et al. 2003; Echenique et al. 2004; Jin and Pancholi 2006). The *S. pyogenes* STK and STP (SP-STK and SP-STP) were characterized as functional manganese-dependent kinase and phosphatase enzymes (Jin and Pancholi 2006). An altered cell-shape, incomplete cell separation, a loosely associated electron-dense layer, and a tendency to settle during growth in laboratory media were the initial phenotypes after mutation of SP-STK. Increases in doubling time and hemolysis activity were noted. For the virulence of *S. pyogenes*, the SP-STK was found to be critical for host cell adherence and resistance to phagocytic killing. As a mediator of SP-STK activity on gene regulation, phosphorylation of a 10 kDa histone-like protein was suggested (Jin and Pancholi 2006). The SP-STP was subsequently shown to be involved in phosphorylation of the *S. pyogenes* VicRK and CovRS TCS (Agarwal et al. 2011). In *S. agalactiae*, mutants defective in STK and STP were attenuated in a neonatal rat sepsis model (Rajagopal et al. 2003). STK expression is vital for resistance to human blood, neutrophils, and oxidative stress. Transcription of the *S. agalactiae* ß-hemolysin is mediated via threonine phosphorylation of the CovR response regulator, which improved CovR repression (Rajagopal et al. 2006). This verified a regulatory connection between TCS and STK/STP systems in these pathogenic streptococcal species. STP mutation in *S. agalactiae* allowed identification of more phenotypes and up to 294 differentially transcribed genes were found. Phosphopeptide enrichment methods allowed identification of 35 STK/STP phosphorylated peptides as part of 27 different proteins (Burnside et al. 2011). In *S. pneumoniae*, pleiotropic functions are under control of a single STK encoded on the chromosome, including virulence, competence, antibiotic resistance, growth and stress response (Echenique et al. 2004), and functional links to active TCS systems were found (Agarwal et al. 2012). Direct STK phosphorylation of target proteins involved in ion transport, cell division, and RNA-polymerization was proven. Currently, the role of the phosphatase is not known. However, it is tempting to speculate that the *S. pneumoniae* phosphatase functionally resembles *S. pyogenes* and *S. agalactiae* phosphatases. Together, these streptococcal serine/threonine kinase–phosphatase systems play an important role in the physiology and virulence of the species discussed here. They apparently have rather conserved functions in all streptococci. Currently, not much is known about the signals that initiate their activity and where in the regulatory hierarchy in conjunction with stand-alone transcriptional regulators and TCS they play their major role. The success story using inhibitors in their eukaryotic counterparts to fight human diseases raises hope that novel inhibitors counteracting bacterial STK/STP systems will be developed and serve as innovative therapies in combating severe bacterial infections.

5 Noncoding Regulatory RNAs

5.1 Cis-Regulatory RNAs

Diverse *cis*-regulatory systems controlling virulence factor genes have been described in gram-positive pathogens including RNAIII in *Staphylococcus aureus* (Novick et al. 1993) and a thermosensor located in the 5'-UTR of the *prfA* mRNA in *Listeria monocytogenes* (Johansson et al. 2002). One of the first examples for a regulatory RNA discovered in *S. pyogenes* was the untranslated 500 bases RNA of the streptococcal pleiotropic effect locus (*pel*). The region contains the structural gene *sagA* coding for a precursor of the streptococcal hemolysin SLS. The locus was initially detected following transposon mutagenesis of a *S. pyogenes* M49 serotype strain. It was found to positively regulate the genes of important streptococcal secreted and surface virulence factors. Loss of the 500 bases transcript decreased transcription of *emm* and *speB* genes and reduced secretion of streptokinase. Whether a protein encoded by the locus or an untranslated RNA was responsible for the regulatory effects was not clear at the time (Li et al. 1999). In a later study, it could be shown that *pel*-dependent virulence factor regulation was mediated by a 459-bases untranslated RNA molecule (Pel). It was demonstrated that regulation by Pel occurs at both transcriptional (e.g. *emm* and *sic*) and post-transcriptional (e.g. SpeB) levels (Mangold et al. 2004). Strain specificity of Pel function is indicated by the fact that in a *sagA*-deficient mutant with an M6 background, *emm* transcription was not affected (Biswas et al. 2001). Similar results have been obtained in *S. pyogenes* M1 and M18 Tn916 *sagA* mutant strains (Betschel et al. 1998). Additionally, *pel* deletion mutant analysis of four M1T1 *S. pyogenes* isolates did not confirm any regulatory function of the Pel sRNA in this serotype (Perez et al. 2009).

Expression of ribosomal protein genes (*rplJ-rplL*) in *E. coli* is regulated by an autogenous control mechanism involving the 5'-untranslated region of the mRNA. Ribosomal proteins L10 and L12 bind to the leader region of the L10 operon and inhibit translation (Johnsen et al. 1982). The secondary structure of the leader RNA is important for the formation of a stable complex (Christensen et al. 1984). A family of putative ribosomal protein leader autoregulatory structures was also found in *B. subtilis* and other Firmicutes. For the genus *Streptococcus*, 393 members of the L10-leader RNA family from 385 species are listed in Rfam (Gardner et al. 2009). Recently, the expression of a L10-leader sRNA, was demonstrated in different growth phases of *S. mutans* (Xia et al. 2012). Growth phase and pH dependency of expression indicate a regulatory role for the L10-leader in the acid adaptive response in this cariogenic bacterium. Expression of L10-leader RNAs and other ribosomal leader RNAs, including L13-, L20-, and L21-leader RNA, was also observed in *S. pyogenes* by tiling array analyses (Perez et al. 2009; Patenge et al. 2012) and in *S. pneumoniae* by tiling array analysis and RNAome sequencing (Kumar et al. 2010; Acebo et al. 2012; Mann et al. 2012).

5.2 FMN-Riboswitch

Riboswitches are regulatory elements found in 5'-untranslated regions of prokaryotic mRNAs. They are acting in *cis* by controlling expression of their downstream genes through a metabolite-induced alteration of their secondary structure. Doing so, riboswitches play a prominent role in bacterial metabolism control (Nudler and Mironov 2004). The flavin mononucleotide (FMN) riboswitch is a metabolite-dependent riboswitch that directly binds FMN. The element controls expression of genes that encode for FMN biosynthesis and transport proteins (Vitreschak et al. 2002). A link to virulence-related gene expression comes from *L. monocytogenes*, where the riboflavin analog roseoflavin targets an FMN-riboswitch and blocks *L. monocytogenes* growth, but also stimulates virulence gene expression and infection (Mansjo and Johansson 2011). Furthermore, two riboswitches acting as noncoding RNAs *in trans* control expression of the virulence regulator PrfA in *L. monocytogenes* (Loh et al. 2009). Metabolite-driven *trans*-regulatory sRNAs form a new class of regulatory noncoding RNAs in bacteria. Expression of FMN-riboswitches has also been detected in streptococci in several recent sRNA expression screens (Perez et al. 2009; Kumar et al. 2010; Patenge et al. 2012; Mann et al. 2012). To date, no involvement of FMN-riboswitches in virulence gene expression has been documented in streptococci. Nevertheless, like other riboswitches, FMN-riboswitches may represent interesting novel drug targets (Blount and Breaker 2006). The natural antibacterial compound roseoflavin binds to FMN-riboswitches and has been shown to inhibit bacterial growth and to regulate gene expression in *B. subtilis* and *L. monocytogenes* (Lee et al. 2009; Mansjo and Johansson 2011). The identification of compounds that trigger FMN-riboswitch function in streptococci may be a promising alley for the development of novel antimicrobial drugs.

5.3 Trans-Antisense sRNAs

sRNAs working by *trans*-antisense binding to their target mRNAs tend to interact with conventional TCS in streptococci. Several examples of sRNA-TCS networks will be described in this chapter.

The small regulatory RNA FasX was one of the first characterized sRNAs in streptococci. It was initially identified during the analysis of the two-component type regulator Fas (fibronectin/fibrinogen binding/hemolytic activity/streptokinase regulator) in *S. pyogenes* (Kreikemeyer et al. 2001). The *fasBCA* operon in *S. pyogenes* encodes two potential sensor kinases and one response regulator. It is expressed in a growth phase-dependent manner and controls the production of several secreted virulence factors such as streptokinase and streptolysin S. Downstream of the *fasBCA* transcriptional unit, an independently transcribed gene, *fasX,* was found to encode a short RNA molecule. Expression of *fasX* depends on the RR FasA. Gene replacement of *fasX* resulted in a phenotype similar to *fasBCA* or *fasA* knock-out mutations with prolonged expression of extracellular matrix-protein-binding adhesins and reduced expression of secreted virulence factors. Complementation of the *fasX* deletion mutant, with *fasX* expressed *in trans* from a plasmid, restored the wild-type *fasBCA* regulation pattern (Kreikemeyer et al. 2001). From these data, it was concluded that FasX, a non-translated RNA, is the main effector molecule of the Fas regulon. Accordingly, *S. pyogenes* carrying a *fasX* deletion induced a reduced response of host epithelial cells in terms of cytokine production, apoptosis, and cytotoxicity parameters (Klenk et al. 2005). FasX enhances streptokinase activity in *S. pyogenes*. Stimulation of streptokinase gene (*ska*) expression by FasX is achieved by binding to the 5' end of the *ska* mRNA and thereby increasing the stability of transcript (Ramirez-Pena et al. 2010). Lack of FasX-*ska*-mRNA interaction in *fasX* mutants led to decreased transcript levels and consequently to a decreased streptokinase protein abundance (Ramirez-Pena et al. 2010). Recently, it has been shown that FasX also controls pilus gene expression by pairing to the extreme 5' end of the pilus biosynthesis operon transcript. In this case, the resulting RNA–RNA interaction reduces the stability of the mRNA, while at the same time inhibiting translation of at least the first gene in the pilus biosynthesis operon. As a consequence of down-regulated pilus expression, adherence to host epithelial cells by *S. pyogenes* was reduced (Liu et al. 2012). Virulence gene regulation by FasX works by classical antisense binding to target mRNAs, but diverse mechanisms could be identified as depicted in Fig. 1: in one example sRNA-mRNA interaction stabilizes the target (*ska* expression, Fig. 1a), in the other it destabilizes the target (pilus biosynthesis control, Fig. 1b). This is particularly striking, because up-regulation of Ska as well as pilus down-regulation represent activities related to the transition of *S. pyogenes* from the colonization stage of infection to the dissemination phase.

Another link between a two-component system and sRNAs became evident recently in *S. pneumoniae*. The *ciaRH* regulatory network plays a major role in the maintenance of cell-wall integrity. Penicillin-binding protein independent resistance to beta-lactam antibiotics is conferred by the *ciaRH* regulatory system (Zahner et al. 2002). Genes controlled by the *ciaRH* system are involved in cell-wall biochemistry. Bacteria carrying mutations in *ciaH* failed to develop genetic competence due to repression of the *comCDE* operon region (Mascher et al. 2003). *ciaRH* mutants were hypersusceptible to a variety of lysis-inducing conditions (Mascher et al. 2006). In two recent studies, a new level of regulation has been introduced to the *ciaRH* network in *S. pneumoniae*. Among the genes regulated by CiaR, five sRNA genes have been identified, designated *cia*-dependent small RNAs (csRNAs). All csRNAs identified so far, show a high sequence similarity and share the same mfold-predicted secondary structure presenting with two stem loops, separated by a stretch of unpaired bases. Deletion mutants lacking csRNA4 and csRNA5, respectively, showed enhanced stationary phase autolysis (Halfmann et al. 2007). The mechanism of csRNA4/5 function is not known yet, but it does not seem to involve interference with LytA and LytC production. Genes for

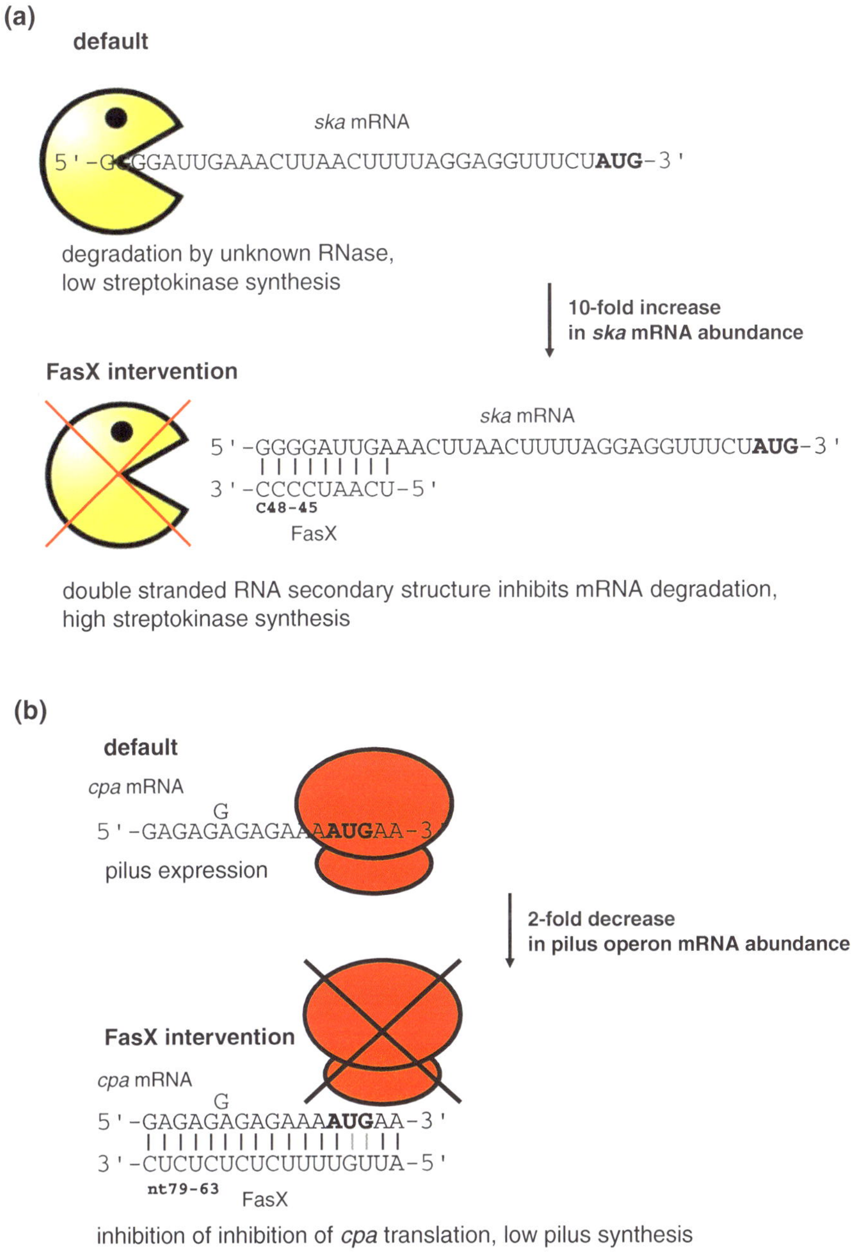

Fig. 1 Schematic of the differential outcome of FasX binding. The start codons of the respective target mRNAs are indicated with bold letters. **a** Stabilization of *ska* transcripts by interaction with FasX, pac man: unknown RNAse molecule. **b** Destabilization of pilus transcripts and inhibition of *cpa*-mRNA translation following interaction with FasX, mushroom: ribosome

csRNAs have been detected on the basis of CiaR binding site presence over a wide range of streptococcal genomes, including *S. mitis, S. oralis, S. sanguinis*, and *S. pyogenes*. Expression of csRNA genes has been demonstrated by Northern blot analyses and suggests that genes for csRNAs belong to the regulon of the response regulator CiaR in all streptococcal species (Marx et al. 2010).

Another excellent example of a regulatory cascade that involves the function of an sRNA is the *rivRX* (RofA-like protein IV regulator R/X) operon in *S. pyogenes*. The global transcriptional regulator CovR represses *mga* and *rivRX*. RivR activates *mga* expression and its mechanism has been studied in a *covR*$^-$ background (Roberts and Scott 2007). RivR belongs to the RofA family of transcriptional regulators (described in Sect. 2.4) and contains a DNA-binding site. The protein seems to interact with Mga in order to enhance *mga* expression. The genes *rivR* and *rivX* are co-transcribed but mediate distinct pathways of Mga regulon activation. The *rivX* sequence contains two putative ORFs. Nonsense mutations in the potential start sites revealed that the *rivX* transcript rather than peptides encoded by *rivX* are responsible for the regulatory phenotype. From primer extension analysis, it was concluded that RNA processing could lead to the production of RivX following co-transcription of *rivR and rivX* explaining the independent function of the two genes. RivX stimulation of *mga* expression and Mga-activated gene expression depended on the presence of Mga protein. There was no stabilization of *mga* transcript observed. The mechanism of RivX function seems to involve a direct or indirect interaction with the *mga* transcript leading to enhancement of Mga translation (Roberts and Scott 2007). The RivR/X system works as integrator of the signals provided by the global CovR and Mga regulatory networks. This cross-talk allows the pathogen to respond to a broad variety of external stimuli with the fine-tuned expression of appropriate virulence factors.

5.4 sRNA Interaction with Proteins

Clustered, regularly interspaced, short palindromic repeat (CRISPR) loci are well-known to provide an adaptive RNA-based immune system in bacteria and archaea. CRISPR protects the cells from horizontal gene transfer originating from phage and plasmid DNA (Marraffini and Sontheimer 2010). Differential RNA sequencing identified a CRISPR/Cas locus in *S. pyogenes* SF370 (M1 serotype) (Deltcheva et al. 2011) encoded by the system II (Nmeni/CASS4 subtype) (Haft et al. 2005). This locus encodes the *trans*-activating CRISPR RNA (tracrRNA), which is, in concert with RNAase III and the CRISPR-associated Csn1 protein, responsible for the maturation of CRISPR RNA (crRNA) (Deltcheva et al. 2011). Following crRNA maturation, the cleavage of substrate DNA needs to be initiated. Very recently, it could be shown that the DNA endonuclease Cas9 from the type II CRISPR system in *S. pyogenes* was guided by a dual RNA molecule to its target DNA. The RNA–RNA complex consisted of the activating tracrRNA and the targeting crRNA, which contained a sequence complementary to the DNA substrate (Jinek et al. 2012). In an independent study, it could be shown that the Cas9-crRNA complex of the *S. thermophilus* CRISPR/Cas system was able to cleave in vitro an artificial DNA substrate containing a sequence complementary to crRNA (Gasiunas et al. 2012). Bacterial CRISPR/Cas systems seem to be functionally conserved over a wide range of phyla. It could be shown that the *S. thermophilus* CRISPR/Cas system provided immunity in *E. coli* (Sapranauskas et al. 2011). An excellent description of the function of the CRISPR system in streptococci is included in a recent review about sRNAs by Le Rhun and Charpentier (Le Rhun and Charpentier 2012).

Formerly regarded as a house keeping RNA, the 4.5S RNA, a component of the bacterial signal recognition particle (SRP), represents another untranslated RNA with influence on streptococcal virulence (Trevino et al. 2010). While the 4.5S RNA gene is not essential for streptococcal growth under laboratory culture conditions, it proved to be essential for *S. pyogenes* to cause lethal infections in a murine bacteraemia model of infection. Mutation of the 4.5S RNA gene resulted in an altered secretome, including a reduction in secretion of the hemolysin streptolysin O and the SpeB protease. Moreover, remodeling of the *S. pyogenes* transcriptome was observed following 4.5S RNA mutation. More detailed assessment of gene expression upon loss of the 4.5S RNA gene revealed that differences in abundance of *grab, speB, spy0430*, and *slo* were *covS*-dependent. A further link of the 4.5S RNA gene to virulence was the strong reduction of growth in human saliva of *S. pyogenes* mutants affected in the gene and decreased virulence of these strains in a murine soft tissue infection model.

5.5 Bioinformatics Prediction Tools for sRNAs

Novel bioinformatics tools and whole-genome expression analyses employing tiling arrays or next generation sequencing helped to study the function of sRNAs in gram-positive pathogens (Mraheil et al. 2010). As knowledge about the role of regulatory RNAs in gram-positive bacteria is rising, new tools are being developed for the analysis of RNA structure and function, e.g., a database focusing on sRNA data from gram-positive bacteria (Pischimarov et al. 2012).

One of the most prominent bioinformatics prediction tools invented for sRNAs was the *sRNA identification protocol using high throughput technology* (SIPHT) tool, which has been used for many bacterial species (Livny et al. 2005; Livny and Waldor 2007). However, comparison of the prediction results with the actual in vivo expression of sRNAs, often revealed a low overlap between the different screening methods (Perez et al. 2009; Arnvig and Young 2009; Mraheil et al. 2011). This phenomenon is due to a combination of limitations of the prediction programs and the fact that not all sRNAs are expressed under all conditions. Today, the development of sRNA prediction software with improved properties is still ongoing. Several recently published bioinformatics tools have been used for the identification of putative sRNAs in streptococci (Raasch et al. 2010; Sridhar et al. 2010; Pichon et al. 2012).

5.6 Whole-Genome sRNA Expression Screens

In recent years, whole-genome sRNA screens in streptococci employing either tiling array or next generation sequencing approaches, revealed an unexpected number of potential sRNAs (Tsui et al. 2010; Kumar et al. 2010; Mraheil et al. 2011; Beaume et al. 2011; Chen et al. 2011). Based on a previous bioinformatic prediction of putative sRNAs (Livny et al. 2006) expression of 40 putative sRNAs was tested in *S. pneumoniae* strain D39 by Northern analysis (Tsui et al. 2010). Nine new pneumococcal sRNAs were identified and the previously reported CcnA sRNA (Halfmann et al. 2007) was confirmed. However, functional characterization of deletion mutants and ectopic overexpression constructs of three of the candidate sRNA genes did not reveal strong effects on the phenotypes tested in this study. In a whole-genome approach, a total of 50 sRNAs from the intergenic regions of *S. pneumoniae* TIGR4 were identified using high-resolution genome tiling arrays (Kumar et al. 2010). In a deep sequencing approach, 88 regulatory sRNAs were identified in the TIGR4 strain of *S. pneumoniae* (Acebo et al. 2012). Of those, three housekeeping sRNAs were detected, several riboswitches and other *cis*-regulatory RNAs and 68 novel sRNAs. One of the novel candidates seems to be involved in the modulation of competence regulation in *S. pneumoniae* (Acebo et al. 2012). In another recent RNAseq study in *S. pneumoniae* TIGR4, 89 sRNAs were detected, 56 of which were novel. Tn-seq analysis testing relative fitness of bacterial mutants during infection predicted a high number of sRNAs involved in pneumococcal pathogenesis (Mann et al. 2012). Attenuated fitness was predicted during infection of the nasopharynx for 26 sRNAs, in the lung for 28 sRNAs, and in the bloodstream for 18 candidates. Fitness predictions were confirmed by individual targeted deletions in a subset of sRNAs. The high number of sRNA genes involved in pathogenesis underlines the overall importance of sRNAs in streptococcal virulence.

A whole-genome intergenic tiling array screen of *S. pyogenes* M1T1 identified approximately 40 sRNAs that were expressed during the exponential growth phase in cells cultivated in THY complex medium (Perez et al. 2009). There was a high conservation of sRNA genes among *S. pyogenes* strains, but the expression of the individual genes was found to be strain dependent. A targeted deletion within the *pel* region did not confirm the phenotype reported from other *S. pyogenes* strains. Expression of 16 candidate genes was confirmed by Northern blot analyses. Together with a former bioinformatics prediction (Livny et al. 2006), the number of putative sRNAs in *S. pyogenes* amounts to 75. In a whole-genome intergenic tiling array screen in *S. pyogenes* M49, 55 putative sRNAs were identified. A total of 42 sRNAs were novel, whereas 13 RNAs had been described before. The sequences of most of the candidates were conserved over streptococcal genomes.

However, comparison of the sRNA expression data to the above-mentioned analysis of the M1T1 strain and to two in silico screening methods revealed a low overlap between the different approaches (Patenge et al. 2012). Thus, the investigation of several conditions and the combination of screening tools will be necessary to gain a comprehensive understanding of the abundance of sRNAs in *S. pyogenes*. It is to be expected that further analyses of *S. pyogenes* genomes by RNAseq will lead to the identification of more sRNA genes.

6 Conclusions

The existing and still increasing richness of information on streptococcal transcriptional regulators, two-component signal transduction systems, eukaryotic-like serine/threonine kinase/phosphatase systems, and noncoding regulatory RNAs summarized in this review underscores the importance to decipher the regulatory networks acting in these pathogenic bacteria. A detailed understanding of virulence and pathogenicity mechanisms is critical for the development of novel antibacterial therapies. This review highlighted the presence of many orthologous/homologous regulators and regulatory mechanisms across streptococcal species, of which many have similar functions and are central for the pathogenesis of all the species. In parallel, many have different functions or have undergone a pathogen-specific alteration, including strain- or serotype-specific adaptation. This raises the question whether developing interference strategies targeting regulators and regulatory networks is a promising goal that should be pursued further. In the opinion of the authors, this question can be answered with "yes". What is required to develop such interference strategies/therapies? (I) The complete operons/regulons of interesting candidates need to be identified. (II) Knowledge needs to be transferred from in vitro to in vivo experiments. (III) Particularly the species-specific regulators/regulatory mechanisms should be kept in focus, as they allow a selective targeting without harmful effects on related species or the residual bacterial flora. In the opinion of the authors the goal of interference therapy should be species-specific virulence/pathogen attenuation rather than killing, as such a strategy allows the immune system of the host to cope with infections and develop immunity. Such novel agents need to be accompanied by available antibiotic approaches. Particular noncoding regulatory RNAs provide a wealth of potential targets, which fulfill above criteria. What are possible means to interfere with their function? In the author's lab antisense-PNA (peptide nucleic acids) technology is currently explored for specific sRNA-activity interference.

References

Acebo P, Martin-Galiano AJ, Navarro S, Zaballos A, Amblar M (2012) Identification of 88 regulatory small RNAs in the TIGR4 strain of the human pathogen *Streptococcus pneumoniae*. RNA 18:530–546

Agarwal S, Agarwal S, Pancholi P, Pancholi V (2011) Role of serine/threonine phosphatase (SP-STP) in *Streptococcus pyogenes* physiology and virulence. J Biol Chem 286:41368–41380

Agarwal S, Agarwal S, Pancholi P, Pancholi V (2012) Strain-specific regulatory role of eukaryote-like serine/threonine phosphatase in pneumococcal adherence. Infect Immun 80:1361–1372

Ahmed NA, Petersen FC, Scheie AA (2008) Biofilm formation and autoinducer-2 signaling in *Streptococcus intermedius*: role of thermal and pH factors. Oral Microbiol Immunol 23: 492–497

Almengor AC, McIver KS (2004) Transcriptional activation of sclA by Mga requires a distal binding site in *Streptococcus pyogenes*. J Bacteriol 186:7847–7857

Almengor AC, Walters MS, McIver KS (2006) Mga is sufficient to activate transcription in vitro of sof-sfbX and other Mga-regulated virulence genes in the group A *Streptococcus*. J Bacteriol 188:2038–2047

Almengor AC, Kinkel TL, Day SJ, McIver KS (2007) The catabolite control protein CcpA binds to Pmga and influences expression of the virulence regulator Mga in the group A *Streptococcus*. J Bacteriol 189:8405–8416

Antunes LC, Ferreira RB, Buckner MM, Finlay BB (2010) Quorum sensing in bacterial virulence. Microbiology 156:2271–2282

Arnvig KB, Young DB (2009) Identification of small RNAs in *Mycobacterium tuberculosis*. Mol Microbiol 73:397–408

Aziz RK, Kansal R, Aronow BJ, Taylor WL, Rowe SL, Kubal M, Chhatwal GS, Walker MJ, Kotb M (2010) Microevolution of group A streptococci in vivo: capturing regulatory networks engaged in sociomicrobiology, niche adaptation, and hypervirulence. PLoS One 5:e9798

Basset A, Turner KH, Boush E, Sayeed S, Dove SL, Malley R (2011) Expression of the type 1 pneumococcal pilus is bistable and negatively regulated by the structural component RrgA. Infect Immun 79:2974–2983

Basset A, Turner KH, Boush E, Sayeed S, Dove SL, Malley R (2012) An epigenetic switch mediates bistable expression of the type 1 pilus genes in *Streptococcus pneumoniae*. J Bacteriol 194:1088–1091

Bastet L, Dube A, Masse E, Lafontaine DA (2011) New insights into riboswitch regulation mechanisms. Mol Microbiol 80:1148–1154

Beaume M, Hernandez D, Docquier M, Delucinge-Vivier C, Descombes P, Francois P (2011) Orientation and expression of methicillin-resistant *Staphylococcus aureus* small RNAs by direct multiplexed measurements using the nCounter of NanoString technology. J Microbiol Methods 84:327–334

Beckert S, Kreikemeyer B, Podbielski A (2001) Group A streptococcal *rofA* gene is involved in the control of several virulence genes and eukaryotic cell attachment and internalization. Infect Immun 69:534–537

Bernish B, van de Rijn I (1999) Characterization of a two-component system in *Streptococcus pyogenes* which is involved in regulation of hyaluronic acid production. J Biol Chem 274:4786–4793

Bessen DE, Hollingshead SK (1994) Allelic polymorphism of emm loci provides evidence for horizontal gene spread in group A streptococci. Proc Natl Acad Sci U S A 91:3280–3284

Bessen DE, Lizano S (2010) Tissue tropisms in group A streptococcal infections. Future Microbiol 5:623–638

Bessen DE, Manoharan A, Luo F, Wertz JE, Robinson DA (2005) Evolution of transcription regulatory genes is linked to niche specialization in the bacterial pathogen *Streptococcus pyogenes*. J Bacteriol 187:4163–4172
Betschel SD, Borgia SM, Barg NL, Low DE, De Azavedo JC (1998) Reduced virulence of group A streptococcal Tn916 mutants that do not produce streptolysin S. Infect Immun 66:1671–1679
Biswas I, Germon P, McDade K, Scott JR (2001) Generation and surface localization of intact M protein in *Streptococcus pyogenes* are dependent on sagA. Infect Immun 69:7029–7038
Blehert DS, Palmer RJ Jr, Xavier JB, Almeida JS, Kolenbrander PE (2003) Autoinducer 2 production by *Streptococcus gordonii* DL1 and the biofilm phenotype of a luxS mutant are influenced by nutritional conditions. J Bacteriol 185:4851–4860
Blount KF, Breaker RR (2006) Riboswitches as antibacterial drug targets. Nat Biotechnol 24:1558–1564
Brantl S (2009) Bacterial chromosome-encoded small regulatory RNAs. Future Microbiol 4: 85–103
Bryan J, Liles R, Cvek U, Trutschl M, Shelver D (2008) Global transcriptional profiling reveals *Streptococcus agalactiae* genes controlled by the MtaR transcription factor. BMC Genomics 9:607
Burnside K, Rajagopal L (2011) Aspects of eukaryotic-like signaling in gram-positive cocci: a focus on virulence. Future Microbiol 6:747–761
Burnside K, Lembo A, Harrell MI, Gurney M, Xue L, BinhTran NT, Connelly JE, Jewell KA, Schmidt BZ, de los RM, Tao WA, Doran KS, Rajagopal L (2011) Serine/threonine phosphatase Stp1 mediates post-transcriptional regulation of hemolysin, autolysis, and virulence of group B *Streptococcus*. J Biol Chem 286:44197–44210
Cao M, Feng Y, Wang C, Zheng F, Li M, Liao H, Mao Y, Pan X, Wang J, Hu D, Hu F, Tang J (2011) Functional definition of LuxS, an autoinducer-2 (AI-2) synthase and its role in full virulence of *Streptococcus suis* serotype 2. J Microbiol 49:1000–1011
Caparon MG, Scott JR (1987) Identification of a gene that regulates expression of M protein, the major virulence determinant of group A streptococci. Proc Natl Acad Sci U S A 84: 8677–8681
Carapetis JR, Steer AC, Mulholland EK, Weber M (2005) The global burden of group A streptococcal diseases. Lancet Infect Dis 5:685–694
Carvalho SM, Kloosterman TG, Kuipers OP, Neves AR (2011) CcpA ensures optimal metabolic fitness of *Streptococcus pneumoniae*. PLoS One 6:e26707
Caymaris S, Bootsma HJ, Martin B, Hermans PW, Prudhomme M, Claverys JP (2010) The global nutritional regulator CodY is an essential protein in the human pathogen *Streptococcus pneumoniae*. Mol Microbiol 78:344–360
Chen C, Tang J, Dong W, Wang C, Feng Y, Wang J, Zheng F, Pan X, Liu D, Li M, Song Y, Zhu X, Sun H, Feng T, Guo Z, Ju A, Ge J, Dong Y, Sun W, Jiang Y, Wang J, Yan J, Yang H, Wang X, Gao GF, Yang R, Wang J, Yu J (2007) A glimpse of streptococcal toxic shock syndrome from comparative genomics of S. *suis* 2 Chinese isolates. PLoS One 2:e315
Chen Y, Indurthi DC, Jones SW, Papoutsakis ET (2011) Small RNAs in the genus *Clostridium*. MBio 2:00340–0010
Christensen T, Johnsen M, Fiil NP, Friesen JD (1984) RNA secondary structure and translation inhibition: analysis of mutants in the rplJ leader. EMBO J 3:1609–1612
Churchward G (2007) The two faces of Janus: virulence gene regulation by CovR/S in group A streptococci. Mol Microbiol 64:34–41
Cunningham MW (2000) Pathogenesis of group A streptococcal infections. Clin Microbiol Rev 13:470–511
Dramsi S, Dubrac S, Konto-Ghiorghi Y, Da C, V, Couve E, Glaser P, Caliot E, Debarbouille M, Bellais S, Trieu-Cuot P, Mistou MY (2012) Rga, a RofA-like regulator, is the major transcriptional activator of the PI-2a pilus in *Streptococcus agalactiae*. Microb Drug Resist 18:286–297

Dalton TL, Collins JT, Barnett TC, Scott JR (2006) RscA, a member of the MDR1 family of transporters, is repressed by CovR and required for growth of *Streptococcus pyogenes* under heat stress. J Bacteriol 188:77–85

Danne C, Entenza JM, Mallet A, Briandet R, Debarbouille M, Nato F, Glaser P, Jouvion G, Moreillon P, Trieu-Cuot P, Dramsi S (2011) Molecular characterization of a *Streptococcus gallolyticus* genomic island encoding a pilus involved in endocarditis. J Infect Dis 204:1960–1970

De Angelis G, Moschioni M, Muzzi A, Pezzicoli A, Censini S, Delany I, Lo SM, Sinisi A, Donati C, Masignani V, Barocchi MA (2011) The *Streptococcus pneumoniae* pilus-1 displays a biphasic expression pattern. PLoS One 6:e21269

Deltcheva E, Chylinski K, Sharma CM, Gonzales K, Chao Y, Pirzada ZA, Eckert MR, Vogel J, Charpentier E (2011) CRISPR RNA maturation by trans-encoded small RNA and host factor RNase III. Nature 471:602–607

Dramsi S, Caliot E, Bonne I, Guadagnini S, Prevost MC, Kojadinovic M, Lalioui L, Poyart C, Trieu-Cuot P (2006) Assembly and role of pili in group B streptococci. Mol Microbiol 60:1401–1413

Echenique J, Kadioglu A, Romao S, Andrew PW, Trombe MC (2004) Protein serine/threonine kinase StkP positively controls virulence and competence in *Streptococcus pneumoniae*. Infect Immun 72:2434–2437

Federle MJ (2009) Autoinducer-2-based chemical communication in bacteria: complexities of interspecies signaling. Contrib Microbiol 16:18–32

Federle MJ, Bassler BL (2003) Interspecies communication in bacteria. J Clin Invest 112: 1291–1299

Fiedler T, Sugareva V, Patenge N, Kreikemeyer B (2010a) Insights into *Streptococcus pyogenes* pathogenesis from transcriptome studies. Future Microbiol 5:1675–1694

Fiedler T, Kreikemeyer B, Sugareva V, Redanz S, Arlt R, Standar K, Podbielski A (2010b) Impact of the *Streptococcus pyogenes* Mga regulator on human matrix protein binding and interaction with eukaryotic cells. Int J Med Microbiol 300:248–258

Fittipaldi N, Takamatsu D, de la Cruz Dominguez-Punaro, Lecours MP, Montpetit D, Osaki M, Sekizaki T, Gottschalk M (2010) Mutations in the gene encoding the ancillary pilin subunit of the *Streptococcus suis* srtF cluster result in pili formed by the major subunit only. PLoS One 5:e8426

Fittipaldi N, Segura M, Grenier D, Gottschalk M (2012) Virulence factors involved in the pathogenesis of the infection caused by the swine pathogen and zoonotic agent *Streptococcus suis*. Future Microbiol 7:259–279

Froehlich BJ, Bates C, Scott JR (2009) *Streptococcus pyogenes* CovRS mediates growth in iron starvation and in the presence of the human cationic antimicrobial peptide LL-37. J Bacteriol 191:673–677

Frohlich KS, Vogel J (2009) Activation of gene expression by small RNA. Curr Opin Microbiol 12:674–682

Gamez G, Hammerschmidt S (2012) Combat pneumococcal infections: adhesins as candidates for protein- based vaccine development. Curr Drug Targets 13:323–337

Gao J, Gusa AA, Scott JR, Churchward G (2005) Binding of the global response regulator protein CovR to the sag promoter of *Streptococcus pyogenes* reveals a new mode of CovR-DNA interaction. J Biol Chem 280:38948–38956

Gardner PP, Daub J, Tate JG, Nawrocki EP, Kolbe DL, Lindgreen S, Wilkinson AC, Finn RD, Griffiths-Jones S, Eddy SR, Bateman A (2009) Rfam: updates to the RNA families database. Nucleic Acids Res 37:D136–D140

Gasiunas G, Barrangou R, Horvath P, Siksnys V (2012) Cas9-crRNA ribonucleoprotein complex mediates specific DNA cleavage for adaptive immunity in bacteria. Proc Natl Acad Sci U S A 109:E2579–E2586

Giammarinaro P, Paton JC (2002) Role of RegM, a homologue of the catabolite repressor protein CcpA, in the virulence of *Streptococcus pneumoniae*. Infect Immun 70:5454–5461

Glaser P, Rusniok C, Buchrieser C, Chevalier F, Frangeul L, Msadek T, Zouine M, Couve E, Lalioui L, Poyart C, Trieu-Cuot P, Kunst F (2002) Genome sequence of *Streptococcus agalactiae*, a pathogen causing invasive neonatal disease. Mol Microbiol 45:1499–1513
Gogol EB, Rhodius VA, Papenfort K, Vogel J, Gross CA (2011) Small RNAs endow a transcriptional activator with essential repressor functions for single-tier control of a global stress regulon. Proc Natl Acad Sci U S A 108:12875–12880
Gorke B, Vogel J (2008) Noncoding RNA control of the making and breaking of sugars. Genes Dev 22:2914–2925
Gottesman S and Storz G (2011) Bacterial small RNA regulators: versatile roles and rapidly evolving variations. Cold Spring Harb Perspect Biol 3:a003798
Gottesman S, McCullen CA, Guillier M, Vanderpool CK, Majdalani N, Benhammou J, Thompson KM, FitzGerald PC, Sowa NA, FitzGerald DJ (2006) Small RNA regulators and the bacterial response to stress. Cold Spring Harb Symp Quant Biol 71:1–11
Graham MR, Virtaneva K, Porcella SF, Barry WT, Gowen BB, Johnson CR, Wright FA, Musser JM (2005) Group A *Streptococcus* transcriptome dynamics during growth in human blood reveals bacterial adaptive and survival strategies. Am J Pathol 166:455–465
Granok AB, Parsonage D, Ross RP, Caparon MG (2000) The RofA binding site in *Streptococcus pyogenes* is utilized in multiple transcriptional pathways. J Bacteriol 182:1529–1540
Gryllos I, Grifantini R, Colaprico A, Jiang S, Deforce E, Hakansson A, Telford JL, Grandi G, Wessels MR (2007) Mg(2 +) signalling defines the group A streptococcal CsrRS (CovRS) regulon. Mol Microbiol 65:671–683
Guedon E, Sperandio B, Pons N, Ehrlich SD, Renault P (2005) Overall control of nitrogen metabolism in *Lactococcus lactis* by CodY, and possible models for CodY regulation in Firmicutes. Microbiology 151:3895–3909
Guenzi E, Gasc AM, Sicard MA, Hakenbeck R (1994) A two-component signal-transducing system is involved in competence and penicillin susceptibility in laboratory mutants of *Streptococcus pneumoniae*. Mol Microbiol 12:505–515
Gusa AA, Scott JR (2005) The CovR response regulator of group A streptococcus (GAS) acts directly to repress its own promoter. Mol Microbiol 56:1195–1207
Gusa AA, Gao J, Stringer V, Churchward G, Scott JR (2006) Phosphorylation of the group A Streptococcal CovR response regulator causes dimerization and promoter-specific recruitment by RNA polymerase. J Bacteriol 188:4620–4626
Gutekunst H, Eikmanns BJ, Reinscheid DJ (2003) Analysis of RogB-controlled virulence mechanisms and gene expression in *Streptococcus agalactiae*. Infect Immun 71:5056–5064
Haanes EJ, Heath DG, Cleary PP (1992) Architecture of the vir regulons of group A streptococci parallels opacity factor phenotype and M protein class. J Bacteriol 174:4967–4976
Haft DH, Selengut J, Mongodin EF, Nelson KE (2005) A guild of 45 CRISPR-associated (Cas) protein families and multiple CRISPR/Cas subtypes exist in prokaryotic genomes. PLoS Comput Biol 1:e60
Halfmann A, Kovacs M, Hakenbeck R, Bruckner R (2007) Identification of the genes directly controlled by the response regulator CiaR in *Streptococcus pneumoniae*: five out of 15 promoters drive expression of small non-coding RNAs. Mol Microbiol 66:110–126
Halfmann A, Schnorpfeil A, Muller M, Marx P, Gunzler U, Hakenbeck R, Bruckner R (2011) Activity of the two-component regulatory system CiaRH in *Streptococcus pneumoniae* R6. J Mol Microbiol Biotechnol 20:96–104
Han XG, Lu CP (2009) Detection of autoinducer-2 and analysis of the profile of luxS and pfs transcription in *Streptococcus suis* serotype 2. Curr Microbiol 58:146–152
Han H, Liu C, Wang Q, Xuan C, Zheng B, Tang J, Yan J, Zhang J, Li M, Cheng H, Lu G, Gao GF (2012) The two-component system Ihk/Irr contributes to the virulence of *Streptococcus suis* serotype 2 strain 05ZYH33 through alteration of the bacterial cell metabolism. Microbiology 158:1852–1866
Hava DL, Camilli A (2002) Large-scale identification of serotype 4 *Streptococcus pneumoniae* virulence factors. Mol Microbiol 45:1389–1406

Hava DL, Hemsley CJ, Camilli A (2003) Transcriptional regulation in the *Streptococcus pneumoniae* rlrA pathogenicity islet by RlrA. J Bacteriol 185:413–421
Heath A, DiRita VJ, Barg NL, Engleberg NC (1999) A two-component regulatory system, CsrR-CsrS, represses expression of three *Streptococcus pyogenes* virulence factors, hyaluronic acid capsule, streptolysin S, and pyrogenic exotoxin B. Infect Immun 67:5298–5305
Heidrich N, Chinali A, Gerth U, Brantl S (2006) The small untranslated RNA SR1 from the *Bacillus subtilis* genome is involved in the regulation of arginine catabolism. Mol Microbiol 62:520–536
Hemsley C, Joyce E, Hava DL, Kawale A, Camilli A (2003) MgrA, an orthologue of Mga, acts as a transcriptional repressor of the genes within the rlrA pathogenicity islet in *Streptococcus pneumoniae*. J Bacteriol 185:6640–6647
Hendriksen WT, Bootsma HJ, Estev + úo S, Hoogenboezem T, de Jong A, de Groot R, Kuipers OP, Hermans PWM (2008) CodY of *Streptococcus pneumoniae*: link between nutritional gene regulation and colonization. J Bacteriol 190:590–601
Hertzen E, Johansson L, Kansal R, Hecht A, Dahesh S, Janos M, Nizet V, Kotb M, Norrby-Teglund A (2012) Intracellular *Streptococcus pyogenes* in human macrophages display an altered gene expression profile. PLoS One 7:e35218
Hoch JA (2000) Two-component and phosphorelay signal transduction. Curr Opin Microbiol 3:165–170
Hollingshead SK, Bessen DE (1995) Evolution of the emm gene family: virulence gene clusters in group A streptococci. Dev Biol Stand 85:163–168
Hollingshead SK, Readdy TL, Yung DL, Bessen DE (1993) Structural heterogeneity of the emm gene cluster in group A streptococci. Mol Microbiol 8:707–717
Hondorp ER, McIver KS (2007) The Mga virulence regulon: infection where the grass is greener. Mol Microbiol 66:1056–1065
Hoskins J, Alborn WE Jr, Arnold J, Blaszczak LC, Burgett S, DeHoff BS, Estrem ST, Fritz L, Fu DJ, Fuller W, Geringer C, Gilmour R, Glass JS, Khoja H, Kraft AR, Lagace RE, LeBlanc DJ, Lee LN, Lefkowitz EJ, Lu J, Matsushima P, McAhren SM, McHenney M, McLeaster K, Mundy CW, Nicas TI, Norris FH, O'Gara M, Peery RB, Robertson GT, Rockey P, Sun PM, Winkler ME, Yang Y, Young-Bellido M, Zhao G, Zook CA, Baltz RH, Jaskunas SR, Rosteck PR Jr, Skatrud PL, Glass JI (2001) Genome of the bacterium *Streptococcus pneumoniae* strain R6. J Bacteriol 183:5709–5717
Iyer R, Baliga NS, Camilli A (2005) Catabolite control protein A (CcpA) contributes to virulence and regulation of sugar metabolism in *Streptococcus pneumoniae*. J Bacteriol 187:8340–8349
Jiang SM, Cieslewicz MJ, Kasper DL, Wessels MR (2005) Regulation of virulence by a two-component system in group B streptococcus. J Bacteriol 187:1105–1113
Jiang SM, Ishmael N, Dunning HJ, Puliti M, Tissi L, Kumar N, Cieslewicz MJ, Tettelin H, Wessels MR (2008) Variation in the group B *Streptococcus* CsrRS regulon and effects on pathogenicity. J Bacteriol 190:1956–1965
Jin H, Pancholi V (2006) Identification and biochemical characterization of a eukaryotic-type serine/threonine kinase and its cognate phosphatase in *Streptococcus pyogenes*: their biological functions and substrate identification. J Mol Biol 357:1351–1372
Jinek M, Chylinski K, Fonfara I, Hauer M, Doudna JA, Charpentier E (2012) A programmable dual-RNA-guided DNA endonuclease in adaptive bacterial immunity. Science 337:816–821
Johansson J, Mandin P, Renzoni A, Chiaruttini C, Springer M, Cossart P (2002) An RNA thermosensor controls expression of virulence genes in *Listeria monocytogenes*. Cell 110:551–561
Johnsen M, Christensen T, Dennis PP, Fiil NP (1982) Autogenous control: ribosomal protein L10–L12 complex binds to the leader sequence of its mRNA. EMBO J 1:999–1004
Joyce EA, Kawale A, Censini S, Kim CC, Covacci A, Falkow S (2004) LuxS is required for persistent pneumococcal carriage and expression of virulence and biosynthesis genes. Infect Immun 72:2964–2975

Kaufman GE, Yother J (2007) CcpA-dependent and -independent control of beta-galactosidase expression in *Streptococcus pneumoniae* occurs via regulation of an upstream phosphotransferase system-encoding operon. J Bacteriol 189:5183–5192

Kazmierczak KM, Wayne KJ, Rechtsteiner A, Winkler ME (2009) Roles of relSpn in stringent response, global regulation and virulence of serotype 2 *Streptococcus pneumoniae* D39. Mol Microbiol 72:590–611

Kietzman CC, Caparon MG (2010) CcpA and LacD.1 affect temporal regulation of *Streptococcus pyogenes* virulence genes. Infect Immun 78:241–252

Kietzman CC, Caparon MG (2011) Distinct time-resolved roles for two catabolite-sensing pathways during *Streptococcus pyogenes* infection. Infect Immun 79:812–821

Kihlberg BM, Cooney J, Caparon MG, Olsen A, Bjorck L (1995) Biological properties of a *Streptococcus pyogenes* mutant generated by Tn916 insertion in mga. Microb Pathog 19:299–315

Kinkel TL, McIver KS (2008) CcpA-mediated repression of streptolysin s expression and virulence in the group A *Streptococcus*. Infect Immun 76:3451–3463

Klenk M, Koczan D, Guthke R, Nakata M, Thiesen HJ, Podbielski A, Kreikemeyer B (2005) Global epithelial cell transcriptional responses reveal *Streptococcus pyogenes* Fas regulator activity association with bacterial aggressiveness. Cell Microbiol 7:1237–1250

Klinkert B, Narberhaus F (2009) Microbial thermosensors. Cell Mol Life Sci 66:2661–2676

Kratovac Z, Manoharan A, Luo F, Lizano S, Bessen DE (2007) Population genetics and linkage analysis of loci within the FCT region of *Streptococcus pyogenes*. J Bacteriol 189:1299–1310

Kreikemeyer B, Boyle MD, Buttaro BA, Heinemann M, Podbielski A (2001) Group A streptococcal growth phase-associated virulence factor regulation by a novel operon (Fas) with homologies to two-component-type regulators requires a small RNA molecule. Mol Microbiol 39:392–406

Kreikemeyer B, McIver KS, Podbielski A (2003) Virulence factor regulation and regulatory networks in *Streptococcus pyogenes* and their impact on pathogen-host interactions. Trends Microbiol 11:224–232

Kreikemeyer B, Nakata M, Koller T, Hildisch H, Kourakos V, Standar K, Kawabata S, Glocker MO, Podbielski A (2007) The *Streptococcus pyogenes* serotype M49 Nra-Ralp3 transcriptional regulatory network and its control of virulence factor expression from the novel eno ralp3 epf sagA pathogenicity region. Infect Immun 75:5698–5710

Kreikemeyer B, Gamez G, Margarit I, Giard JC, Hammerschmidt S, Hartke A, Podbielski A (2011) Genomic organization, structure, regulation and pathogenic role of pilus constituents in major pathogenic Streptococci and Enterococci. Int J Med Microbiol 301:240–251

Kreth J, Chen Z, Ferretti J, Malke H (2011) Counteractive balancing of transcriptome expression involving CodY and CovRS in *Streptococcus pyogenes*. J Bacteriol 193:4153–4165

Kumar R, Shah P, Swiatlo E, Burgess SC, Lawrence ML, Nanduri B (2010) Identification of novel non-coding small RNAs from *Streptococcus pneumoniae* TIGR4 using high-resolution genome tiling arrays. BMC Genomics 11:350

Lamy MC, Zouine M, Fert J, Vergassola M, Couve E, Pellegrini E, Glaser P, Kunst F, Msadek T, Trieu-Cuot P, Poyart C (2004) CovS/CovR of group B streptococcus: a two-component global regulatory system involved in virulence. Mol Microbiol 54:1250–1268

Lanie JA, Ng WL, Kazmierczak KM, Andrzejewski TM, Davidsen TM, Wayne KJ, Tettelin H, Glass JI, Winkler ME (2007) Genome sequence of Avery's virulent serotype 2 strain D39 of *Streptococcus pneumoniae* and comparison with that of unencapsulated laboratory strain R6. J Bacteriol 189:38–51

Le Rhun A, Charpentier E (2012) Small RNAs in streptococci. RNA Biol 9:414–426

Leday TV, Gold KM, Kinkel TL, Roberts SA, Scott JR, McIver KS (2008) TrxR, a new CovR-repressed response regulator that activates the Mga virulence regulon in group A *Streptococcus*. Infect Immun 76:4659–4668

Lee ER, Blount KF, Breaker RR (2009) Roseoflavin is a natural antibacterial compound that binds to FMN riboswitches and regulates gene expression. RNA Biol 6:187–194

Levin JC, Wessels MR (1998) Identification of csrR/csrS, a genetic locus that regulates hyaluronic acid capsule synthesis in group A *Streptococcus*. Mol Microbiol 30:209–219

Li Z, Sledjeski DD, Kreikemeyer B, Podbielski A, Boyle MD (1999) Identification of pel, a *Streptococcus pyogenes* locus that affects both surface and secreted proteins. J Bacteriol 181:6019–6027

Li N, Wang F, Niu S, Cao J, Wu K, Li Y, Yin N, Zhang X, Zhu W, Yin Y (2009) Discovery of novel inhibitors of *Streptococcus pneumoniae* based on the virtual screening with the homology-modeled structure of histidine kinase (VicK). BMC Microbiol 9:129

Li W, Liu L, Qiu D, Chen H, Zhou R (2010) Identification of *Streptococcus suis* serotype 2 genes preferentially expressed in the natural host. Int J Med Microbiol 300:482–488

Li J, Tan C, Zhou Y, Fu S, Hu L, Hu J, Chen H, Bei W (2011) The two-component regulatory system CiaRH contributes to the virulence of *Streptococcus suis* 2. Vet Microbiol 148:99–104

Liu M, Hanks TS, Zhang J, McClure MJ, Siemsen DW, Elser JL, Quinn MT, Lei B (2006) Defects in ex vivo and in vivo growth and sensitivity to osmotic stress of group A *Streptococcus* caused by interruption of response regulator gene vicR. Microbiology 152: 967–978

Liu Z, Trevino J, Ramirez-Pena E, Sumby P (2012) The small regulatory RNA FasX controls pilus expression and adherence in the human bacterial pathogen group A *Streptococcus*. Mol Microbiol 86:140–154

Livny J, Waldor MK (2007) Identification of small RNAs in diverse bacterial species. Curr Opin Microbiol 10:96–101

Livny J, Fogel MA, Davis BM, Waldor MK (2005) sRNAPredict: an integrative computational approach to identify sRNAs in bacterial genomes. Nucleic Acids Res 33:4096–4105

Livny J, Brencic A, Lory S, Waldor MK (2006) Identification of 17 *Pseudomonas aeruginosa* sRNAs and prediction of sRNA-encoding genes in 10 diverse pathogens using the bioinformatic tool sRNAPredict2. Nucleic Acids Res 34:3484–3493

Loh E, Dussurget O, Gripenland J, Vaitkevicius K, Tiensuu T, Mandin P, Repoila F, Buchrieser C, Cossart P, Johansson J (2009) A trans-acting riboswitch controls expression of the virulence regulator PrfA in *Listeria monocytogenes*. Cell 139:770–779

Loughman JA, Caparon MG (2006) A novel adaptation of aldolase regulates virulence in *Streptococcus pyogenes*. EMBO J 25:5414–5422

Loughman JA, Caparon MG (2007) Comparative functional analysis of the lac operons in *Streptococcus pyogenes*. Mol Microbiol 64:269–280

Luo F, Lizano S, Banik S, Zhang H, Bessen DE (2008) Role of Mga in group A streptococcal infection at the skin epithelium. Microb Pathog 45:217–224

Lyon WR, Madden JC, Levin JC, Stein JL, Caparon MG (2001) Mutation of luxS affects growth and virulence factor expression in *Streptococcus pyogenes*. Mol Microbiol 42:145–157

Malke H, Ferretti JJ (2007) CodY-affected transcriptional gene expression of *Streptococcus pyogenes* during growth in human blood. J Med Microbiol 56:707–714

Malke H, Steiner K, McShan WM, Ferretti JJ (2006) Linking the nutritional status of *Streptococcus pyogenes* to alteration of transcriptional gene expression: the action of CodY and RelA. Int J Med Microbiol 296:259–275

Manetti AG, Koller T, Becherelli M, Buccato S, Kreikemeyer B, Podbielski A, Grandi G, Margarit I (2010) Environmental acidification drives S. *pyogenes* pilus expression and microcolony formation on epithelial cells in a FCT-dependent manner. PLoS One 5:e13864

Mangold M, Siller M, Roppenser B, Vlaminckx BJ, Penfound TA, Klein R, Novak R, Novick RP, Charpentier E (2004) Synthesis of group A streptococcal virulence factors is controlled by a regulatory RNA molecule. Mol Microbiol 53:1515–1527

Mann B, van OT, Wang J, Obert C, Wang YD, Carter R, McGoldrick DJ, Ridout G, Camilli A, Tuomanen EI, Rosch JW (2012) Control of Virulence by Small RNAs in *Streptococcus pneumoniae*. PLoS Pathog 8:e1002788

Mansjo M, Johansson J (2011) The riboflavin analog roseoflavin targets an FMN-riboswitch and blocks *Listeria monocytogenes* growth, but also stimulates virulence gene-expression and infection. RNA Biol 8:674–680

Marouni MJ, Sela S (2003) The luxS Gene of *Streptococcus pyogenes* regulates expression of genes that affect internalization by epithelial cells. Infect Immun 71:5633–5639

Marraffini LA, Sontheimer EJ (2010) Self versus non-self discrimination during CRISPR RNA-directed immunity. Nature 463:568–571
Marx P, Nuhn M, Kovacs M, Hakenbeck R, Bruckner R (2010) Identification of genes for small non-coding RNAs that belong to the regulon of the two-component regulatory system CiaRH in *Streptococcus*. BMC Genomics 11:661
Mascher T, Zahner D, Merai M, Balmelle N, de Saizieu AB, Hakenbeck R (2003) The *Streptococcus pneumoniae* cia regulon: CiaR target sites and transcription profile analysis. J Bacteriol 185:60–70
Mascher T, Heintz M, Zahner D, Merai M, Hakenbeck R (2006) The CiaRH system of *Streptococcus pneumoniae* prevents lysis during stress induced by treatment with cell wall inhibitors and by mutations in pbp2x involved in beta-lactam resistance. J Bacteriol 188:1959–1968
McIver KS (2009) Stand-alone response regulators controlling global virulence networks in *Streptococcus pyogenes*. Contrib Microbiol 16:103–119
McIver KS, Scott JR (1997) Role of mga in growth phase regulation of virulence genes of the group A *Streptococcus*. J Bacteriol 179:5178–5187
McIver KS, Heath AS, Green BD, Scott JR (1995) Specific binding of the activator Mga to promoter sequences of the emm and scpA genes in the group A *Streptococcus*. J Bacteriol 177:6619–6624
McIver KS, Thurman AS, Scott JR (1999) Regulation of mga transcription in the group A *Streptococcus*: specific binding of mga within its own promoter and evidence for a negative regulator. J Bacteriol 181:5373–5383
Mistou MY, Dramsi S, Brega S, Poyart C, Trieu-Cuot P (2009) Molecular dissection of the secA2 locus of group B *Streptococcus* reveals that glycosylation of the Srr1 LPXTG protein is required for full virulence. J Bacteriol 191:4195–4206
Mitchell AM, Mitchell TJ (2010) *Streptococcus pneumoniae*: virulence factors and variation. Clin Microbiol Infect 16:411–418
Mraheil MA, Billion A, Kuenne C, Pischimarov J, Kreikemeyer B, Engelmann S, Hartke A, Giard JC, Rupnik M, Vorwerk S, Beier M, Retey J, Hartsch T, Jacob A, Cemic F, Hemberger J, Chakraborty T, Hain T (2010) Comparative genome-wide analysis of small RNAs of major Gram-positive pathogens: from identification to application. Microb Biotechnol 3:658–676
Mraheil MA, Billion A, Mohamed W, Mukherjee K, Kuenne C, Pischimarov J, Krawitz C, Retey J, Hartsch T, Chakraborty T, Hain T (2011) The intracellular sRNA transcriptome of *Listeria monocytogenes* during growth in macrophages. Nucleic Acids Res 39:4235–4248
Muller M, Marx P, Hakenbeck R, Bruckner R (2011) Effect of new alleles of the histidine kinase gene ciaH on the activity of the response regulator CiaR in *Streptococcus pneumoniae* R6. Microbiology 157:3104–3112
Nakata M, Podbielski A, Kreikemeyer B (2005) MsmR, a specific positive regulator of the *Streptococcus pyogenes* FCT pathogenicity region and cytolysin-mediated translocation system genes. Mol Microbiol 57:786–803
Nakata M, Koller T, Moritz K, Ribardo D, Jonas L, McIver KS, Sumitomo T, Terao Y, Kawabata S, Podbielski A, Kreikemeyer B (2009) Mode of expression and functional characterization of FCT-3 pilus region-encoded proteins in *Streptococcus pyogenes* serotype M49. Infect Immun 77:32–44
Narberhaus F, Vogel J (2009) Regulatory RNAs in prokaryotes: here, there and everywhere. Mol Microbiol 74:261–269
Novick RP, Ross HF, Projan SJ, Kornblum J, Kreiswirth B, Moghazeh S (1993) Synthesis of staphylococcal virulence factors is controlled by a regulatory RNA molecule. EMBO J 12:3967–3975
Nudler E, Mironov AS (2004) The riboswitch control of bacterial metabolism. Trends Biochem Sci 29:11–17
Okada N, Geist RT, Caparon MG (1993) Positive transcriptional control of mry regulates virulence in the group A streptococcus. Mol Microbiol 7:893–903

Ou YQ, Ma WL, Liu CH, Shi R, Zheng WL (2005) [Detection of quorum-sensing pathway and construction of LuxS gene deletion mutants of group B *Streptococcus*]. Di Yi Jun Yi Da Xue Xue Bao 25:1135–1139

Ouyang Q, Ma WL, Liu CH, Shi R, Zheng WL (2006) [Phenotypic analysis of luxS gene deletion mutants and its application in virulence regulation research in group B *Streptococcus*]. Nan Fang Yi Ke Da Xue Xue Bao 26:117–121

Pan X, Ge J, Li M, Wu B, Wang C, Wang J, Feng Y, Yin Z, Zheng F, Cheng G, Sun W, Ji H, Hu D, Shi P, Feng X, Hao X, Dong R, Hu F, Tang J (2009) The orphan response regulator CovR: a globally negative modulator of virulence in Streptococcus suis serotype 2. J Bacteriol 191:2601–2612

Papenfort K, Vogel J (2010) Regulatory RNA in bacterial pathogens. Cell Host Microbe 8: 116–127

Park SE, Jiang S, Wessels MR (2012) CsrRS and environmental pH regulate group B *Streptococcus* adherence to human epithelial cells and extracellular matrix. Infect Immun 80:3975–3984

Patenge N, Billion A, Raasch P, Normann J, Wisniewska-Kucper A, Retey J, Boisguérin V, Hartsch T, Hain T, Kreikemeyer B (2012) Identification of novel growth phase- and media-dependent small non-coding RNAs in *Streptococcus pyogenes* M49 using intergenic tiling arrays. BMC Genomics Genomics 13:550

Paterson GK, Blue CE, Mitchell TJ (2006) Role of two-component systems in the virulence of *Streptococcus pneumoniae*. J Med Microbiol 55:355–363

Perez N, Trevino J, Liu Z, Ho SC, Babitzke P, Sumby P (2009) A genome-wide analysis of small regulatory RNAs in the human pathogen group A *Streptococcus*. PLoS One 4:e7668

Perez-Casal JF, Dillon HF, Husmann LK, Graham B, Scott JR (1993) Virulence of two *Streptococcus pyogenes* strains (types M1 and M3) associated with toxic-shock-like syndrome depends on an intact mry-like gene. Infect Immun 61:5426–5430

Petersen FC, Ahmed NA, Naemi A, Scheie AA (2006) LuxS-mediated signalling in *Streptococcus anginosus* and its role in biofilm formation. Antonie Van Leeuwenhoek 90:109–121

Pichon C, du ML, Caliot ME, Trieu-Cuot P, Le BC (2012) An in silico model for identification of small RNAs in whole bacterial genomes: characterization of antisense RNAs in pathogenic *Escherichia coli* and *Streptococcus agalactiae* strains. Nucleic Acids Res 40:2846–2861

Pischimarov J, Kuenne C, Billion A, Hemberger J, Cemic F, Chakraborty T, Hain T (2012) sRNAdb: A small non-coding RNA database for gram-positive bacteria. BMC Genomics 13:384

Podbielski A, Flosdorff A, Weber-Heynemann J (1995) The group A streptococcal virR49 gene controls expression of four structural vir regulon genes. Infect Immun 63:9–20

Podbielski A, Woischnik M, Leonard BA, Schmidt KH (1999) Characterization of nra, a global negative regulator gene in group A streptococci. Mol Microbiol 31:1051–1064

Podkaminski D, Vogel J (2010) Small RNAs promote mRNA stability to activate the synthesis of virulence factors. Mol Microbiol 78:1327–1331

Price CE, Zeyniyev A, Kuipers OP, Kok J (2011) From meadows to milk to mucosa—adaptation of *Streptococcus* and *Lactococcus* species to their nutritional environments. FEMS Microbiol Rev 36:949–971

Quach D, van Sorge NM, Kristian SA, Bryan JD, Shelver DW, Doran KS (2009) The CiaR response regulator in group B *Streptococcus* promotes intracellular survival and resistance to innate immune defenses. J Bacteriol 191:2023–2032

Raasch P, Schmitz U, Patenge N, Vera J, Kreikemeyer B, Wolkenhauer O (2010) Non-coding RNA detection methods combined to improve usability, reproducibility and precision. BMC Bioinform 11:491

Rajagopal L (2009) Understanding the regulation of Group B Streptococcal virulence factors. Future Microbiol 4:201–221

Rajagopal L, Clancy A, Rubens CE (2003) A eukaryotic type serine/threonine kinase and phosphatase in *Streptococcus agalactiae* reversibly phosphorylate an inorganic pyrophosphatase and affect growth, cell segregation, and virulence. J Biol Chem 278:14429–14441

Rajagopal L, Vo A, Silvestroni A, Rubens CE (2006) Regulation of cytotoxin expression by converging eukaryotic-type and two-component signalling mechanisms in *Streptococcus agalactiae*. Mol Microbiol 62:941–957

Ramirez-Pena E, Trevino J, Liu Z, Perez N, Sumby P (2010) The group A *Streptococcus* small regulatory RNA FasX enhances streptokinase activity by increasing the stability of the ska mRNA transcript. Mol Microbiol 78:1332–1347

Redanz S, Standar K, Podbielski A, Kreikemeyer B (2012) Heterologous expression of sahH reveals that biofilm formation is autoinducer-2 independent in *Streptococcus sanguinis*, but is associated with an intact AMC. J Biol Chem 287:36111–36122

Riani C, Standar K, Srimuang S, Lembke C, Kreikemeyer B, Podbielski A (2007) Transcriptome analyses extend understanding of *Streptococcus pyogenes* regulatory mechanisms and behavior toward immunomodulatory substances. Int J Med Microbiol 297:513–523

Ribardo DA, McIver KS (2003) amrA encodes a putative membrane protein necessary for maximal exponential phase expression of the Mga virulence regulon in *Streptococcus pyogenes*. Mol Microbiol 50:673–685

Ribardo DA, McIver KS (2006) Defining the Mga regulon: comparative transcriptome analysis reveals both direct and indirect regulation by Mga in the group A *Streptococcus*. Mol Microbiol 62:491–508

Roberts SA, Scott JR (2007) RivR and the small RNA RivX: the missing links between the CovR regulatory cascade and the Mga regulon. Mol Microbiol 66:1506–1522

Romao S, Memmi G, Oggioni MR, Trombe MC (2006) LuxS impacts on LytA-dependent autolysis and on competence in *Streptococcus pneumoniae*. Microbiology 152:333–341

Rosch JW, Mann B, Thornton J, Sublett J, Tuomanen E (2008) Convergence of regulatory networks on the pilus locus of *Streptococcus pneumoniae*. Infect Immun 76:3187–3196

Samen U, Heinz B, Boisvert H, Eikmanns BJ, Reinscheid DJ, Borges F (2011) Rga is a regulator of adherence and pilus formation in *Streptococcus agalactiae*. Microbiology 157:2319–2327

Santi I, Grifantini R, Jiang SM, Brettoni C, Grandi G, Wessels MR, Soriani M (2009) CsrRS regulates group B *Streptococcus* virulence gene expression in response to environmental pH: a new perspective on vaccine development. J Bacteriol 191:5387–5397

Sapranauskas R, Gasiunas G, Fremaux C, Barrangou R, Horvath P, Siksnys V (2011) The *Streptococcus thermophilus* CRISPR/Cas system provides immunity in *Escherichia coli*. Nucleic Acids Res 39:9275–9282

Schauder S, Shokat K, Surette MG, Bassler BL (2001) The LuxS family of bacterial autoinducers: biosynthesis of a novel quorum-sensing signal molecule. Mol Microbiol 41:463–476

Scott JR, Zahner D (2006) Pili with strong attachments: Gram-positive bacteria do it differently. Mol Microbiol 62:320–330

Sebert ME, Patel KP, Plotnick M, Weiser JN (2005) Pneumococcal HtrA protease mediates inhibition of competence by the CiaRH two-component signaling system. J Bacteriol 187:3969–3979

Sharma CM, Hoffmann S, Darfeuille F, Reignier J, Findeiss S, Sittka A, Chabas S, Reiche K, Hackermuller J, Reinhardt R, Stadler PF, Vogel J (2010) The primary transcriptome of the major human pathogen *Helicobacter pylori*. Nature 464:250–255

Sharma CM, Papenfort K, Pernitzsch SR, Mollenkopf HJ, Hinton JC, Vogel J (2011) Pervasive post-transcriptional control of genes involved in amino acid metabolism by the Hfq-dependent GcvB small RNA. Mol Microbiol 81:1144–1165

Shelburne SA III, Sumby P, Sitkiewicz I, Granville C, Deleo FR, Musser JM (2005) Central role of a bacterial two-component gene regulatory system of previously unknown function in pathogen persistence in human saliva. Proc Natl Acad Sci U S A 102:16037–16042

Shelburne SA, Keith D, Horstmann N, Sumby P, Davenport MT, Graviss EA, Brennan RG, Musser JM (2008) A direct link between carbohydrate utilization and virulence in the major human pathogen group A *Streptococcus*. Proc Natl Acad Sci 105:1698–1703

Shelburne SA, Olsen RJ, Suber B, Sahasrabhojane P, Sumby P, Brennan RG, Musser JM (2010) A combination of independent transcriptional regulators shapes bacterial virulence gene expression during infection. PLoS Pathog 6:e1000817

Shelver D, Rajagopal L, Harris TO, Rubens CE (2003) MtaR, a regulator of methionine transport, is critical for survival of group B *Streptococcus* in vivo. J Bacteriol 185:6592–6599

Siller M, Janapatla R, Pirzada Z, Hassler C, Zinkl D, Charpentier E (2008) Functional analysis of the group A streptococcal luxS/AI-2 system in metabolism, adaptation to stress and interaction with host cells. BMC Microbiol 8:188

Solano-Collado V, Espinosa M, Bravo A (2012) Activator role of the pneumococcal mga-like virulence transcriptional regulator. J Bacteriol 194:4197–4207

Sonenshein AL (2005) CodY, a global regulator of stationary phase and virulence in gram-positive bacteria. Curr Opin Microbiol 8:203–207

Sridhar J, Sambaturu N, Sabarinathan R, Ou HY, Deng Z, Sekar K, Rafi ZA, Rajakumar K (2010) sRNAscanner: a computational tool for intergenic small RNA detection in bacterial genomes. PLoS One 5:e11970

Stenz L, Francois P, Whiteson K, Wolz C, Linder P, Schrenzel J (2011) The CodY pleiotropic repressor controls virulence in gram-positive pathogens. FEMS Immunol Med Microbiol 62:123–139

Stock AM, Robinson VL, Goudreau PN (2000) Two-component signal transduction. Annu Rev Biochem 69:183–215

Storz G, Vogel J, Wassarman KM (2011) Regulation by small RNAs in bacteria: expanding frontiers. Mol Cell 43:880–891

Stroeher UH, Paton AW, Ogunniyi AD, Paton JC (2003) Mutation of luxS of *Streptococcus pneumoniae* affects virulence in a mouse model. Infect Immun 71:3206–3212

Sugareva V, Arlt R, Fiedler T, Riani C, Podbielski A, Kreikemeyer B (2010) Serotype- and strain- dependent contribution of the sensor kinase CovS of the CovRS two-component system to *Streptococcus pyogenes* pathogenesis. BMC Microbiol 10:34

Sumby P, Whitney AR, Graviss EA, DeLeo FR, Musser JM (2006) Genome-wide analysis of group A Streptococci reveals a mutation that modulates global phenotype and disease specificity. PLoS Pathog 2:e5

Surette MG, Miller MB, Bassler BL (1999) Quorum sensing in *Escherichia coli*, *Salmonella typhimurium*, and *Vibrio harveyi*: a new family of genes responsible for autoinducer production. Proc Natl Acad Sci U S A 96:1639–1644

Takamatsu D, Nishino H, Ishiji T, Ishii J, Osaki M, Fittipaldi N, Gottschalk M, Tharavichitkul P, Takai S, Sekizaki T (2009) Genetic organization and preferential distribution of putative pilus gene clusters in *Streptococcus suis*. Vet Microbiol 138:132–139

Telford JL, Barocchi MA, Margarit I, Rappuoli R, Grandi G (2006) Pili in gram-positive pathogens. Nat Rev Microbiol 4:509–519

Tettelin H, Nelson KE, Paulsen IT, Eisen JA, Read TD, Peterson S, Heidelberg J, Deboy RT, Haft DH, Dodson RJ, Durkin AS, Gwinn M, Kolonay JF, Nelson WC, Peterson JD, Umayam LA, White O, Salzberg SL, Lewis MR, Radune D, Holtzapple E, Khouri H, Wolf AM, Utterback TR, Hansen CL, McDonald LA, Feldblyum TV, Angiuoli S, Dickinson T, Hickey EK, Holt IE, Loftus BJ, Yang F, Smith HO, Venter JC, Dougherty BA, Morrison DA, Hollingshead SK, Fraser CM (2001) Complete genome sequence of a virulent isolate of *Streptococcus pneumoniae*. Science 293: 498–506

Tettelin H, Masignani V, Cieslewicz MJ, Eisen JA, Peterson S, Wessels MR, Paulsen IT, Nelson KE, Margarit I, Read TD, Madoff LC, Wolf AM, Beanan MJ, Brinkac LM, Daugherty SC, Deboy RT, Durkin AS, Kolonay JF, Madupu R, Lewis MR, Radune D, Fedorova NB, Scanlan D, Khouri H, Mulligan S, Carty HA, Cline RT, Van Aken SE, Gill J, Scarselli M, Mora M, Iacobini ET, Brettoni C, Galli G, Mariani M, Vegni F, Maione D, Rinaudo D, Rappuoli R, Telford JL, Kasper DL, Grandi G, Fraser CM (2002) Complete genome sequence and comparative genomic analysis of an emerging human pathogen, serotype V *Streptococcus agalactiae*. Proc Natl Acad Sci U S A 99:12391–12396

Thomason MK, Storz G (2010) Bacterial antisense RNAs: how many are there, and what are they doing? Annu Rev Genet 44:167–188

Trappetti C, Potter AJ, Paton AW, Oggioni MR, Paton JC (2011) LuxS mediates iron-dependent biofilm formation, competence, and fratricide in *Streptococcus pneumoniae*. Infect Immun 79:4550–4558

Trevino J, Perez N, Ramirez-Pena E, Liu Z, Shelburne SA III, Musser JM, Sumby P (2009) CovS simultaneously activates and inhibits the CovR-mediated repression of distinct subsets of group A *Streptococcus* virulence factor-encoding genes. Infect Immun 77:3141–3149

Trevino J, Perez N, Sumby P (2010) The 4.5S RNA component of the signal recognition particle is required for group A *Streptococcus* virulence. Microbiology 156:1342–1350

Tsui HC, Mukherjee D, Ray VA, Sham LT, Feig AL, Winkler ME (2010) Identification and characterization of noncoding small RNAs in *Streptococcus pneumoniae* serotype 2 strain D39. J Bacteriol 192:264–279

Vahling CM, McIver KS (2006) Domains required for transcriptional activation show conservation in the mga family of virulence gene regulators. J Bacteriol 188:863–873

Vanderpool CK, Gottesman S (2005) Noncoding RNAs at the membrane. Nat Struct Mol Biol 12:285–286

Vendeville A, Winzer K, Heurlier K, Tang CM, Hardie KR (2005) Making 'sense' of metabolism: autoinducer-2, LuxS and pathogenic bacteria. Nat Rev Microbiol 3:383–396

Vidal JE, Ludewick HP, Kunkel RM, Zähner D, Klugman KP (2011) The LuxS-dependent quorum-sensing system regulates early biofilm formation by *Streptococcus pneumoniae* strain D39. Infect Immun 79:4050–4060

Vitreschak AG, Rodionov DA, Mironov AA, Gelfand MS (2002) Regulation of riboflavin biosynthesis and transport genes in bacteria by transcriptional and translational attenuation. Nucleic Acids Res 30:3141–3151

Voyich JM, Sturdevant DE, Braughton KR, Kobayashi SD, Lei B, Virtaneva K, Dorward DW, Musser JM, Deleo FR (2003) Genome-wide protective response used by group A *Streptococcus* to evade destruction by human polymorphonuclear leukocytes. Proc Natl Acad Sci U S A 100:1996–2001

Wang Y, Zhang W, Wu Z, Zhu X, Lu C (2011) Functional analysis of luxS in *Streptococcus suis* reveals a key role in biofilm formation and virulence. Vet Microbiol 152:151–160

Waters LS, Storz G (2009) Regulatory RNAs in bacteria. Cell 136:615–628

Willenborg J, Fulde M, de Greeff A, Rohde M, Smith HE, Valentin-Weigand P, Goethe R (2011) Role of glucose and CcpA in capsule expression and virulence of *Streptococcus suis*. Microbiology 157:1823–1833

Winkler ME, Hoch JA (2008) Essentiality, bypass, and targeting of the YycFG (VicRK) two-component regulatory system in gram-positive bacteria. J Bacteriol 190:2645–2648

Xia L, Xia W, Li S, Li W, Liu J, Ding H, Li J, Li H, Chen Y, Su X, Wang W, Sun L, Wang C, Shao N, Chu B (2012) Identification and expression of small non-coding RNA, L10-Leader, in different growth phases of *Streptococcus mutans*. Nucleic Acid Ther 22:177–186

Yoshida A, Ansai T, Takehara T, Kuramitsu HK (2005) LuxS-based signaling affects *Streptococcus mutans* biofilm formation. Appl Environ Microbiol 71:2372–2380

Zahner D, Kaminski K, van der Linden M, Mascher T, Meral M, Hakenbeck R (2002) The ciaR/ciaH regulatory network of *Streptococcus pneumoniae*. J Mol Microbiol Biotechnol 4: 211–216

Zahner D, Gandhi AR, Yi H, Stephens DS (2011) Mitis group streptococci express variable pilus islet 2 pili. PLoS One 6:25124

Zhu J, Patel R, Pei D (2004) Catalytic mechanism of S-ribosylhomocysteinase (LuxS): stereochemical course and kinetic isotope effect of proton transfer reactions. Biochemistry 43:10166–10172

Host–Pathogen Interactions in Streptococcal Immune Sequelae

D. Patric Nitsche-Schmitz and Gursharan S. Chhatwal

Abstract Otherwise uncomplicated infections with *Streptococcus pyogenes* can cause two insidious immune sequelae known as post-streptococcal glomerulonephritis (PSGN) and acute rheumatic fever (ARF). These diseases follow with a latency of a few weeks or months after primary infection and are responsible for high mortality and morbidity. PSGN has also been linked to infections with group C streptococci of the species *S. equi* ssp. *zooepidemicus* (SESZ). Moreover, there are some indications that infection with group C and G streptococci (GCGS) of the subspecies *Streptococcus dysgalactiae* ssp. *equisimilis* (SDSE) leads to ARF. Despite decades of research, the picture of the molecular pathogenesis of streptococcal immune sequelae resembles a jigsaw puzzle. Herein we try to put some of the puzzle bits together that have been collected till date.

Contents

G. S. Chhatwal
Department of Medical Microbiology, Helmholtz Centre for Infection Research, D-38124 Braunschweig, Germany

D. P. Nitsche-Schmitz (✉)
Helmholtz Centre for Infection Research, Inhoffenstraße 7 D-38124 Braunschweig, Germany
e-mail: Patric.Nitsche@helmholtz-hzi.de

Current Topics in Microbiology and Immunology (2013) 368: 155–171
DOI: 10.1007/82_2012_296

Published Online: 6 December 2012

1 Introduction

More than 15 million people worldwide suffer from consequences of the severe immune sequelae post-streptococcal glomerulonephritis (PSGN) and acute rheumatic fever (ARF). They are caused by mild but inadequately treated infections with pyogenic streptococci. ARF and PSGN claim 350,000 human lives every year (Carapetis et al. 2005a, b). Approaches to combat these immune sequelae include the development of streptococcal vaccines for prevention of the causative infections and of diagnostic tests for early detection of these infections. For both approaches, profound information on the causative streptococci is indispensable. A wide-ranging knowledge of the factors and processes that trigger and drive the pathogenesis of PSGN and ARF is a further support and may deliver novel ideas for preventive and therapeutic approaches. This article is a summary of the current knowledge of the etiology and pathogenesis of the aforementioned diseases.

2 Post-streptococcal Glomerulonephritis

2.1 Characteristics of PSGN

The immune sequela PSGN follows with a delay of 1–3 weeks after skin infections, pharyngitis, or scarlet fever caused by *S. pyogenes* (Dick and Dick 1983; Futcher 1940; Lyttle et al. 1938). Infections with SESZ typically occur as outbreaks. Such outbreaks can be followed by a wave of PSGN cases, which proves the nephritogenic potential of this group C streptococcal species (Balter et al. 2000; Barnham et al. 1983; Duca et al. 1969; Francis et al. 1993). The molecular pathogenesis of PSGN is still elusive. Deposition of immune complexes that contain streptococcal components, aberrant complement reactions, autoimmunity, and other injurious cellular immune responses are suspected causes of this severe disease that affects the kidney (Chhatwal and Graham 2008).

Examination of the diseased glomeruli by electron microscopy revealed characteristic electron-dense structures on the luminal site of the glomerular capillary basement membrane, called sub-epithelial "humps" (Cochrane 1971; Fish et al. 1966). Streptococcal antigens, complement factors, and immunoglobulin have been localized in these humps. This indicates formation of immune complexes, hence crucial contributions of the host immune system to PSGN pathogenesis (Andres et al. 1966; Batsford et al. 2005). Potential causes for glomerular damage comprise autoimmunity against components of the basement membrane of capillary walls, deposition of immune complexes due to anti-IgG autoantibodies, and antibodies that react against streptococcal proteins. Moreover, it has been suggested that plasminogen binding and activating proteins may take part in the pathogenesis of glomerulonephritis (Rodriguez-Iturbe and Batsford 2007).

2.2 Role of Complement in PSGN

Concentrations of complement factors are strongly reduced in the blood of patients with PSGN. This condition, called hypocomplementemia, indicates the activation and consumption of the complement system during disease (Dedeoglu et al. 1996). Mice that lack complement factor 1 subunit q (C1q) k.o.-mice develop glomerulonephritis that resembles PSGN, pathohistologically. Proliferation of cells in the Bowman's space and electron-dense sub-endothelial and sub-epithelial depositions occur in the affected glomeruli (Botto et al. 1998). Observation of sub-epithelial depositions in the C1q-deficient mice suggests that deficiencies in the complement system cause the sub-epithelial humps in PSGN and contribute to the pathogenesis of this immune disease. This is contradictory to the localization of complement factors in the humps, which indicates activity of the complement system in the affected tissue rather than diminished response (Rodriguez-Iturbe and Batsford 2007). Similar to sub-epithelial humps in PSGN, glomerular depositions in the C1q-deficient mice contained complement factor 3 (Botto et al. 1998), indicating activity of the lectin or alternative pathway of complement. The precise role of complement reactions in the pathogenesis of PSGN remains a matter of investigation.

2.3 Factors Involved in PSGN

Immunohistochemistry has localized a variety of different streptococcal antigens in the glomeruli of PSGN patients (Batsford et al. 2005; Michael et al. 1966; Ohkuni et al. 1983; Seegal et al. 1965; Treser et al. 1970; Villarreal et al. 1979). The foci of streptococcal infections that cause PSGN are distant from the kidney. Therefore, glomerular deposition of streptococcal antigens is evidence for blood borne bacteria or bacterial components. These foreign antigens are not efficiently cleared by the immune system and may be the triggers of PSGN pathogenesis. Notably, intact bacteria have never been found in renal tissue affected by PSGN.

The glyceraldehyde-3-phosphate-dehydrogenase of *S. pyogenes* has been suggested as a candidate nephritogenic protein, under the name nephritis-associated plasminogen receptor (NAPlr). As a major glycolytic enzyme, it is present in all strains and species of the genus *Streptococcus*. Although having an important intracellular metabolic function, a fraction of NAPlr is released into the extracellular space. There it binds back to the streptococcal surface and serves as a receptor for plasmin(ogen) (Pancholi and Fischetti 1993). Although NAPlr is a trigger of the alternative complement pathway (Yoshizawa et al. 2004) it does not co-localize with complement and immunoglobulin deposits in glomeruli of PSGN patients. Therefore, activation of plasmin(ogen) by NAPlr is a more likely mode of action in PSGN pathogenesis than activation of the complement system by this bacterial protein (Oda et al. 2005). Homologs of NAPlr are also present on the surface of commensal oral streptococci, which also bind and activate plasmin(ogen) (Itzek et al. 2010; Jones and Holt 2004). Transient bacteremia with oral streptococci is frequent in humans (Daly et al. 1997; Roberts et al. 1997; Westling et al. 2002), but not associated with PSGN. The features that distinguish NAPlr of *S. pyogenes* from other streptococcal glyceraldehyde-3-phosphate-dehydrogenases remain to be identified and may shed light on a possible role of NAPlr in PSGN.

Only a few *S. pyogenes* strains lack the streptococcal cysteine proteinase SpeB (Darenberg et al. 2007; Schmitz et al. 2003; Vlaminckx et al. 2003), which is another suspected nephritogenic factor. SpeB and its zymogen zSpeB interact with plasminogen. Other than NAPlr, SpeB co-localizes with complement factor 3 and IgG in the sub-epithelial humps of glomeruli that are affected by PSGN (Batsford et al. 2005; Rodriguez-Iturbe and Batsford 2007). The cationic character of SpeB and zSpeB is thought to allow the transmigration of these molecules from the blood through the glomerular basement membrane, which is followed by deposition of the proteins on the sub-epithelial side and development of nephritis (Vogt et al. 1983). On one hand this hypothesis is supported by examinations on kidney biopsies and anti-SpeB serum titers of PSGN patients (Cu et al. 1998; Parra et al. 1998). On the other hand a SESZ strain that caused an epidemic of PSGN in Nova Serrana, Brazil (Balter et al. 2000; Beres et al. 2008), lacks SpeB. Hence, alternative pathogenicity factors exist that cause PSGN. The chain of evidence for SpeB as a nephritogenic molecule is not yet closed.

2.4 Correlation of M Types with PSGN

The nephritogenic potential of *S. pyogenes* strains correlates with certain M-types such as M1, 2, 4, 12, 18, 25, 49, 55, 57, and 60 (Stollerman 1969). As one of the type-specific genes the M protein itself may be involved in the pathogenesis of PSGN. Released from the bacterial surface by the actions of bacterial and host proteases like SpeB or neutrophil elastase, M protein forms insoluble aggregates with the plasma protein fibrinogen (Berge and Björck 1995; Herwald et al. 2004). When injected into mice, the M protein binds and precipitates in the glomeruli (Kantor 1965; Kaplan 1958) and causes renal lesions that could be nephritogenic (Humair et al. 1969). Notably, M-like proteins exist in SESZ, including the outbreak associated strain MGCS10565 (Beres et al. 2008) and may add to the nephritogenic potential of this species.

Serological examinations on PSGN patients suggest protein SIC (streptococcal inhibitor of complement-lysis) as a nephritogenic factor (Skattum et al. 2006; Sriprakash et al. 2002). Presence of genes that code for protein SIC or its variant DRS is limited to *S. pyogenes* strains of nephritogenic M-types like M1, 12, 55, and 57 (Hartas and Sriprakash 1999), which supports the notion that they are nephritogenic factors. SIC binds to complement factors that form the membrane attack complex, preventing its assembly (Åkesson et al. 1996). However, it is not yet known how this property or other host-interactions of SIC or DRS may cause PSGN. SIC or similar proteins have not been found in SESZ.

Of the variety of potential nephritogenic factors that have been under investigation, so far there is none that can be considered as the sole nephritogenic factor. Despite the fact that many of the suspected nephritogens accumulate in the kidney of laboratory animals, leading to glomerular injury, none of the factors alone were capable to evoke a disorder that resembled PSGN, sufficiently. Moreover, some of the factors do not occur in SESZ. The repertoire of host–pathogen interactions that leads to PSGN is not yet uncovered.

3 Acute Rheumatic Fever and Rheumatic Heart Disease

3.1 Characterization of ARF

Autoimmune responses caused by an untreated or insufficiently treated *S. pyogenes* pharyngitis cause acute rheumatic fever (ARF) with a latency period of 1–4 weeks. The molecular and immunological processes that drive the pathogenesis of this disorder and subsequent rheumatic heart disease (RHD) have been partially discovered. (Carapetis et al. 2005a, b; Cunningham 2012; Dinkla et al. 2003; Guilherme et al. 2006; McCarty 1956; Stollerman 1969; Tontsch et al. 2000; Wannamaker 1973). The diagnostic criteria for ARF are referred to as the major Jones criteria and include arthritis, erythema, carditis, and neurological dysfunctions (chorea) (Keitzer 2005). About a third of the patients develop RHD, leading to irreversible damage to the heart valves that requires surgical intervention. Most of the patients with ARF are children, adolescents, and young adults. Globally, about 1 per 1,000 people is suffering from ARF or RHD with some geographic regions and specific communities being particularly affected (Carapetis et al. 2005a, b; Chhatwal and Graham 2008; Cunningham 2000). Certain humans appear particularly prone to contract ARF as indicated by associations with genetic markers such as major histocompatibility complex (MHC) antigen phenotypes (for references see: Cunningham 2000). In addition to host susceptibility, differences in medical care and the socioeconomic conditions are influencing the epidemiology of ARF.

3.2 Role of Humoral and Cellular Immunity in ARF

Elevated titers of autoantibodies in the sera of ARF/RHD patients (Carapetis et al. 2005a, b; Dinkla et al. 2003, 2007; Guilherme et al. 2006) and deposition of immunoglobulin and complement in the affected hearts (Kaplan et al. 1964) indicate that humoral immunity is crucially involved in ARF pathogenesis. Moreover, the early processes in rheumatic carditis involve cellular immune responses. Elevated numbers of T-lymphocytes and macrophages occur in the cellular infiltrates that are observed in myocardial lesions of affected hearts (Cunningham 2003; Guilherme et al. 2006; Roberts et al. 2001). The inflammation that takes place in those patients and is characterized by increased numbers of mature circulating T helper cells (CD4 + T cells), elevated levels of cytokines such as IL-1 and IL-2, IFNγ, TNF-α and higher amounts of receptors for these cytokines (Guilherme et al. 2006). In summary, ARF is caused by destructive immune responses and inflammation that are not sufficiently suppressed after infection.

3.3 Streptococcal Factors in ARF

Several streptococcal factors such as superantigens, group A carbohydrate, hyaluronic acid capsule, and M proteins have been suspected to trigger the autoimmunity in ARF. During acute infection superantigens act as toxins that contribute to streptococcal toxic shock-like syndrome because they are triggers of exaggerated inflammation (Eriksson et al. 1999; Kotb 1995; Sriskandan et al. 1996; Stevens et al. 1989; Yu and Ferretti 1989). Moreover, they exert mitogenic activity on T-lymphocytes that may crucially influence the development of destructive autoimmunity in response to streptococcal infection. By means of specific interactions, superantigens bridge the MHC class II proteins of antigen presenting cells with the variable Vβ-region of the T-cell receptor independent of a presented antigen. In response to that, a T-cell population that carries matching Vβ-regions, will be activated and release pro-inflammatory cytokines. Interaction of co-receptors CD28 or LFA-1 on T-cells with the suiting ligands B7 or ICAM-1, respectively, on antigen presenting cells stimulate a polyclonal expansion of the activated T-cells. In the absence of these co-stimulating interactions, the activated T-cells fall into anergy or apoptosis. As a consequence selection for a certain T-cell population takes place. Depending on further co-stimulatory interactions, activated T-cells can stimulate B-cell maturation to plasma cells and ensuing secretion of a set of polyclonal immunoglobulins. In summary, superantigens can expand certain polyclonal populations of lymphocytes while driving others into anergy or apoptosis. If this combines with a break of tolerance the patient may produce defined populations of auto-reactive cells (for reference see (Kotb 1995; Proft and Fraser 2003). If and how this is generating the autoimmunity against articular, neural, and cardiac tissue that is seen in ARF/RHD is not yet studied in depth.

As indicated by the deposition of immunoglobulin and complement factors in the diseased heart, humoral immune reactions participate in ARF pathogenesis (Kaplan et al. 1964). Autoimmune responses against several host proteins such as cardiac myosin, heart valve glycoproteins, and antigens of the sarcolemma membrane occur in ARF or RHD patients (Carapetis et al. 2005a, b; Goldstein et al. 1968; Guilherme et al. 2006; Tontsch et al. 2000; Wannamaker 1973). Cross-reactivity of the autoantibodies with streptococcal cell wall carbohydrates, the cell membrane or the M protein was demonstrated (Kaplan and Suchy 1964; Zabriskie and Freimer 1966). In 1968, Goldstein and colleagues isolated glycoproteins from the heart valve that contained N-acetylglucosamine (GlcNAc) and shared epitopes with the streptococcal group A carbohydrate that contains the same monosaccharide unit (Goldstein et al. 1968). Antibodies that reacted with group A carbohydrate circulated in patients with rheumatic valvular disease implicating their role in the pathogenesis of this disorder (Dudding and Ayoub 1968). Notably, monoclonal antibodies against GlcNAc could be isolated from ARF that cross-reacted between human and streptococcal peptide antigens (Adderson et al. 1998). As a monosaccharide subunit in N-glycosylated proteins and glycosaminoglycans, such as the hyaluronic acid, GlcNAc is ubiquitous in the host's extracellular matrix and part of the surrounding matrix (calix) of virtually all cells. It is also located on the surface of erythrocytes. Thus, autoimmune responses against GlcNAc may be causative for the early, less specific symptoms in ARF.

3.4 Role of M protein and Collagen in ARF

S. pyogenes isolates of M type 18, which were thickly encapsulated with hyaluronic acid, caused an ARF outbreak in Utah (Veasy et al. 2004). Hyaluronic acid is not only produced by the bacteria. It is abundant in the host's extracellular matrix. Therefore, the *S. pyogenes* capsule was suspected to cause autoimmunity. Some polysaccharide antigens may activate specific B cells, because their repetitive epitopes can cross-link B-cell receptors (Vos et al. 2000). Although it remains speculative, hyaluronic acid may be one of those polymeric antigens. In connection with a break of tolerance, which could be caused by streptococcal triggers of inflammation such as superantigens, the clonal expansion and maturation of the B-lymphocytes may lead to prolonged autoimmune responses against this carbohydrate and maybe disease. More recent work suggests that the hyaluronic acid capsule binds and aggregates collagen on the streptococcal surface, evoking the production of collagen autoantibodies (Dinkla et al. 2003). Hyaluronic acid is also produced in high quantities by SESZ (Chong et al. 2005; Woolcock 1974), but there is no evidence for a rheumatogenic potential of this species. The isolated hyaluronic acid alone, either from animal tissues or from SESZ, seems not sufficient to trigger disease, since it is widely used as cosmetic filler with a low risk for immune complications.

Some of the vaccines that are currently under development are designed to exploit the immune responses against M protein that protect humans against reinfection with streptococci of the same or a cross-reacting M-type (Bauer et al. 2012; Dale et al. 2011; Steer et al. 2009a, b). This is complicated by severe side effects of this streptococcal surface antigen that seem to be caused by autoimmunity. After a vaccination trial with M protein type 3 three out of 21 human vaccinees developed ARF or ARF-like symptoms (Massell et al. 1969). M3 is a collagen-binding M protein (Dinkla et al. 2003).

M proteins oligomerize via long α-helical coiled-coil regions to form extended thread-like molecules on the streptococcal surface (Fischetti et al. 1990; Phillips et al. 1981). This structural feature is also found in several host proteins such as α-keratin, human heart myosin, and laminin and gives rise to immunological cross-reactivity known as molecular mimicry (Anonymous 1979; Blank et al. 2007). This mimicry is considered as one of the triggers for the destructive immune responses in ARF (Cunningham 2012; Guilherme and Kalil 2010). As compared to healthy individuals, the blood of ARF/RHD patients contains higher concentrations of these cross-reactive antibodies (McCormack et al. 1993). Notably, the autoantibodies from ARF patients differ in reactivity from anti-cardiac antibodies in postcardiotomy or heart failure patients. This may reflect that the autoimmunity in ARF is the cause for the cardiologic symptoms, while in the other two patient groups autoantibodies are a sequel of cardiac damage (Zabriskie 1967). Monoclonal antibodies derived from ARF/RHD patients were poly-reactive, binding to M-protein, aforementioned human coiled-coil proteins, as well as GlcNAc (Cunningham 2000). Antibodies with a similar reactivity against those human and bacterial antigens were raised in mice by injection of *S. pyogenes* (Cunningham et al. 1984; Cunningham and Russell 1983). Human and murine poly-reactive antibodies with myosin cross-reactivity, as suggested by their cytotoxicity and tissue localization, appear to contribute to the pathogenesis of carditis in ARF (Cunningham 2000). Myosin cross-reactive antibodies recognize epitopes in M proteins M5, 6, and 19 (Bronze et al. 1988; Cunningham et al. 1989; Dale and Beachey 1985a, b, 1986). Further support for a role of molecular mimicry of M protein in ARF is the isolation of reactive T-cell clones that cross-react between M5 protein and cardiac myosin from myocardial and valvular lesions of RHD patients (Fae et al. 2006). Taken together there is a considerable body of evidence for molecular mimicry as a cause for autoimmunity that leads to ARF (Cunningham 2012; Guilherme and Kalil 2010).

3.5 Octapeptide PARF and Its Role in ARF

Alternatively, M proteins trigger autoimmune response by a mechanism that requires binding of collagen (Barroso et al. 2009; Dinkla et al. 2003, 2007). The rheumatogenic M protein of type 3 (Stollerman 1969) that caused ARF and ARF-like symptoms in vaccinees (Massell et al. 1969), binds collagens to the streptococcal surface and causes their aggregation (Dinkla et al. 2003). This interaction, which occurs with collagen of types I–IV, has been observed in different types of M proteins including M proteins of SDSE. It depends on an (A/T/E)XYLXX(L/F)N octapeptide motif that is located in the N-terminal type-specific part of the M protein (Fig. 1) (Barroso et al. 2009; Dinkla et al. 2007; Reißmann et al. 2012) and contributes to acute infections by facilitating streptococcal colonization of the ECM (Dinkla et al. 2003; Nitsche et al. 2006). When injected into mice, M3 protein and other collagen-binding M proteins that carry the (A/T/E)XYLXX(L/F)N motif trigger production of anti-collagen IV antibodies. Collagen IV, as a major component of basement membranes, is localized underneath the endothelium that lines the heart valves. Thus, it is a potential target of the autoimmune responses that lead to or aggravate valvular damage. These collagen autoantibodies were also found in increased levels in ARF and RHD patients as compared to healthy individuals (Dinkla et al. 2003, 2007). Therefore, the collagen binding octapeptide motif was named PARF, which is the acronym for peptide associated with rheumatic fever (Dinkla et al. 2007). The increase in collagen autoimmunity occurs early during acute pharyngitis before the onset of ARF, which indicates that the collagen autoimmune response is causative for the sequela rather than a consequence of the tissue destruction in ARF and RHD (Dinkla et al. 2003) (see Fig. 1).

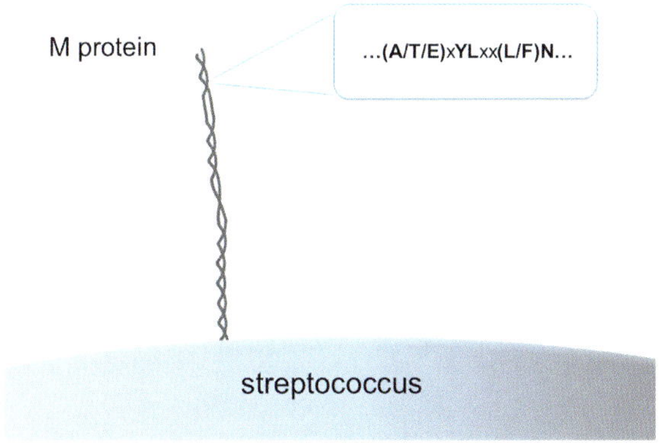

Fig. 1 Schematic representation of an M protein with PARF motif. The dimeric thread-like M protein is covalently bound to the peptidoglycan of the streptococcal surface and protrudes from the bacterium. The PARF motif (*box*) located in the N-terminal end of all known M proteins confers to the streptococci a high binding capacity for collagen IV (Reißmann et al. 2012)

Collagen autoantibodies produced in response to PARF positive collagen binding M proteins do not cross-react with the M protein. This indicates that the immune reactions against collagen in ARF/RHD, other than myosin-directed autoimmunity, are not caused by molecular mimicry. The collagen-binding M protein alters the presentation of the autoantigen in a hitherto uncharacterized way, leading to a break of tolerance (Dinkla et al. 2003, 2007). The altered presentation of collagen to the immune system could be caused by the interaction with streptococcal M protein; the underlying principle of PARF-dependent collagen autoimmunity may be a "conformeropathy" such as the Goodpastuer's syndrome, a collagen IV autoimmune disease that affects the lungs and kidneys (Chan et al. 2011). In Goodpasture's syndrome a conformational change in the non-collagenous domain NC1 of collagen IV leads to exposure of epitopes that evokes the production autoantibodies against the glomerular basement membrane. This causes an autoimmune glomerulonephritis with histological features that differ from PSGN (Chan et al. 2011; Rodriguez-Iturbe and Batsford 2007).

Globally, streptococcal *emm*-types that possess collagen-binding M proteins with PARF motif have a high epidemiological relevance. Based on a world-spanning meta-analysis of *emm*-typing studies (Steer et al. 2009a, b) type 3 is of particular relevance, amounting to 7 % of the *S. pyogenes* isolates (Reißmann et al. 2012). Only M proteins with the PARF motif are shown to endow streptococci with a high binding capacity for a variety of collagens, including collagen IV. However, high affinity interaction has also been observed between M1 protein and collagens I and VI. This potential adhesin had only low affinity for the basement membrane collagen IV (Bober et al. 2011) and binds collagen VI via the non-collagenous region of this connective tissue protein. This suggests principal mechanistic differences as compared to the interaction between collagen and PARF. Thus, it remains unexamined whether M1 and other PARF-negative collagen-binding M proteins are able to trigger a collagen autoimmune response. If they do, the epidemiological relevance of collagen-binding M proteins and collagen autoimmunity in the pathogenesis of ARF and RHD may be even higher than indicated by the previous estimates (Reißmann et al. 2012).

M protein-based vaccines generate serotype-specific immune responses that protect the vaccinee against streptococcal infection and in spite of the ARF-like side effects that were observed in trials with M3 proteins (Massell et al. 1969), such vaccines are still under development (Bauer et al. 2012; Dale et al. 2011; Steer et al. 2009a, b). Available data give rise to concerns. Vaccine antigens that contain a PARF motif may cause autoimmune disease (Dinkla et al. 2003, 2007). This obstacle may be overcome by well-chosen point mutations that inactivate the collagen binding motif, but do not affect the vaccine epitopes, which induce protective immune responses against streptococcal infection.

Continued research on the various processes and factors that cause the destructive immune responses will not only improve our understanding of ARF pathogenesis but may also pave the way for a safe vaccine against the causative streptococci.

4 On the Role of *S. dysgalactiae* ssp. *equisimilis* in Acute Rheumatic Fever

In most if not all of the studies that determined the streptococcal subspecies in GCGS infections in humans, SDSE was the prevalent subspecies (Broyles et al. 2009; Reißmann et al. 2010). There are conceivable indications that infection with SDSE can cause ARF. In contrast, *S. equi* ssp. *zooepidemicus* infections are clearly associated with PSGN (Balter et al. 2000; Barnham et al. 1983; Duca et al. 1969; Francis et al. 1993), but not with ARF.

Early epidemiological data from the U.S. linked ARF to *S. pyogenes* pharyngitis only (McCarty 1956; Wannamaker 1973). In many regions of the world *S. pyogenes* is the predominant streptococcus in sore throat. However, in certain geographic regions with a high prevalence of ARF and RHD, β-hemolytic streptococci have an unusual epidemiology. For instance, in a study from 2006 more than 60 % of the throat isolates from Indigenous Australians in the Northern Territory of Australia were SDSE, exceeding the isolation rate of *S. pyogenes* (McDonald et al. 2006). In 1978 about 75 % of the pharyngitis cases that occurred in Lagos, Nigeria were caused by GCGS (Ogunbi et al. 1978). Both, in Lagos and in the Northern Territory *S. pyogenes* was predominant in skin infections. This is challenging the previous picture of the etiology of ARF. Speculations about a contribution of *S. pyogenes* skin infections in causing ARF have been raised (McDonald et al. 2004). However, experimental evidence is missing that suppurative skin infections can cause autoimmunity or ARF. Another possible explanation for the high prevalence of ARF and RHD in the aforementioned high incidence regions could be an unrecognized or underestimated rheumatogenic potential of SDSE pharyngitis. The vast majority of SDSE strains that colonize or infect humans have M proteins like *S. pyogenes*. This explains why pharyngeal SDSE isolates from a community of Indigenous Australians, evoked anti-myosin responses in mice. Most likely, this is reflecting a rheumatogenic potential of SDSE strains that is responsible for the high prevalence of ARF in this community (Haidan et al. 2000).

Regionally and temporarily, rates of isolates with collagen binding M proteins with PARF motif that trigger collagen autoimmunity exceed 10 % in different regions of the globe both, for SDSE or *S. pyogenes* (Reißmann et al. 2012). More than 10 % of clinical SDSE isolates collected in Southern India bore such collagen-binding M proteins. This region is also known to suffer from a high prevalence of ARF (Reißmann et al. 2012) as well as high rates of SDSE throat carriage and pharyngitis (Bramhachari et al. 2010).

Despite the fact that SDSE isolates carry the factors that trigger ARF-associated autoimmunity, the relevance of this in ARF requires further epidemiological and experimental examination. However, for the time being, these bacteria must not be neglected as a cause for this severe immune sequela.

5 Outlook

Pinpointing the factors that trigger the pathogenesis of streptococcal sequelae will contribute to the identification of the streptococci that cause these immune diseases. Such an ability to detect strains with a nephritogenic or rheumatogenic potential facilitates identification of patients who are at high risk of contracting these immune diseases and who benefit from an intensified preventive treatment. Moreover, our growing knowledge of the spectrum of etiological bacteria may have a substantial impact on the design of vaccines that aim at preventing ARF. For instance, the successful vaccine may have to protect against a broader range of streptococci that includes SDSE. The risk of side effects is another obstacle in the development of streptococcal vaccines. This obstacle may be overcome by excluding streptococcal factors that induce detrimental immune reactions such as molecular mimicry motifs or PARF from vaccine formulations. However, we should be aware that the current knowledge of the streptococcal inducers of immune disease maybe is not yet complete.

References

Adderson EE, Shikhman AR, Ward KE, Cunningham MW (1998) Molecular analysis of polyreactive monoclonal antibodies from rheumatic carditis: human anti-N-acetylglucosamine/anti-myosin antibody V region genes. J Immunol 161(4):2020–2031

Åkesson P, Sjöholm AG, Björck L (1996) Protein SIC, a novel extracellular protein of Streptococcus pyogenes interfering with complement function. J Biol Chem 271(2):1081–1088

Andres GA, Accinni L, Hsu KC, Zabriskie JB, Seegal BC (1966) Electron microscopic studies of human glomerulonephritis with ferritin-conjugated antibody. Localization of antigen-antibody complexes in glomerular structures of patients with acute glomerulonephritis. J Exp Med 123(2):399–412

Anonymous (1979) Klebsiella and ankylosing spondylitis–molecular mimicry? Lancet 1(8124):1012–1013

Balter S, Benin A, Pinto SW, Teixeira LM, Alvim GG, Luna E, Jackson D, LaClaire L, Elliott J, Facklam R, Schuchat A (2000) Epidemic nephritis in Nova Serrana Brazil. Lancet 355(9217):1776–1780

Barnham M, Thornton TJ, Lange K (1983) Nephritis caused by *Streptococcus zooepidemicus* (lancefield group C). Lancet 1(8331):945–948

Barroso V, Rohde M, Davies MR, Gillen CM, Nitsche-Schmitz DP, Dinkla K, Chhatwal GS (2009) Identification of active variants of PARF in human pathogenic group C and group G streptococci leads to an amended description of its consensus motif. Int J Med Microbiol 299(8):547–553

Batsford SR, Mezzano S, Mihatsch M, Schiltz E, Rodriguez-Iturbe B (2005) Is the nephritogenic antigen in post-streptococcal glomerulonephritis pyrogenic exotoxin B (SPE B) or GAPDH? Kidney Int 68(3):1120–1129

Bauer MJ, Georgousakis MM, Vu T, Henningham A, Hofmann A, Rettel M, Hafner LM, Sriprakash KS, McMillan DJ (2012) Evaluation of novel Streptococcus pyogenes vaccine candidates incorporating multiple conserved sequences from the C-repeat region of the M-protein. Vaccine 30(12):2197–2205

Beres SB, Sesso R, Pinto SW, Hoe NP, Porcella SF, Deleo FR, Musser JM (2008) Genome sequence of a lancefield group C Streptococcus zooepidemicus strain causing epidemic nephritis: new information about an old disease. PLoS One 3(8):e3026

Berge A, Björck L (1995) Streptococcal cysteine proteinase releases biologically active fragments of streptococcal surface proteins. J Biol Chem 270(17):9862–9867

Blank M, Barzilai O, Shoenfeld Y (2007) Molecular mimicry and auto-immunity. Clin Rev Allergy Immunol 32(1):111–118

Bober M, Mörgelin M, Olin AI, von Pawel-Rammingen U, Collin M (2011) The membrane bound LRR lipoprotein Slr, and the cell wall-anchored M1 protein from Streptococcus pyogenes both interact with type I collagen. PLoS One 6(5):e20345

Botto M, Dell'Agnola C, Bygrave AE, Thompson EM, Cook HT, Petry F, Loos M, Pandolfi PP, Walport MJ (1998) Homozygous C1q deficiency causes glomerulonephritis associated with multiple apoptotic bodies. Nat Genet 19(1):56–59

Bramhachari PV, Kaul SY, McMillan DJ, Shaila MS, Karmarkar MG, Sriprakash KS (2010) Disease burden due to *Streptococcus dysgalactiae* subsp. equisimilis (group G and C streptococcus) is higher than that due to Streptococcus pyogenes among Mumbai school children. J Med Microbiol 59(Pt 2):220–223

Bronze MS, Beachey EH, Dale JB (1988) Protective and heart-crossreactive epitopes located within the NH2 terminus of type 19 streptococcal M protein. J Exp Med 167(6):1849–1859

Broyles LN, Van Beneden C, Beall B, Facklam R, Shewmaker PL, Malpiedi P, Daily P, Reingold A, Farley MM (2009) Population-based study of invasive disease due to beta-hemolytic streptococci of groups other than A and B. Clin Infect Dis 48(6):706–712

Carapetis JR, McDonald M, Wilson NJ (2005a) Acute rheumatic fever. Lancet 366(9480):155–168

Carapetis JR, Steer AC, Mulholland EK, Weber M (2005b) The global burden of group A streptococcal diseases. Lancet Infect Dis 5(11):685–694

Chan AL, Louie S, Leslie KO, Juarez MM, Albertson TE (2011) Cutting edge issues in Goodpasture's disease. Clin Rev Allergy Immunol 41(2):151–162

Chhatwal GS, Graham RM (2008) Streptococcal diseases. In: Encyclopedia of public health 2008, vol 6, Academic Press, San Diego, pp 231–241

Chong BF, Blank LM, McLaughlin R, Nielsen LK (2005) Microbial hyaluronic acid production. Appl Microbiol Biotechnol 66(4):341–351

Cochrane CG (1971) Mechanisms involved in the deposition of immune complexes in tissues. J Exp Med 134(3):75–89

Cu GA, Mezzano S, Bannan JD, Zabriskie JB (1998) Immunohistochemical and serological evidence for the role of streptococcal proteinase in acute post-streptococcal glomerulonephritis. Kidney Int 54(3):819–826

Cunningham MW (2000) Pathogenesis of group A streptococcal infections. Clin Microbiol Rev 13(3):470–511

Cunningham MW (2003) Autoimmunity and molecular mimicry in the pathogenesis of post-streptococcal heart disease. Front Biosci 8:533–543

Cunningham MW (2012) Streptococcus and rheumatic fever. Curr Opin Rheumatol 24(4):408–416

Cunningham MW, Russell SM (1983) Study of heart-reactive antibody in antisera and hybridoma culture fluids against group A streptococci. Infect Immun 42(2):531–538

Cunningham MW, Krisher K, Graves DC (1984) Murine monoclonal antibodies reactive with human heart and group A streptococcal membrane antigens. Infect Immun 46(1):34–41

Cunningham MW, McCormack JM, Fenderson PG, Ho MK, Beachey EH, Dale JB (1989) Human and murine antibodies cross-reactive with streptococcal M protein and myosin recognize the sequence GLN-LYS-SER-LYS-GLN in M protein. J Immunol 143(8):2677–2683

Dale JB, Beachey EH (1985a) Epitopes of streptococcal M proteins shared with cardiac myosin. J Exp Med 162(2):583–591

Dale JB, Beachey EH (1985b) Multiple, heart-cross-reactive epitopes of streptococcal M proteins. J Exp Med 161(1):113–122

Dale JB, Beachey EH (1986) Sequence of myosin-crossreactive epitopes of streptococcal M protein. J Exp Med 164(5):1785–1790

Dale JB, Penfound TA, Chiang EY, Walton WJ (2011) New 30-valent M protein-based vaccine evokes cross-opsonic antibodies against non-vaccine serotypes of group A streptococci. Vaccine 29(46):8175–8178

Daly C, Mitchell D, Grossberg D, Highfield J, Stewart D (1997) Bacteraemia caused by periodontal probing. Aust Dent J 42(2):77–80

Darenberg J, Luca-Harari B, Jasir A, Sandgren A, Pettersson H, Schalen C, Norgren M, Romanus V, Norrby-Teglund A, Normark BH (2007) Molecular and clinical characteristics of invasive group A streptococcal infection in Sweden. Clin Infect Dis 45(4):450–458

Dedeoglu IO, Springate JE, Waz WR, Stapleton FB, Feld LG (1996) Prolonged hypocomplementemia in post streptococcal acute glomerulonephritis. Clin Nephrol 46(5):302–305

Dick GF, Dick GH (1983) Landmark article Jan 26, 1924: the etiology of scarlet fever. By George F. Dick and Gladys Henry Dick. JAMA 250(22):3096

Dinkla K, Rohde M, Jansen WT, Kaplan EL, Chhatwal GS, Talay SR (2003) Rheumatic fever-associated Streptococcus pyogenes isolates aggregate collagen. J Clin Invest 111(12):1905–1912

Dinkla K, Nitsche-Schmitz DP, Barroso V, Reissmann S, Johansson HM, Frick IM, Rohde M, Chhatwal GS (2007) Identification of a streptococcal octapeptide motif involved in acute rheumatic fever. J Biol Chem 282(26):18686–18693

Duca E, Teodorovici G, Radu C, Vita A, Talasman-Niculescu P, Bernescu E, Feldi C, Rosca V (1969) A new nephritogenic streptococcus. J Hyg (Lond) 67(4):691–698

Dudding BA, Ayoub EM (1968) Persistence of streptococcal group A antibody in patients with rheumatic valvular disease. J Exp Med 128(5):1081–1098

Eriksson BK, Andersson J, Holm SE, Norgren M (1999) Invasive group A streptococcal infections: T1M1 isolates expressing pyrogenic exotoxins A and B in combination with selective lack of toxin-neutralizing antibodies are associated with increased risk of streptococcal toxic shock syndrome. J Infect Dis 180(2):410–418

Fae KC, da Silva DD, Oshiro SE, Tanaka AC, Pomerantzeff PM, Douay C, Charron D, Toubert A, Cunningham MW, Kalil J, Guilherme L (2006) Mimicry in recognition of cardiac myosin peptides by heart-intralesional T cell clones from rheumatic heart disease. J Immunol 176(9):5662–5670

Fischetti VA, Pancholi V, Schneewind O (1990) Conservation of a hexapeptide sequence in the anchor region of surface proteins from gram-positive cocci. Mol Microbiol 4(9):1603–1605

Fish AJ, Michael AF, Vernier RL, Good RA (1966) Acute serum sickness nephritis in the rabbit. An immune deposit disease. Am J Pathol 49(6):997–1022

Francis AJ, Nimmo GR, Efstratiou A, Galanis V, Nuttall N (1993) Investigation of milk-borne *Streptococcus zooepidemicus* infection associated with glomerulonephritis in Australia. J Infect 27(3):317–323

Futcher PH (1940) Glomerular nephritis following skin infections. Arch Intern Med 65:1192–1210

Goldstein I, Rebeyrotte P, Parlebas J, Halpern B (1968) Isolation from heart valves of glycopeptides which share immunological properties with Streptococcus haemolyticus group A polysaccharides. Nature 219(5156):866–868

Guilherme L, Kalil J (2010) Rheumatic fever and rheumatic heart disease: cellular mechanisms leading autoimmune reactivity and disease. J Clin Immunol 30(1):17–23

Guilherme L, Kalil J, Cunningham M (2006) Molecular mimicry in the autoimmune pathogenesis of rheumatic heart disease. Autoimmunity 39(1):31–39

Haidan A, Talay SR, Rohde M, Sriprakash KS, Currie BJ, Chhatwal GS (2000) Pharyngeal carriage of group C and group G streptococci and acute rheumatic fever in an Aboriginal population. Lancet 356(9236):1167–1169

Hartas J, Sriprakash KS (1999) Streptococcus pyogenes strains containing emm12 and emm55 possess a novel gene coding for distantly related SIC protein. Microb Pathog 26(1):25–33
Herwald H, Cramer H, Mörgelin M, Russell W, Sollenberg U, Norrby-Teglund A, Flodgaard H, Lindbom L, Björck L (2004) M protein, a classical bacterial virulence determinant, forms complexes with fibrinogen that induce vascular leakage. Cell 116(3):367–379
Humair L, Potter EV, Kwaan HC (1969) The role of fibrinogen in renal disease. I. Production of experimental lesions in mice. J Lab Clin Med 74(1):60–71
Itzek A, Gillen CM, Fulde M, Friedrichs C, Rodloff AC, Chhatwal GS, Nitsche-Schmitz DP (2010) Contribution of plasminogen activation toward the pathogenic potential of oral streptococci. PLoS One 5(11):e13826
Jones MN, Holt RG (2004) Activation of plasminogen by Streptococcus mutans. Biochem Biophys Res Commun 322(1):37–41
Kantor FS (1965) Fibrogen precipitation by streptococcal M protein. Ii. Renal lesions induced by intravenous injection of M protein into mice and rats. J Exp Med 121:861–872
Kaplan MH (1958) Localization of streptococcal antigens in tissues. I. Histologic distribution and persistence of M protein, types 1, 5, 12, and 19 in the tissues of the mouse. J Exp Med 107(3):341–352
Kaplan MH, Suchy ML (1964) Immunologic relation of streptococcal and tissue antigens. Ii. Cross-reaction of antisera to mammalian heart tissue with a cell wall constituent of certain strains of group a streptococci. J Exp Med 119:643–650
Kaplan MH, Bolande R, Rakita L, Blair J (1964) Presence of bound immunoglobulins and complement in the myocardium in acute rheumatic fever. Association with cardiac failure. N Engl J Med 271:637–645
Keitzer R (2005) Acute rheumatic fever (ARF) and poststreptococcal reactive arthritis (PSRA)–an update. Z Rheumatol 64(5):295–307
Kotb M (1995) Bacterial pyrogenic exotoxins as superantigens. Clin Microbiol Rev 8(3):411–426
Lyttle JD, Seegal D, Loeb EN, Jost EL (1938) The serum antistreptolysin titer in acute glomerulonephritis. J Clin Invest 17(5):631–639
Massell BF, Honikman LH, Amezcua J (1969) Rheumatic fever following streptococcal vaccination. Report of three cases. Jama 207(6):1115–1119
McCarty M (1956) Nature of rheumatic fever. Circulation 14(6):1138–1143
McCormack JM, Crossley CA, Ayoub EM, Harley JB, Cunningham MW (1993) Poststreptococcal anti-myosin antibody idiotype associated with systemic lupus erythematosus and Sjogren's syndrome. J Infect Dis 168(4):915–921
McDonald M, Currie BJ, Carapetis JR (2004) Acute rheumatic fever: a chink in the chain that links the heart to the throat? Lancet Infect Dis 4(4):240–245
McDonald MI, Towers RJ, Andrews RM, Benger N, Currie BJ, Carapetis JR (2006) Low rates of streptococcal pharyngitis and high rates of pyoderma in Australian aboriginal communities where acute rheumatic fever is hyperendemic. Clin Infect Dis 43(6):683–689
Michael AF Jr, Drummond KN, Good RA, Vernier RL (1966) Acute poststreptococcal glomerulonephritis: immune deposit disease. J Clin Invest 45(2):237–248
Nitsche DP, Johansson HM, Frick IM, Mörgelin M (2006) Streptococcal protein FOG, a novel matrix adhesin interacting with collagen I in vivo. J Biol Chem 281(3):1670–1679
Oda T, Yamakami K, Omasu F, Suzuki S, Miura S, Sugisaki T, Yoshizawa N (2005) Glomerular plasmin-like activity in relation to nephritis-associated plasmin receptor in acute poststreptococcal glomerulonephritis. J Am Soc Nephrol 16(1):247–254
Ogunbi O, Fadahunsi HO, Ahmed I, Animashaun A, Daniel SO, Onuoha DU, Ogunbi LQ (1978) An epidemiological study of rheumatic fever and rheumatic heart disease in Lagos. J Epidemiol Community Health 32(1):68–71
Ohkuni H, Friedman J, van de Rijn I, Fischetti VA, Poon-King T, Zabriskie JB (1983) Immunological studies of post-streptococcal sequelae: serological studies with an extracellular protein associated with nephritogenic streptococci. Clin Exp Immunol 54(1):185–193

Pancholi V, Fischetti VA (1993) Glyceraldehyde-3-phosphate dehydrogenase on the surface of group A streptococci is also an ADP-ribosylating enzyme. Proc Natl Acad Sci U S A 90(17):8154–8158

Parra G, Rodriguez-Iturbe B, Batsford S, Vogt A, Mezzano S, Olavarria F, Exeni R, Laso M, Orta N (1998) Antibody to streptococcal zymogen in the serum of patients with acute glomerulonephritis: a multicentric study. Kidney Int 54(2):509–517

Phillips GN Jr, Flicker PF, Cohen C, Manjula BN, Fischetti VA (1981) Streptococcal M protein: alpha-helical coiled-coil structure and arrangement on the cell surface. Proc Natl Acad Sci U S A 78(8):4689–4693

Proft T, Fraser JD (2003) Bacterial superantigens. Clin Exp Immunol 133(3):299–306

Reißmann S, Friedrichs C, Rajkumari R, Itzek A, Fulde M, Rodloff AC, Brahmadathan KN, Chhatwal GS, Nitsche-Schmitz DP (2010) Contribution of Streptococcus anginosus to infections caused by groups C and G streptococci, southern India. Emerg Infect Dis 16(4):656–663

Reißmann S, Gillen CM, Fulde M, Bergmann R, Nerlich A, Rajkumari R, Brahmadathan KN, Chhatwal GS, Nitsche-Schmitz DP (2012) Region specific and worldwide distribution of collagen-binding M proteins with PARF motifs among human pathogenic streptococcal isolates. PLoS One 7(1):e30122

Roberts GJ, Holzel HS, Sury MR, Simmons NA, Gardner P, Longhurst P (1997) Dental bacteremia in children. Pediatr Cardiol 18(1):24–27

Roberts S, Kosanke S, Terrence Dunn S, Jankelow D, Duran CM, Cunningham MW (2001) Pathogenic mechanisms in rheumatic carditis: focus on valvular endothelium. J Infect Dis 183(3):507–511

Rodriguez-Iturbe B, Batsford S (2007) Pathogenesis of poststreptococcal glomerulonephritis a century after Clemens von Pirquet. Kidney Int 71(11):1094–1104

Schmitz FJ, Beyer A, Charpentier E, Normark BH, Schade M, Fluit AC, Hafner D, Novak R (2003) Toxin-gene profile heterogeneity among endemic invasive European group A streptococcal isolates. J Infect Dis 188(10):1578–1586

Seegal BC, Andres GA, Hsu KC, Zabriskie JB (1965) Studies on the pathogenesis of acute and progressive glomerulonephritis in man by immunofluorescein and immunoferritin techniques. Fed Proc 24:100–108

Skattum L, Åkesson P, Truedsson L, Sjöholm AG (2006) Antibodies against four proteins from a Streptococcus pyogenes serotype M1 strain and levels of circulating mannan-binding lectin in acute poststreptococcal glomerulonephritis. Int Arch Allergy Immunol 140(1):9–19

Sriprakash KS, Hartas J, White A (2002) Antibodies to streptococcal inhibitor of complement function and M peptides in a post-streptococcal glomerulonephritis endemic region of Australia. J Med Microbiol 51(7):589–594

Sriskandan S, Moyes D, Cohen J (1996) Detection of circulating bacterial superantigen and lymphotoxin-alpha in patients with streptococcal toxic-shock syndrome. Lancet 348(9037):1315–1316

Steer AC, Batzloff MR, Mulholland K, Carapetis JR (2009a) Group A streptococcal vaccines: facts versus fantasy. Curr Opin Infect Dis 22(6):544–552

Steer AC, Law I, Matatolu L, Beall BW, Carapetis JR (2009b) Global emm type distribution of group A streptococci: systematic review and implications for vaccine development. Lancet Infect Dis 9(10):611–616

Stevens DL, Tanner MH, Winship J, Swarts R, Ries KM, Schlievert PM, Kaplan E (1989) Severe group A streptococcal infections associated with a toxic shock-like syndrome and scarlet fever toxin A. N Engl J Med 321(1):1–7

Stollerman GH (1969) Nephritogenic and rheumatogenic group A streptococci. J Infect Dis 120(2):258–263

Tontsch D, Pankuweit S, Maisch B (2000) Autoantibodies in the sera of patients with rheumatic heart disease: characterization of myocardial antigens by two-dimensional immunoblotting and N-terminal sequence analysis. Clin Exp Immunol 121(2):270–274

Treser G, Semar M, Ty A, Agel I, Franklin MA, Lange K (1970) Partial characterization of antigenic streptococcal plasma membrane components in acute glomerulonephritis. J Clin Invest 49(4):762–768

Veasy LG, Tani LY, Daly JA, Korgenski K, Miner L, Bale J, Kaplan EL, Musser JM, Hill HR (2004) Temporal association of the appearance of mucoid strains of Streptococcus pyogenes with a continuing high incidence of rheumatic fever in Utah. Pediatrics 113(3 Pt 1):168–172

Villarreal H Jr, Fischetti VA, van de Rijn I, Zabriskie JB (1979) The occurrence of a protein in the extracellular products of streptococci isolated from patients with acute glomerulonephritis. J Exp Med 149(2):459–472

Vlaminckx BJ, Mascini EM, Schellekens J, Schouls LM, Paauw A, Fluit AC, Novak R, Verhoef J, Schmitz FJ (2003) Site-specific manifestations of invasive group a streptococcal disease: type distribution and corresponding patterns of virulence determinants. J Clin Microbiol 41(11):4941–4949

Vogt A, Batsford S, Rodriguez-Iturbe B, Garcia R (1983) Cationic antigens in poststreptococcal glomerulonephritis. Clin Nephrol 20(6):271–279

Vos Q, Lees A, Wu ZQ, Snapper CM, Mond JJ (2000) B-cell activation by T-cell-independent type 2 antigens as an integral part of the humoral immune response to pathogenic microorganisms. Immunol Rev 176:154–170

Wannamaker LW (1973) The chain that links the heart to the throat. Circulation 48(1):9–18

Westling K, Ljungman P, Thalme A, Julander I (2002) Streptococcus viridans septicaemia: a comparison study in patients admitted to the departments of infectious diseases and haematology in a university hospital. Scand J Infect Dis 34(4):316–319

Woolcock JB (1974) The capsule of *Streptococcus equi*. J Gen Microbiol 85(2):372–375

Yoshizawa N, Yamakami K, Fujino M, Oda T, Tamura K, Matsumoto K, Sugisaki T, Boyle MD (2004) Nephritis-associated plasmin receptor and acute poststreptococcal glomerulonephritis: characterization of the antigen and associated immune response. J Am Soc Nephrol 15(7):1785–1793

Yu CE, Ferretti JJ (1989) Molecular epidemiologic analysis of the type A streptococcal exotoxin (erythrogenic toxin) gene (speA) in clinical Streptococcus pyogenes strains. Infect Immun 57(12):3715–3719

Zabriskie JB (1967) Mimetic relationships between group A streptococci and mammalian tissues. Adv Immunol 7:147–188

Zabriskie JB, Freimer EH (1966) An immunological relationship between the group. A streptococcus and mammalian muscle. J Exp Med 124(4):661–678

Immunopathogenesis of Streptococcal Deep Tissue Infections

Linda Johansson and Anna Norrby-Teglund

Abstract *Streptococcus pyogenes* is an important human pathogen that can cause a variety of diseases in immunocompetent individuals ranging from uncomplicated superficial infections to severe life-threatening infections including rapidly progressing deep tissue infections, such as necrotizing fasciitis (NF) and severe cellulitis. The pathogenesis of these infections is complex and multifactorial involving numerous virulence factors expressed by the bacteria. Here, we review data from epidemiologic, pathogenomic, and pathogenesis studies that have provided insight into the host–pathogen interactions that contribute to *S. pyogenes* tissue infections. The role of tissue-specific streptococcal types, intracellular bacterial persistence, and other immune evasion strategies resulting in massive bacterial load at the tissue site, as well as virulence factors contributing to a local hyperinflammatory response are highlighted. A particular focus is on in vivo findings in patients that provide insight into host and bacterial factors that are expressed at the infected tissue site, and the mechanisms underlying tissue pathology.

Contents

L. Johansson · A. Norrby-Teglund (✉)
Karolinska Institutet, Center for Infectious Medicine, F59,
Karolinska University Hospital Huddinge, S-141 86 Stockholm, Sweden
e-mail: Anna.Norrby-Teglund@ki.se

L. Johansson
e-mail: Linda.Johansson.2@ki.se

Current Topics in Microbiology and Immunology (2013) 368: 173–188
DOI: 10.1007/82_2012_282

Published Online: 9 December 2012

1 *Streptococcus pyogenes* Soft Tissue Infections

Streptococcus species, most frequently *Streptococcus pyogenes* but also group B, C, and G streptococci, are common causes of skin and soft tissue infections (Tognetti et al. 2012; Wong et al. 2003). The infections vary in their location and severity from superficial uncomplicated infections, i.e., impetigo and erysipelas, to fulminant deep tissue infections such as NF and pyomyositis. The former are frequent conditions that typically respond well to antibiotic therapy whereas the latter are rare infections that are associated with substantial morbidity and mortality (Bisno and Stevens 1996).

Several prospective population-based studies of invasive streptococcal infections have reported skin and soft tissue as the dominating foci of infection (Davies et al. 1996; Lamagni et al. 2008; Moses et al. 2002). In a joint European surveillance study, Strep-Euro, over 5,000 cases of invasive *S. pyogenes* were identified in 11 European countries (Lamagni et al. 2008). In this patient cohort, skin lesions/wounds were the most common predisposing factor, reported in 25 % of cases; and skin and soft tissue were the most common foci of infection, with 32 % of patients having cellulitis and 8 % NF.

A prospective population-based surveillance of NF caused by *S. pyogenes* was conducted in Canada during 1991–1997 (Kaul et al. 1997). Among the 77 NF patients identified, 46 % were bacteremic and 47 % developed streptococcal toxic shock syndrome. The overall case fatality rate was 36 % and outcome was associated with increasing age, presence of hypotension, and bacteremia. NF patients who also developed streptococcal toxic shock syndrome had a mortality of 67 % as compared to 4.9 % in patients who did not. Common risk factors of NF are blunt trauma and varicella infections (Kaul et al. 1997; Stevens et al. 1989). A case-control study confirmed that NF cases were sixfold more likely to have had a recent blunt trauma as compared to controls (Nuwayhid et al. 2007). In a murine model of myonecrosis, muscle injury was found to be associated with enhanced bacterial colonization at the site of injury and more severe tissue pathology (Bryant et al. 2006). Also increased vimentin expression was evident on injured muscle cells, and the authors proposed that vimentin may tether circulating bacteria from the circulation thereby providing a molecular basis for the association with blunt trauma and streptococcal severe tissue infections. This is an interesting model considering that approximately half of all *S. pyogenes* NF or pyomyositis cases have no known portal of entry; yet, for unknown reasons, infection becomes established at the site of a prior, nonpenetrating minor trauma, such as a muscle strain.

Among children, the most significant risk factor is varicella, which was reported to be associated with a 58-fold increased risk of acquisition of invasive *S. pyogenes* infections (Laupland et al. 2000). This seems to be particularly relevant for NF, as several studies have shown higher NF rates in children with preceding varicella infections (Imöhl et al. 2011; Laupland et al. 2000; Minodier et al. 2009; Patel et al. 2004). The underlying mechanisms to this association remains unknown, but it has been proposed that the full-thickness skin lesion of chickenpox may serve as a portal of entry for the organism or that the viral infections cause a transient immunologic derangement that predisposes to secondary bacterial infection (Stevens 1992).

2 Tissue-Specific *S. pyogenes* Strains

The throat and skin of the human host are the principal reservoirs for *S. pyogenes*, and asymptomatic carriage is generally noted in 2–5 % of the population. It has long been recognized that tissue-specific subpopulations of *S. pyogenes* exist (Wannamaker 1970), and epidemiologic studies identified defined genetic markers within the *emm* genes that distinguished between many isolates derived from the throat or skin (Bessen et al. 1996). Five basic *emm* chromosomal patterns (A through E) were identified, and whereas the isolates of *emm* pattern A–C were disproportionately associated with the nasopharynx, *emm* pattern D isolates were most often isolated from skin and impetigo lesions (Bessen et al. 1996). Organisms of a third pattern group, *emm* pattern E, were found at both tissue sites.

Similarly, certain *S. pyogenes emm* types are over-represented in specific disease manifestations. Since the late 1980s, there are numerous reports of the M1 and M3 strains causing large outbreaks of invasive *S. pyogenes* infections, including both streptococcal toxic shock syndrome and NF (reviewed in (Aziz and Kotb 2008; Cunningham 2000; Olsen and Musser 2010)). However, it is important to note that this association is far from exclusive, and also other types are significant causes of these severe invasive manifestations (Luca-Harari et al. 2009; Tyrrell et al. 2010). This was clearly shown in a recent European wide surveillance study that identified 5,521 cases with invasive *S. pyogenes* isolates (Strep-Euro), in which 50 % of the STSS cases and 55 % of the NF cases were caused by types other than *emm*1 or *emm*3 (Luca-Harari et al. 2009). Another recent example is the report of *emm*59 that went from being an extremely rare serotype to the most common type causing invasive infections in Canada, with a predominance of severe tissue infections (Tyrrell et al. 2010).

Pathogenomic studies of streptococcal epidemics have revealed new insights in the molecular pathogenesis of these infections, including basis for tissue-specificity and hypervirulence as reviewed in the literature (Musser and Shelbourne 2009). Of particular interest to this review, is the full genome dissection of the Canadian *emm*59 epidemic clone (Fittipaldi et al. 2012). The analyses revealed that it was genetically distinct from other *emm*59 *S. pyogenes* strains, either historic strains or strains causing sporadic infections in other countries (Fittipaldi et al. 2012). Furthermore, the epidemic *emm*59 strain was found to be significantly more virulent in experimental models of invasive skin infection and NF as compared to historic *emm*59 strains (Fittipaldi et al. 2012).

Also highly relevant are findings in *emm*3 strains in which a naturally occurring single-nucleotide mutation in the *mts*R gene was epidemiologically associated with significantly decreased number of human NF cases (Beres et al. 2006; Olsen et al. 2010). The mutation results in a stop codon and premature termination of the translation of MtsR, which alters a multiple gene virulence axis and results in increased expression of PrsA. This in turn leads to reduced activity of the cysteine protease SpeB and attenuated NF capacity. Isogenic mutant strains that overexpress *prs*A or lack speB had decreased secreted protease activity in vivo and recapitulated the NF-negative phenotype of the Delta*mts*R mutant strain in mice and monkeys. In addition, strains with this *mts*R mutation caused normal levels of other human invasive infection (Olsen et al. 2010).

3 Massive *S. pyogenes* Load and Intracellular Persistence at the Tissue Site

NF is characterized by widespread host-tissue destruction arising from the action of bacterial products as well as activated host immune cells. In fact, these infected areas are burdened with bacteria. In an analysis based on snap-frozen biopsies from tissue samples collected from patients with NF or severe cellulitis caused by *S. pyogenes* of varying serotypes, Thulin et al. (2006) revealed that viable *S. pyogenes* were readily detectable in tissue collected from the epicenter of infection as well as in distal, seemingly unaffected, tissue. In situ imaging demonstrated that the bacterial load was significantly associated with severity of soft tissue inflammation, with over 70 % of the tissue from severely involved areas having high bacterial load whereas distal tissue commonly had low bacterial load (Fig. 1a). Of clinical concern was the finding that viable bacteria were detected even in biopsies collected as late as up to 20 days after diagnosis of infection and initiation of intravenous antibiotic (i.e. a β-lactam in combination with clindamycin) (Fig. 1a). Thus, indicating an antibiotic eradication failure at the tissue site of infection.

Streptococcus pyogenes is classically known as an extracellular pathogen; however, intracellular bacteria have been found in phagocytic cells in infected tissue from patients with NF or severe cellulitis (Fig. 1b) (Thulin et al. 2006). Taken together with the persistent bacterial source at the tissue site, it was proposed that intracellular persistence may represent a mechanism by which the bacteria can avoid antibiotic clearance in infected soft tissue. This has previously been suggested in recurrent tonsillitis where intracellular cocci were identified in human tonsillar epithelium (Osterlund et al. 1997). Several different cell types have been shown to harbor intracellular *S. pyogenes* in in vitro cell cultures, including epithelial cells (LaPenta et al. 1994; Rohde et al. 2003) and neutrophils (Medina et al. 2003; Staali et al. 2003). Analyses of patient tissue biopsies implicated macrophages as the predominant host cell in infected soft tissue, although some neutrophils that harbored *S. pyogenes* could also be detected (Thulin et al. 2006). Interestingly, while the

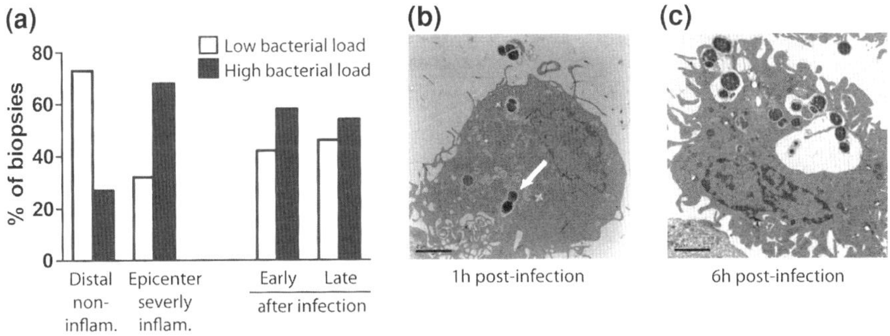

Fig. 1 Massive bacterial burden in tissue biopsies from patients with *Streptococcus pyogenes* infection. **a** Bacteria were detected by immunohistochemical staining of cryosections of tissue biopsies collected from patients with NF or cellulitis caused by *S. pyogenes*. Percentage of biopsies with low or high bacterial load defined by their in situ imaging value, in groups of noninflamed distal tissue or severely involved epicenter tissue. Original data have previously been presented (Thulin et al. 2006). **b** and **c** show transmission electron microscopy images of human monocyte derived macrophages infected with a clinical M1T1 isolate after one 1- and 6-h postinfection (Images: Dr. Matthias Mörgelin, Sweden). The *arrow* indicates bacteria residing in intracellular vacuoles

extracellular streptococci dominated in biopsies collected from highly inflamed epicenter areas with severe tissue pathology, the intracellular localization was most established in the noninflamed parts of the infected tissue. Not only do the bacteria persist within human macrophages but there is also evidence of an intracellular replication (Hertzén et al. 2010). Moreover, after a prolonged time inside the macrophage the bacteria egress from the cell and are now capable of reinfecting new nearby located phagocytic cells, thereby creating a vicious cycle that results in a massive bacterial reservoir at the tissue site that contributes to continued infection and tissue injury (Fig. 1c) (Hertzén et al. 2010).

4 Host and Bacterial Factors Contributing to Inflammation and Tissue Pathology

In studies of patients with *S. pyogenes* sepsis and toxic shock syndrome, there is convincing evidence that the proinflammatory cytokine response in circulation is strongly linked to systemic severity (Kotb et al. 2002; Norrby-Teglund et al. 2000). This is influenced by host immunogenetic variations in the HLA class II locus that determines the cytokine response to streptococcal superantigens (Kotb et al. 2002; Nooh et al. 2011). Presentation of superantigens by the risk HLA class II haplotypes augmented T cell proliferation whereas the response was attenuated with the protective HLA class II haplotype. In addition, the risk haplotype resulted in a polarization toward a proinflammatory profile, whereas the protective HLA types in higher ratios of anti- to proinflammatory cytokines (Kotb et al. 2002; Nooh et al. 2011).

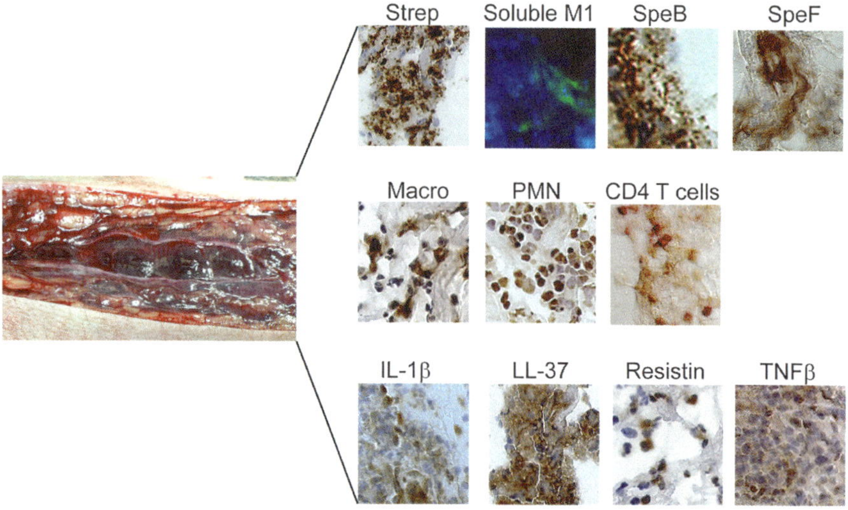

Fig. 2 Illustration of immunostainings for bacterial and host factors in cryosections of epicenter tissue of infected patients. The large image shows a photo of the tissue lesion in a patient with necrotizing fasciitis in the arm, note the extensive areas of necrosis (Photo: Dr. Pontus Thulin, Sweden). The small images show immunostainings of indicated host and bacterial factors in cryosections of severely involved tissue. Strep, *Streptococcus pyogenes*; Macro, macrophages; PMN, neutrophils; HBP, heparin-binding protein

A similar association between excessive inflammation and disease severity has been reported at the tissue site of infection in patients with severe soft tissue *S. pyogenes* infections (Johansson et al. 2009, 2008; Norrby-Teglund et al. 2001). In these studies, direct assessment of inflammatory responses at the tissue site was achieved by microscopical analyses of patient tissue biopsies, which revealed massive cell infiltrates consisting of predominantly macrophages, neutrophils, and T cells, as well as high levels of proinflammatory cytokines including IL-1, IL-6, IL-8, TNFα, TNFβ and IFNγ (Fig. 2). The cytokine profile at the tissue site was characterized by increasing levels of IL-1 and significantly higher frequencies of TNFβ and INFγ, i.e., Th1 cytokines, producing cells in biopsies with more severe tissue infection. As the Th1 cytokines are hallmark cytokines of a superantigen response, the cytokine profile at the tissue site resembled that of a typical superantigen response (Andersson et al. 1992; Norrby-Teglund et al. 1997). This finding together with the detection of the superantigen SpeF provided strong support for the direct action of superantigens at the tissue site (Norrby-Teglund et al. 2001).

Also SpeB has been emphasized as a critical virulence factors in *S. pyogenes* infections, particularly with tissue focus of infections. This protein has multiple functions, mainly as a proteinase that modulates several host defense systems including the cytokine, kallikrein-kinin, coagulation, and complement system. Known substrates of SpeB include, among others, IL1β-precursor (IL1β convertase) (Kapur et al. 1993a), metalloproteases (Burns et al. 1996), the extracellular matrix proteins vitronectin and fibronectin (Kapur et al. 1993b), and kininogens (Herwald et al. 1996). This results in enhanced inflammation and increased vascular permeability through generation of bioactive proinflammatory IL1β and TNFα, as well as release of kinins. SpeB not only interacts with human proteins but it also modulates a number of streptococcal virulence factors, such as degradation of superantigens, streptokinase, and DNases, or by releasing cell wall-attached factors such as M-protein, which results in attenuation of virulence, reviewed in the literature (Nelson et al. 2011).

SpeB can thus affect bacterial virulence in opposing ways, and it has been proposed that these proteolytic events differ temporally and spatially during distinct stages of infections (Rasmussen and Bjorck 2002). The importance of SpeB to the streptococcus is underscored by its tightly regulated expression involving many different layers of control, reviewed in Ref. (Carroll and Musser 2011). The impact on bacterial virulence is illustrated by the findings that loss of SpeB is strongly associated with facilitated bacterial transmission and systemic infection due to enhanced resistance to neutrophil killing through the action of DNases (further discussed below) (Cole et al. 2006; Walker et al. 2007), as well as through accumulation of surface plasmin activity that triggers systemic spread (Cole et al. 2006). On the other hand, lack of protective acute-phase antibodies against SpeB was identified as a risk factor for development of invasive *S. pyogenes* infections, indicating an important role in these infections (Basma et al. 1999; Norrby-Teglund et al. 1994). Furthermore, SpeB was shown to be highly expressed in vivo in snap-frozen tissue biopsies of patients with NF or erysipelas with the highest expression found in severely involved tissue with a combination of intra- and extracellular bacteria and dense cell infiltration (Thulin et al. 2006). In addition, SpeB has been shown in numerous experimental models to contribute to severe tissue injury and to be required for the bacteria to cause NF (Kapur et al. 1994; Kuo et al. 1998; Olsen et al. 2010; Svensson et al. 2000).

Another factor that has emerged as a proinflammatory molecule is the M-protein, classically known as an anti-phagocytic factor on the cell surface of the bacteria, but lately recognized as a multifunctional immune activator (Herwald et al. 2004; Påhlman et al. 2006, 2007). Herwald et al. (2004) demonstrated that soluble M-protein, resulting from either the action of SpeB or host proteases, forms complexes with fibrinogen. These complexes bind to β2-integrins on neutrophils heparin-binding protein, which results in activation and release of massive amounts of the granule protein, which in murine models were directly responsible for induction of vascular leakage and acute lung damage (Herwald et al. 2004; Kahn et al. 2008; Soehnlein et al. 2008). Soluble M1-protein and M1-protein/fibrinogen complexes could also be demonstrated in patient biopsies (Herwald et al. 2004; Kahn et al. 2008; Soehnlein et al. 2008), where neutrophils represent one the dominant cell populations and the degree of infiltration correlate with severity of tissue infection (Thulin et al. 2006). The potential pathophysiological significance of these complexes generated during infection was further substantiated by the presence of neutrophil effector proteins at the infected tissue site, including heparin-binding protein, IL-8, resistin, and LL-37, all of which are likely to contribute to the hyperinflammatory state that characterizes these infections (Fig. 2) (Herwald et al. 2004; Johansson et al. 2009, 2008; Norrby-Teglund et al. 2001).

The cytolytic toxin streptolysin S (SLS) is yet another streptococcal factor which has been implicated in the pathogenesis of NF (Ginsburg 1999). This was highlighted in a paper of Humar et al. (2002), who reported on three patients with severe necrotizing soft tissue infections due to β-haemolytic group G streptococcus. In this paper, they used targeted mutagenesis of the putative SLS structural gene sagA in group G streptococcus as well as *S. pyogenes*, which eliminated β-haemolytic activity. Whereas subcutaneous infection with wildtype *S. pyogenes* or group G streptococcus in mice resulted in inflammatory lesions with high bacterial counts, marked neutrophil infiltration, and histopathological evidence of diffuse tissue necrosis, such changes were not evident in mice infected with the isogenic SLS-negative mutants.

5 Immune Evasion Strategies of *S. pyogenes*

Phagocytic cells and antimicrobial peptides have a pivotal role in bacterial clearance at the site of infection. However, the invading *S. pyogenes* is equipped with an impressive array of products that by various mechanisms counteract the host immune defense; thereby promoting bacterial escape from immune clearance and efficient infection of the host (Table 1). As illustrated in Table 1, there is a striking redundancy with many bacterial factors affecting the same defense mechanisms. M-proteins are, as mentioned above, primary streptococcal virulence determinants that possess anti-phagocytic activities. This effect is achieved via interference with the complement system and inhibitory regulators of the complement system (reviewed in Refs. (Metzgar and Zampolli 2011; Smeesters et al. 2010)). On the other hand, M1 protein has been shown to play an important role in prevention of phagolysosomal maturation in both human macrophages and neutrophils, thereby promoting intracellular persistence (Hertzén et al. 2010; Staali et al. 2006). Thus, M-protein has two completely different mechanisms by which it impairs the ability of phagocytic cells to clear the bacterial infections. On one hand, it aids the bacteria to avoid phagocytosis and on the other, it modulates the phagocyte into a hospitable host in which bacterial persistence and replication can take place (Fig. 3).

Antimicrobial peptides are essential components of the innate immune defense through their ability to eliminate microorganisms by disruption of their cell membranes (Cederlund et al. 2011). The two main classes in humans are cathelicidins (LL-37) and defensins (α-defensin and β-defensin), both important in controlling bacterial infections. However, many bacterial pathogens have evolved mechanisms to counteract these peptides (Schmidtchen et al. 2002). *S. pyogenes* secretes at least two factors capable of inactivating LL-37 and defensins in vitro, namely, SpeB (Schmidtchen et al. 2002) and the streptococcal inhibitor of complement (SIC) (Frick et al. 2003) (Table 1). As mentioned above, due to the many

Table 1 Immune evasion strategies of *Streptococcus pyogenes*

Immune evasion	Main virulence factors involved
Resistance to phagocytosis	Hyaluronic acid capsule
	M-protein
	M-like-protein
Intracellular persistence in macrophages	M1 protein
Intracellular persistence in neutrophils	M-protein
Degradation of NETs	Sda1
	SpnA
Inhibition of AMP	M1 protein
	SIC
	SpeB
Cleavage of IgG	Endo S
	Mac/IdeS
	SpeB
Inactivation of complement	C5a peptidase (ScpA)
	SpeB
Degradation of IL-8	SpyCEP (ScpC)

AMP, antimicrobial peptides; Endo S, endoglycosidase S; IdeS, IgG-degrading enzyme of *S. pyogenes*; NETs, neutrophilic extracellular traps; ScpA, streptococcal C5a peptidase; ScpC/SpyCEP, streptococcal cell envelope serine proteinase; Sda1, extracellular streptodornase D; SIC, streptococcal inhibitor of complement; SpeB, streptococcal pyrogenic exotoxin B; SpnA, *Streptococcus pyogenes* nuclease

The table is based on the following reviews and original papers (Chang et al. 2011; Cole et al. 2011; Collin and Olsén 2001; Cunningham 2000; Hertzén et al. 2010; Johansson et al. 2010; Metzgar and Zampolli 2011)

different functions of SpeB including cleavage of its own virulence factors, its expression is tightly controlled. One sophisticated regulatory mechanism is exerted at the bacterial surface (Rasmussen et al. 1999). This is achieved through expression of the surface-attached G-related α2-macroglobulin binding (GRAB) protein, which binds α2-macroglobulin, a major protease inhibitor of human plasma. Once secreted, SpeB becomes entrapped in the GRAB/α2-macroglobulin cage were it remains proteolytically active against LL-37, thereby achieving an accumulation of active SpeB around the bacteria (Nyberg et al. 2004) (Fig. 3).

The cathelicidin LL-37 has been ascribed a central role in protection against murine necrotic skin infections caused by *S. pyogenes* (Nizet et al. 2001). Subsequent studies in *S. pyogenes* infected soft tissue biopsies from patients revealed that the active form of LL-37 was highly expressed in the infected tissue and in fact showed a strong positive correlation to bacterial load (Johansson et al. 2008). However, despite LL-37 being present at the tissue site, the high bacterial load and the fact that there are viable bacteria in the tissue during a prolonged time, suggest a lack of significant antimicrobial effect of LL-37 in these patients (Johansson et al. 2008; Thulin et al. 2006). An explanation to this lack of antimicrobial activity was provided by the finding that coccus-like structures positive for both GRAB and SpeB are present in the patient tissue, indicating that this regulation occurs in vivo (Johansson et al. 2008). Recently, another interesting finding

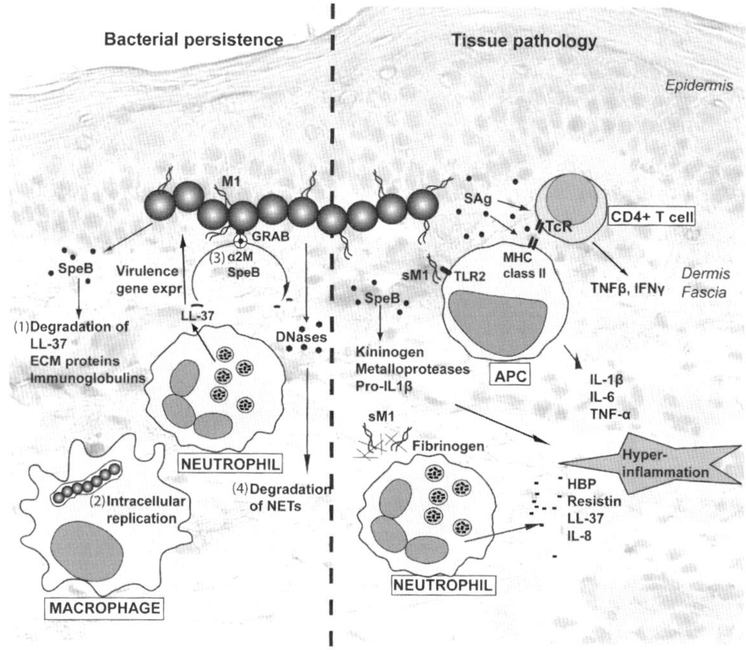

Fig. 3 Schematic model of host–pathogen interactions at the tissue site of infection during severe *Streptococcus pyogenes* deep tissue infections. *S. pyogenes* have evolved several immune evasion mechanisms that contribute to the bacterial persistence that characterizes severely involved tissue infections. Immune evasion strategies include *1* proteolytic degradation of host immune effector mediators including LL-37, immunoglobulins; *2* intracellular persistence and replication within phagocytic cells predominantly macrophages; *3* protection against antimicrobial peptides by SpeB entrapped in GRAB/α-2 macroglobulin (α2 m) complexes on the bacterial surface; and *4* degradation of neutrophil extracellular traps (NETs) by bacterial DNases such as streptodornase A. Dissemination of infection and tissue injury is contributed by proteolytic events, such as SpeB-mediated degradation of extracellular matrix (ECM) proteins, as well as induction of excessive inflammatory responses mediated largely by superantigens (SAg) and soluble M1 protein (sM1) that activate T cells, antigen presenting cells (APC) and neutrophils. This activation results in release of pathologic levels of proinflammatory mediators as well as mediators affecting vascular leakage, in particular neutrophil-derived heparin-binding protein (HBP). The figure is modified based on the literature (Johansson et al. 2010)

regarding LL-37 was reported demonstrating that subinhibitory concentrations of LL-37 activates the expression of several CsrRS(CovRS) controlled streptococcal virulence factors including the capsule operon, *mac/IdeS*(Mac/IgG protease), *spyCEP* (IL-8 protease), *ska* (streptokinase), *slo* (streptolysin O), *nga* (NAD-glycohydrolase), *speA* (pyrogenic exotoxin A) and *sda1* (DNase) (Gryllos et al. 2008; Tran-Winkler et al. 2011). Moreover, expression of several genes was also downregulated by the presence of LL-37. These genes included *metB* (putative cystathionine *β*-lyase), *Spy1414* (putative cation (potassium) transport protein), *grab* (protein G-related α_2-macroglobulin protein) and *speB* (cysteine protease). All together, these changes in gene expression promote the conversion of the bacteria from a colonizing to an invasive phenotype with increased resistance to killing by phagocytes in response to sensing LL-37 in its surrounding environment.

Neutrophils are one of the dominant phagocytic cell populations found in tissue from patients with *S. pyogenes* infection, and degree of infiltration is associated with bacterial burden (Thulin et al. 2006). One recently described mechanism used by neutrophils to eliminate bacteria is the formation of neutrophil extracellular traps (NETs) (Brinkmann et al. 2004). NETs are actively released from neutrophils through a cell death process called NETosis (Fuchs et al. 2007) and consist of extracellular chromatin fibers covered with histones and granule-derived antimicrobial peptides and enzymes such as neutrophil elastase, cathepsin G, myeloperoxidase (MPO) and LL-37 (Brinkmann et al. 2004). This web-like DNA structure entraps and kills pathogenic bacteria through the high local concentration of the antimicrobial components present in the NETs. However, *S. pyogenes* are capable of producing DNases, such as Sda1 and SpnA, that have been shown to degrade the DNA backbone in vitro and in mouse models and thereby the bacteria avoids being trapped within these NET formations (Buchanan et al. 2006; Chang et al. 2011). DNases (i.e., SpeF) has been detected in *S. pyogenes* infected tissue biopsies, but analyses have not included relation to presence or absence of NETs (Norrby-Teglund et al. 2001).

6 Conclusion

The pathogenesis of NF caused by *S. pyogenes* is multifactorial and involves a plethora of different bacterial virulence factors and host mediators (Fig. 3). The data demonstrate that these infections are hyperinflammatory conditions characterized by massive bacterial load and infiltration of inflammatory cells including neutrophils, macrophages, and T cells. Various immune evasion mechanisms, including resistance to antimicrobial peptides and intracellular bacterial replication, contribute to a persisting bacterial reservoir at the tissue site which is not efficiently cleared despite prolonged intravenous antibiotic treatment. As a result, there is a continuous production of exotoxins, proteases, and cytotoxic bacterial factors that mediate a pathologic inflammatory response and tissue injury at the local infection site. The high mortality and morbidity associated with these infections underscore the need for introduction of improved therapeutic strategies targeting the intracellular bacterial reservoir and dampening the hyperinflammatory state.

Acknowledgments Financial support: The Swedish Research Council (ANT: 12610; LJ: 522-2010-2868), Torsten and Ragnar Söderberg's Foundation (ANT), ESCMID (LJ), Karolinska University Hospital and the Karolinska Institutet.

References

Andersson J, Nagy S, Björk L, Abrams J, Holm S, Andersson U (1992) Bacterial toxin-induced cytokine production studied at the single-cell level. Immunol Rev 127:69–96

Aziz RK, Kotb M (2008) Rise and persistence of global M1T1 clone of *Streptococcus pyogenes*. Emerg Infect Dis 14:1511–1517

Basma H, Norrby-Teglund A, Guedez Y, McGeer A, Low DE, El-Ahmedy O, Schwartz B, Kotb M (1999) Risk factors in the pathogenesis of invasive group A streptococcal infections: Role of protective humoral immunity. Infect Immun 67:1871–1877

Beres SB, Richter EW, Nagiec MJ, Sumby P, Porcella SF, DeLeo FR, Musser JM (2006) Molecular genetic anatomy of inter- and intraserotype variation in the human bacterial pathogen group A Streptococcus. Proc Natl Acad Sci USA 103:7059–7064

Bessen DE, Sotir CM, Readdy TL, Hollingshead SK (1996) Genetic corrrelates of throat and skin isolates of group A streptococci. J Infect Dis 173:896–900

Bisno AL, Stevens DL (1996) Streptococcal infections of skin and soft tissues. N Engl J Med 334:240–245

Brinkmann V, Reichard U, Goosmann C, Fauler B, Uhlemann Y, Weiss DS, Weinrauch Y, Zychlinsky A (2004) Neutrophil extracellular traps kill bacteria. Science 303:1532–1535

Bryant AE, Bayer CR, Huntington JD, Stevens DL (2006) Group A streptococcal myonecrosis: increased vimentin expression after skeletal-muscle injury mediates the binding of *Streptococcus pyogenes*. J Infect Dis 193:1685–1692

Buchanan JT, Simpson AJ, Aziz RK, Liu GY, Kristian SA, Kotb M et al (2006) DNase expression allows the pathogen group A streptococcus to escape killing in neutrophil extracellular traps. Curr Biol 16: 396–400

Burns EH, Marciel AM, Musser JM (1996) Activation of a 66-kilodalton human endothelial cell matrix metalloprotease by *Streptococcus pyogenes* extracellular cysteine protease. Infect Immun 64:4744–4750

Carroll RK, Musser JM (2011) From transcription to activation: how group A streptococcus, the flesh-eating pathogen, regulates SpeB cysteine protease production. Mol Microbiol 81:588–601

Cederlund A, Gudmundsson GH, Agerberth B (2011) Antimicrobial peptides important in innate immunity. FEBS J 278:3942–3951

Chang A, Khemlani A, Kang H, Proft T (2011) Functional analysis of Streptococcus pyogenes nuclease A (SpnA), a novel group A streptococcal virulence factor. Mol Microbiol 79:1629–1642

Cole JN, Barnett TC, Nizet V, Walker MJ (2011) Molecular insight into invasive group A streptococcal disease. Nat Rev Microbiol 9:724–736

Cole JN, McArthur JD, McKay FC, Sanderson-Smith ML, Cork AJ, Ranson M, Rohde M, Itzek A, Sun H, Ginsburg D, Kotb M, Nizet V, Chhatwal GS, Walker MJ (2006) Trigger for group A streptococcal M1T1 invasive disease. FASEB 20:E1139–E1145

Collin M, Olsén A (2001) Effect of SpeB and EndoS from *Streptococcus pyogenes* on human immunoglobulins. Infect Immun 69:7187–7189

Cunningham MW (2000) Pathogenesis of group A streptococcal infections. Clin Microbiol Rev 13:470–511

Davies DH, McGeer A, Schwartz B, Green K, Cann D, Simor AE, Low DE (1996) The Ontario group A streptococcal study group Invasive group A streptococcal infections in Ontario, Canada. N Engl J Med 135:547–554

Fittipaldi N, Beres SB, Olsen RJ, Kapur V, Shea PR, Watkins ME, Cantu CC, Laucirica DR, Jenkins L, Flores AR, Lovgren M, Ardanuy C, Liñares J, Low DE, Tyrrell GJ, Musser JM (2012) Full-genome dissection of an epidemic of severe invasive disease caused by a hypervirulent, recently emerged clone of group A Streptococcus. Am J Pathol 180:1522–1534

Frick IM, Akesson P, Rasmussen M, Schmidtchen A, Bjorck L (2003) SIC, a secreted protein of Streptococcus pyogenes that inactivates antibacterial peptides. J Biol Chem 278:16561–16566

Fuchs TA, Abed U, Goosmann C, Hurwitz R, Schulze I, Wahn V, Weinrauch Y, Brinkmann V, Zychlinsky A (2007) Novel cell death program leads to neutrophil extracellular traps. J Cell Biol 176:231–241

Ginsburg I (1999) Is streptolysin S of group A streptococci a virulence factor? APMIS 107:1051–1059

Gryllos I, Tran-Winkler HJ, Cheng MF, Chung H, Bolcome Rr, Lu W, Lehrer RI, Wessels MR (2008) Induction of group A Streptococcus virulence by a human antimicrobial peptide. Proc Natl Acad Sci USA 105:16755–16760

Hertzén E, Johansson L, Wallin R, Schmidt H, Kroll M, Rehn AP, Kotb M, Mörgelin M, Norrby-Teglund A (2010) M1 protein-dependent intracellular trafficking promotes persistence and replication of *Streptococcus pyogenes* in macrophages. J Innate Immun. 2:534–545

Herwald H, Collin M, Muller-Esterl W, Björck L (1996) Streptococcal cysteine proteinase releases kinins: a virulence mechanism. J Exp Med 184:665–673

Herwald H, Cramer H, Mörgelin M, Russell W, Sollenberg U, Norrby-Teglund A, Flodgaard H, Lindblom L, Björck L (2004) M protein, a classical bacterial virulence determinant, forms complexes with fibrinogen that induce vascular leakage. Cell 116:367–379

Humar D, Datta V, Bast DJ, Beall B, De Azavedo JC, Nizet V (2002) Streptolysin S and necrotising infections produced by group G streptococcus. Lancet 359:124–129

Imöhl M, van der Linden M, Reinert RR, Ritter K (2011) Invasive group A streptococal disease and association with varicella in Germany, 1996–2009. FEMS Immunol Med Microbiol 62:101–109

Johansson L, Linnér A, Sundén-Cullberg J, Haggar A, Herwald H, Loré K, Treutiger C-J, Norrby-Teglund A (2009) Neutrophil-derived hyperresistinemia in severe acute streptococcal infections. J Immunol 183:4047–4054

Johansson L, Thulin P, Low DE, Norrby-Teglund A (2010) Getting under the skin: the immunopathogenesis of *Streptococcus pyogenes* deep tissue infections. Clin Infect Dis 51:58–65

Johansson L, Thulin P, Sendi P, Hertzén E, Linder A, Åkesson P, Low DE, Agerberth B, Norrby-Teglund A (2008) Cathelicidin LL-37 in severe *Streptococcus pyogenes* soft tissue infections in humans. Infect Immun 76:3399–3404

Kahn F, Mörgelin M, Shannon O, Norrby-Teglund A, Herwald H, Olin AI, Björck L (2008) Antibodies against a surface protein of *Streptococcus pyogenes* promote a pathologic inflammatory response. PLoS Pathog 4(9):e1000149

Kapur V, Maffei JT, Greer RS, Li LL, Adams GJ, Musser JM (1994) Vaccination with streptococcal extracellular cysteine protease (interleukin-1 beta convertase) protects mice against challenge with heterologous group A streptococci. Microb Pathog 16:443–450

Kapur V, Majesky MW, Li LL, Black RA, Musser JM (1993a) Cleavage of interleukin 1 beta (IL-1 beta) precursor to produce active IL-1 beta by a conserved extracellular cysteine protease from *Streptococcus pyogenes*. Proc Natl Acad Sci USA 90:7676–7680

Kapur V, Topouzis S, Majesky MW, Li LL, Hamrick MR, Hamill RJ, Patti JM, Musser JM (1993b) A conserved *Streptococcus pyogenes* extracellular cysteine protease cleaves human fibronectin and degrades vitronectin. Microb Pathog 15:327–346

Kaul R, McGeer A, Low DE, Green K, Schwartz B (1997) Ontario group A streptococcal study group. In: Simor E (ed) Population-based surveillance for group A streptococcal necrotizing fasciitis: clinical features, prognostic indicators, and microbiologic analysis of seventy-seven cases. Am J Med 103:18–24

Kotb M, Norrby-Teglund A, McGeer A, Dorak MK, Khurshid A, El-Shirbini H, Green K, Peeples J, Wade J, Thomson G, Schwartz B, Low DE (2002) Immunogenetic and molecular basis for the differences in outcomes of invasive group A streptococcal infections. Nat Med 8:1398–1404

Kuo CF, Wu JJ, Lin KY, Tsai PJ, Lee SC, Jin YT, Lei HY, Lin YS (1998) Role of streptococcal pyrogenic exotoxin B in the mouse model of group A streptococcal infection. Infect Immun 66:3931–3935

Lamagni TL, Darenberg J, Luca-Harari B, Siljander T, Efstratiou A, Henriques-Normark B, Vuopio-Varkila J, Bouvet A, Creti R, Ekelund K, Koliou M, Reinert RR, Stathi A, Strakova L, Ungureanu V, Schalén C (2008) Strep-EURO Study Group. In: Jasir A (ed) Epidemiology of severe *Streptococcus pyogenes* disease in Europe. J Clin Microbiol 46:2359–2367

LaPenta D, Rubens C, Chi E, Cleary PP (1994) Group A streptococci efficiently invade human respiratory epithelial cells. Proc Natl Acad Sci USA 91:12115–12119

Laupland KB, Davies HD, Low DE, Schwartz B, Green K, McGeer A (2000) Invasive group A streptococcal disease in children and association with varicella-zoster virus infection. Ontario Group A Streptococcal Study Group. Pediatrics 105:E60

Luca-Harari B, Darenberg J, Neal S, Siljander T, Strakova L, Tanna A, Creti R, Ekelund K, Koliou M, Tassios PT, van der Linden M, Straut M, Vuopio-Varkila J, Bouvet A, Efstratiou A, Schalén C, Henriques-Normark B, Strep-EURO study group, Jasir A (2009) Clinical and microbiological characteristics of severe *Streptococcus pyogenes* disease in Europe. J Clin Microbiol 47:1155–1165

Medina E, Goldmann O, Toppel AW, Chhatwal GS (2003) Survival of *Streptococcus pyogenes* within host phagocytic cells: a pathogenic mechanism for persistence and systemic invasion. J Infect Dis 187:597–603

Metzgar D, Zampolli A (2011) The M protein of group A Streptococcus is a key virulence factor and a clinically relevant strain identification marker. Virulence 2:402–412

Minodier P, Bidet P, Rallu F, Tapiero B, Bingen E, Ovetchkine P (2009) Clinical and microbiologic characteristics of group A streptococcal necrotizing fasciitis in children. Pediatr Infect Dis J 28:541–543

Moses AE, Goldberg S, Korenman Z, Ravins M, Hanski E, Shapiro M (2002) Invasive group A streptococcal infections, Israel. Emerg Infect Dis 8:421–426

Musser JM, Shelbourne SA (2009) A decade of molecular pathogenomic analysis of group A streptococcus. J Clin Invest 119:2455–2463

Nelson DC, Garbe J, Collin M (2011) Cysteine proteinase SpeB from *Streptococcus pyogenes*—a potent modifier of immunologically important host and bacterial proteins. Biol Chem 392:1077–1088

Nizet V, Ohtake T, Lauth X, Trowbridge J, Rudisill J, Dorschner RA, Pestonjamasp V, Piraino J, Huttner K, Gallo RL (2001) Innate antimicrobial peptide protects the skin from invasive bacterial infection. Nature 414:454–457

Nooh MM, Nookala S, Kansal R, Kotb M (2011) Individual genetic variations directly effect polarization of cytokine responses to superantigens associated with streptococcal sepsis: implications for customized patient care. J Immunol. 186:3156–3163

Norrby-Teglund A, Chatellier S, Low DE, McGeer A, Green K, Kotb M (2000) Host variation in cytokine responses to superantigens determine the severity of invasive group A streptococcal infection. Eur J Immunol 30:3247–3255

Norrby-Teglund A, Lustig R, Kotb M (1997) Differential induction of Th1 versus Th2 cytokines by group A streptococcal toxic shock syndrome isolates. Infect Immun 65:5209–5215

Norrby-Teglund A, Pauksens K, Holm SE, Norgren M (1994) Relation between low capacity of human sera to inhibit streptococcal mitogens and serious manifestation of disease. J Infect Dis 170:585–591

Norrby-Teglund A, Thulin P, Gan BS, Kotb M, McGeer A, Andersson J, Low DE (2001) Evidence for superantigen involvement in severe group A streptococcal tissue infection. J Infect Dis 186:853–860

Nuwayhid ZB, Aronoff DM, Mulla ZD (2007) Blunt trauma as a risk factor for group A streptococcal necrotizing fasciitis. Ann Epidemiol 17:878–881

Nyberg P, Rasmussen M, Björck L (2004) ∝2-Macroglobulin-proteinase complexes protects *Streptococcus pyogenes* from killing by the antimicrobial peptide LL-37. J Biol Chem 279:52820–52823

Olsen RJ, Musser JM (2010) Molecular pathogenesis of necrotizing fasciitis. Ann Rev Pathol Mech Dis 5:1–31

Olsen RJ, Sitkiewicz I, Ayeras AA, Gonulal VE, Cantu C, Beres SB, Green NM, Lei B, Humbird T, Greaver J, Chang E, Ragasa WP, Montgomery CA, Cartwright JJ, McGeer A, Low DE, Whitney AR, Cagle PT, Blasdel TL, DeLeo FR, Musser JM (2010) Decreased necrotizing fasciitis capacity caused by a single nucleotide mutation that alters a multiple gene virulence axis. Proc Natl Acad Sci USA 107:888–893

Osterlund A, Popa R, Nikkila T, Scheynius A, Engstrand L (1997) Intracellular reservoir of *Streptococcus pyogenes* in vivo: a possible explanation for recurrent pharyngotonsillitis. Laryngoscope 107:640–647

Patel RA, Binns HJ, Shulman ST (2004) Reduction in pediatric hospitalizations for varicella-related invasive group A streptococcal infections in the varicella vaccine era. J Pediatr 144:68–74

Påhlman LI, Möreglin M, Eckert J, Johansson L, Russell W, Riesbeck K, Soehnlein O, Lindbom L, Norrby-Teglund A, Schumann RR, Björck L, Herwald H (2006) Streptococcal M protein: a multipotent and powerful inducer of inflammation. J Immunol 177:1221–1228

Påhlman LI, Olin AI, Darenberg J, Mörgelin M, Kotb M, Herwald H, Norrby-Teglund A (2007) Soluble M1 protein of *Streptococcus pyogenes* triggers potent T cell activation. Cell Microbiol 10:404–414

Rasmussen M, Bjorck L (2002) Proteolysis and its regulation at the surface of *Streptococcus pyogenes*. Mol Microbiol 43:537–544

Rasmussen M, Muller HP, Bjorck L (1999) Protein GRAB of *streptococcus pyogenes* regulates proteolysis at the bacterial surface by binding alpha2-macroglobulin. J Biol Chem 28:15336–15344

Rohde M, Muller E, Chhatwal GS, Talay SR (2003) Host cell caveolae act as an entry-port for group A streptococci. Cell Microbiol 5:323–342

Schmidtchen A, Frick IM, Andersson E, Tapper H, Bjorck L (2002) Proteinases of common pathogenic bacteria degrade and inactivate the antibacterial peptide LL-37. Mol Microbiol 46:157–168

Smeesters PR, McMillan DJ, Sriprakash KS (2010) The streptococcal M protein: a highly versatile molecule. Trends Microbiol 18:275–282

Soehnlein O, Oehmcke S, Ma X, Rothfuchs AG, Frithiof R, van Rooijen N, Mörgelin M, Herwald H, Lindbom L (2008) Neutrophil degranulation mediates severe lung damage triggered by streptococcal M1 protein. Eur Respir J 32:405–412

Staali L, Bauer S, Mörgelin M, Björck L, Tapper H (2006) *Streptococcus pyogenes* bacteria modulate membrane traffic in human neutrophils and selectively inhibit azurophilic granule fusion with phagosomes. Cell Microbiol 8:690–703

Staali L, Morgelin M, Bjorck L, Tapper H (2003) *Streptococcus pyogenes* expressing M and M-like surface proteins are phagocytosed but survive inside human neutrophils. Cell Microbiol 5:253–265

Stevens DL (1992) Invasive group A streptococcus infections. Clin Infect Dis 14:2–13

Stevens DL, Tanner MH, Winship J, Swarts R, Ries KM, Schlievert PM, Kaplan E (1989) Severe group A streptococcal infections associated with a toxic shock-like syndrome and scarlet fever toxin A. N Engl J Med 321:1–7

Svensson MD, Scaramuzzino DA, Sjobring U, Olsen A, Frank C, Bessen DE (2000) Role for a secreted cysteine proteinase in the establishment of host tissue tropism by group A streptococci. Mol Microbiol 38:242–253

Thulin P, Johansson L, Low DE, Gan BS, Kotb M, McGeer A, Norrby-Teglund A (2006) Viable group A streptococci in macrophages during acute soft tissue infections. PLoS Med 3:0371–0379

Tognetti L, Martinelli C, Hercogiva J, Lotti T, Leoncini F, Moretti S (2012) Bacterial skin and soft tissue infections: review of the epidemiology, microbioogy, aetiopåathogenesis and treatment. J Eur Acad Dermatol Venerol 26:931–941

Tran-Winkler HJ, Love JF, Gryllos I, Wessels MR (2011) Signal transduction through CsrRS confers an invasive phenotype in group A Streptococcus. PLoS Pathog 7:e1002361

Tyrrell GJ, Lovgren M, St Jean T, Hoang L, Patrick DM, Horsman G, Van Caeseele P, Sieswerda LE, McGeer A, Laurence RA, Bourgault AM, Low DE (2010) Epidemic of group A Streptococcus M/emm59 causing invasive disease in Canada. Clin Infect Dis 51:1290–1297

Walker MJ, Hollands A, Sanderson-Smith ML, Cole JN, Kirk JK, Henningham A, McArthur JD, Dinkla K, Aziz RK, Kansal RG, Simpson AJ, Buchanan JT, Chhatwal GS, Kotb M, Nizet V (2007) DNase SdaI provides selection pressure for a switch to invasive group A streptococcal infection. Nat Med 13:981–985

Wannamaker LW (1970) Differences between streptococcal infections of the throat and of the skin. N Engl J Med 282:23–31

Wong CH, Chang HC, Pasupathy S, Khin LW, Tan JL, Low CO (2003) Necrotizing fasciitis: clinical presentation, microbiology, and determinants of mortality. J Bone Joint Surg Am 85-A:1454–1460

Modulation of the Coagulation System During Severe Streptococcal Disease

Oonagh Shannon, Heiko Herwald and Sonja Oehmcke

Abstract Haemostasis is maintained by a tightly regulated coagulation system that comprises platelets, procoagulant proteins, and anticoagulant proteins. During the local and systemic response to bacterial infection, the coagulation system becomes activated, and contributes to the pathophysiological response to infection. The significant human pathogen, *Streptococcus pyogenes* has multiple strategies to modulate coagulation. This can range from systemic activation of the intrinsic and extrinsic pathway of coagulation to local stimulation of fibrinolysis. Such diverse effects on this host system imply a finely tuned host–bacteria interaction. The molecular mechanisms that underlie this modulation of the coagulation system are discussed in this review.

Contents

O. Shannon (✉)
Division of Infection Medicine, Department of Clinical Sciences,
Biomedical Centre (BMC), B14, Lund University,
SE-22184 Lund, Sweden
e-mail: oonagh.shannon@med.lu.se

H. Herwald · S. Oehmcke
Virology and Hygiene, Medical Faculty, Institute of Medical Microbiology,
University of Rostock, Rostock, Germany

Current Topics in Microbiology and Immunology (2013) 368: 189–205
DOI: 10.1007/82_2012_283

Published Online: 8 December 2012

1 Introduction

Haemostasis is the process that maintains the integrity of the circulatory system. Resting platelets patrol the vasculature and become activated in response to endothelial damage and exposure to subendothelial matrix components. Adhesion and activation allows firm binding between platelets to occur, resulting in the formation of platelet thrombi. The proteins of the coagulation cascade are concomitantly activated and the platelet plug is stabilised by the generation of fibrin. The coagulation cascade can be broken down into an extrinsic (tissue factor driven) and intrinsic pathway (contact activation). Both arms are triggered by limited proteolysis and are amplified, like most other protease-driven effector systems, in a snowball-like manner to generate active thrombin, which initiates formation of a fibrin network (Furie and Furie 1988). Traditionally, it was believed that the coagulation cascade was entirely regulated by acellular components and lacked the involvement of cellular surfaces. However, it is now clear that coagulation is more complex and can be divided into three parts, initiation, amplification and propagation (Smith 2009), of which the amplification step involves the presence of platelets (Hoffman 2003). Activation of the coagulation cascade can also amplify platelet activation since the thrombin generated is also a potent platelet agonist. Both the extrinsic and intrinsic pathways have important functions in clot induction and stabilisation. While clotting is triggered by the exposure of tissue factor, activation of the contact system helps to stabilise the formed clot and also triggers inflammatory reactions via the release of kinins (Oehmcke and Herwald 2010; Renné 2012). As is the case for all protease-driven cascades, the coagulation system is tightly regulated. The anticoagulant fibrinolytic system is initiated when plasminogen is converted to the active protease plasmin, which degrades the fibrin clot. Furthermore, negative feedback systems of anticoagulation, such as the Activated Protein C pathway, are initiated simultaneously with activation of coagulation. The maintenance of haemostasis and avoidance of pathological thrombosis are highly dependent on maintaining the balance between these pro- and anticoagulant systems.

2 Crosstalk Between Coagulation and Inflammation

The close links between innate immunity and coagulation are exemplified by the host defence systems of the horseshoe crab. These ancient invertebrates have developed a defence system that relies on one cellular component, the hemocyte which functions as both platelet and phagocyte. In response to bacterial products, the hemocyte releases a cascade of proteins that generate a clot, wherein bacteria are sequestered and prevented from spreading systemically. During vertebrate evolution, the systems of inflammation and coagulation have diverged; however, there remains considerable crosstalk between the two systems in humans (Opal and Esmon 2003). Activated platelets release procoagulant factors, growth factors and proinflammatory factors including cytokines and chemokines (Semple et al. 2011). Platelets also adhere to and modulate the function of endothelial cells and leucocytes, promoting adherence and activation (Smyth et al. 2009). Activation of the coagulation cascade also plays a fundamental role in sequestering bacteria and limits invasiveness of the pathogen (van der Poll et al. 2011; Loof et al. 2011b).

3 Severe Infectious Disease

The inflammatory response and the coagulation system are simultaneously activated during the response to infection. The systemic activation of the coagulation system in infectious diseases constitutes a serious health threat and is associated with high morbidity and mortality rates (de Jong et al. 2010). In the pathology of severe infectious diseases, there seems to be a common theme of exaggerated activation of host defence by microorganisms, which contributes to the pathogenesis of infection. Bacterial superantigens are perhaps a classic example, whereby a massive cytokine storm and pathologic immune reactions are generated by the bacterial toxin (Brosnahan and Schlievert 2011). The pathophysiology of coagulation abnormalities in severe infectious diseases, such as sepsis or septic shock, is complex and often follows the same principles i.e., pathologic induction of pro- and anti-inflammatory reactions, systemic activation of coagulation, downregulation of anticoagulant pathways and impaired fibrinolysis (Semeraro et al. 2012). Both coagulation and anticoagulation systems have been shown to be activated in septic patients (Lorente et al. 1993; Mavrommatis et al. 2000). This may lead to the formation of circulating microthrombi that can target organs, followed by a consumption of clotting factors and platelets. The latter bears the risk of additional complications such as secondary severe bleeding. In particular, the level of thrombocytopenia present has been reported to correlate with the severity of infection (Gawaz et al. 1997; Vandijck et al. 2010). Due to the rapid progression of the disease, the early recognition of patients with severe infectious diseases is of great importance, and for every hour that proper treatment is delayed, the mortality increases by 7.5 % (Kumar et al. 2006).

Streptococcus pyogenes is an important causative agent of severe infectious disease, which can manifest as Streptococcal Toxic Shock Syndrome (STSS) or necrotising fasciitis. The portal of entry is often the skin and the bacteria invade from this site to cause bacteraemia, which may escalate to sepsis or STSS. The mortality rates remain high despite adequate antibiotic treatment and modern intensive care (Lamagni et al. 2008). The streptococcal surface is surrounded by M proteins that contribute to multiple aspects of bacterial virulence (for a review see (Oehmcke et al. 2010). *S. pyogenes* can be classified into different serotypes based on variations in M protein, and certain serotypes, including the M1 and M3 serotype, are more frequently isolated from patients with invasive disease. Sepsis and invasive *S. pyogenes* infection are often associated with coagulopathy. Histopathological studies of soft tissue from patients suffering from necrotising fasciitis revealed thrombosis and infarction of the tissue, and the amount of thrombosis was higher in acute cases as compared with subacute cases (Barker et al. 1987). In a mouse model of invasive *S. pyogenes* soft tissue infection, the vessels of the subcutaneous tissue were reported to be thrombosed and infarcted (Ashbaugh et al. 1998). Coagulopathy has also been observed in a baboon model of invasive *S. pyogenes* infection (Taylor et al. 1999).

4 Platelet Activation

As early as 1971, it was reported that platelets become activated in response to bacteria, and *S. pyogenes* was one of the bacteria investigated (Clawsson and White 1971). Almost ten years later, Kurpiewski and colleagues also demonstrated that *S. pyogenes* can stimulate platelet aggregation (Kurpiewski et al. 1983). The molecular mechanisms involved were not described until 2002 (Sjöbring et al. 2002). In this study, *S. pyogenes* bacteria were shown to bind plasma fibrinogen via the M protein and adhere to the fibrinogen receptor GPIIb/IIIa on platelets. Platelet activation occurs when specific IgG antibodies against the bacteria are also present and engage the platelet Fc receptor for IgG (Sjöbring et al. 2002). This is in agreement with the mechanism of platelet activation described for other significant gram-positive pathogens, including various species of streptococci and *Staphylococcus aureus* (Cox et al. 2011). Functional *S. pyogenes* M1 protein can be released from the bacterial surface (Berge and Björck 1995; Herwald et al. 2004). In 2007, soluble M1 protein was demonstrated to be a powerful platelet agonist (Shannon et al. 2007). The authors demonstrated that M1 protein forms complexes with fibrinogen and anti-M1 IgG, which mediate platelet activation. An interindividual variation in the ability of platelets to become activated in response to M1 protein was observed and this was correlated to the level of anti-M1 IgG present in the individual's plasma. This implies that certain individuals may have a higher risk of developing platelet-mediated coagulopathy in response to *S. pyogenes*, and it can be speculated that this may contribute to the significant interindividual host susceptibility to *S. pyogenes* infection.

Platelets activated by M1 protein also form complexes with neutrophils and monocytes, resulting in the activation of both cell types (Shannon et al. 2007). A crucial element in the genesis of coagulopathy in sepsis is the generation of tissue factor. The activation of platelets by M1 protein stimulated tissue factor expression on platelet–monocyte aggregates (Shannon et al. 2007). This provides an important positive feedback to thrombin generation, activation of coagulation and further platelet activation. Streptolysin O, secreted by *S. pyogenes*, has also been reported to stimulate the leucoctye–platelet complex formation (Bryant et al. 2005). Platelet thrombi and platelet–leukocyte complexes may occlude the microvasculature, contributing to the vascular dysfunction and organ damage that is a hallmark of STSS. Interestingly, platelet–neutrophil complexes have been detected in the bloodstream of septic patients and their levels decreased in severe sepsis associated with organ failure (Gawaz et al. 1997). Aggregated platelets that co-localised with *S. pyogenes* and M1 protein were detected at the epicentre of infection in patients with severe soft tissue infections caused by *S. pyogenes* M1 isolates (Shannon et al. 2007), supporting the hypothesis that M1 protein-induced platelet activation occurred in vivo. A recent study has shown that bacteria isolated from patients presenting with gram-positive bacteraemia will activate platelets from the infected individual, ex vivo, implying that bacteria-mediated activation may occur during infection (Johansson et al. 2011). Interestingly, the beta haemolytic streptococci were among the most potent platelet activators in this study.

5 Systemic Activation of Coagulation

S. pyogenes has evolved a variety of strategies to cause systemic activation of coagulation. In 2003, Bryant and colleagues reported that heat-killed Streptococci of serotype M1 and M3 trigger tissue factor mediated procoagulant activity in human endothelial cells and monocytes (Bryant et al. 2003). Four years later, Påhlman and colleagues reported that tissue factor expression on monocytes is induced by M proteins (Påhlman et al. 2007). M proteins also evoke the release of proinflammatory cytokines from monocytes by binding to toll-like receptor 2 (Påhlman et al. 2006). It is therefore tempting to speculate that M protein triggered tissue factor expression is also mediated by toll-like receptors.

The interaction between streptococci and the intrinsic pathway was first described in 1991 by de la Cadena et al. (1991). In this study, the injection of cell wall peptidoglycan–polysaccharide polymers from group A streptococci into rats produced a syndrome of relapsing polyarthritis and anaemia. These complications were combined with a depletion of contact factors which led the authors to conclude that contact activation is involved in the pathogenesis of inflammatory processes in infectious diseases (de la Cadena et al. 1991). Nine years later, Sriskandan and colleagues employed a murine model of streptococcal necrotizing fasciitis and reported that infected animals exhibited increased clotting times of the intrinsic pathway of coagulation and depletion of contact factors (Sriskandan et al. 2000). However, no bleeding abnormalities were seen in these animals.

A completely novel mode of coagulation activation was recently described for *S. pyogenes* (Oehmcke et al. 2012). In this study, it was found that soluble M protein triggers the release of procoagulant microparticles from monocytes, which induce both pathways of coagulation. The extrinsic pathway is induced by up-regulation of tissue factor on the microparticle surface, and the intrinsic pathway by translocation of phosphatidylserine to the outer leaflet, facilitating docking of contact factors. Further analyses revealed that both pathways are required to induce clotting (extrinsic pathway) and stabilisation of the formed clot (intrinsic pathway) (Oehmcke et al. 2012).

6 Release of Kinins

In addition to its procoagulant activity, induction of the contact system leads to the generation of small proinflammatory peptides also known as kinins (Renné 2012). Although the half-life of kinins in the circulation is extremely short (<20 s), pathologically high kinin levels can be measured in septic patients (Mattsson et al. 2001) suggesting an important role for these peptides in severe infectious diseases. The binding of contact factors to Streptococci was first reported in 1995 (Ben Nasr et al. 1995). The assembly and activation of contact factors at the surface of group A streptococci triggers the release of bradykinin (Ben Nasr et al. 1997). As previously mentioned, kinins are potent inflammatory mediators and their generation leads to the induction of vascular leakage, oedema formation, hypotension and pain (Leeb-Lundberg et al. 2005). Since many of these complications are also found in patients with severe streptococcal infections, it has been proposed that kinin release is a bacterial strategy to cause pathological inflammatory reactions. Indeed, in a murine model of streptococcal shock, systemic degradation of H-kininogen (the bradykinin precursor protein) was observed, which points to a massive release of kinins (Sriskandan et al. 2000). Similar findings were reported when biopsies from patients with streptococcal erysipelas were analysed (Linder et al. 2010).

Kinins explore their activities by binding to two distinct receptors referred to as B1R and B2R, respectively (Leeb-Lundberg et al. 2005). While bradykinin preferentially binds to B2R, its carboxy terminally truncated derivative (desArg^9 bradykinin) has a higher affinity for B1R. The two receptors have different pharmacological properties of which the B1R is involved in acute inflammatory processes whereas the B2R can trigger chronic reactions (Leeb-Lundberg et al. 2005). The trimming of bradykinin to desArg^9 bradykinin involves the action of endopeptidases. It has recently been shown that *S. pyogenes* recruits and activates TAFI, a plasma carboxypeptidase at the bacterial surface (Bengtson et al. 2009). Bound and activated TAFI can then remove the carboxy-terminal arginine from bradykinin and convert the peptide from a B2R to a B1R agonist. This mechanism may contribute to chronic inflammatory reactions that are often observed in patients with streptococcal skin infection, and an up-regulation of the B1R has been detected in human biopsies from patients with erysipelas and necrotizing fasciitis (Bengtson et al. 2009).

7 Coagulation as a Host Defence Mechanism

In 2006, Frick and colleagues provided direct evidence that the contact system contributes to defence against bacteria. In this study, binding of contact factors to the surface of *S. pyogenes* led to the generation of antimicrobial peptides (Frick et al. 2006). It is now appreciated that the contact system contributes to other immune responses. Neutrophil extracellular traps (NETs) were described eight - years ago as a novel innate defence mechanism (Brinkmann et al. 2004). Neutrophils release a complex network of granule proteins and chromatin, which captures and kills invading pathogens. Recently, M proteins from Streptococci were demonstrated to be potent inducers of NET formation and these NETs present a surface that facilitates the assembly and activation of the contact system (Oehmcke et al. 2009a). These findings suggest that the crosstalk between the contact system and host defence mechanisms may lead to an amplification of the immune response and facilitate elimination of the pathogen. As previously mentioned, contact activation eventually leads to the formation of a fibrin network that is stabilised by thrombin-activated coagulation factor XIII. One of the main functions of this transglutaminase is to covalently crosslink adjacent fibrin chains and renders the newly formed clot insoluble (Muszbek et al. 2011). This mechanism will prevent the efflux of plasma and blood cells into the injured tissue, and also avert the infiltration of invading pathogens into the circulation (Loof et al. 2011a). Loof and colleagues found that initiation of the contact system leads to an activation of factor XIII at the streptococcal surface. This activated factor XIII covalently weaved the bacteria into the fibrin network thereby blocking escape of Streptococci from the clot and facilitating bacterial killing by antimicrobial peptides (Loof et al. 2011a). Taken together a novel concept is emerging, whereby the contact system participates in the early recognition and elimination of invading pathogens.

The significance of fibrin clot formation for host defence against *S. pyogenes* has also been demonstrated by Sun and colleagues (Sun et al. 2009). In this study, mice with genetic alterations in several coagulation factors were investigated for their susceptibility to *S. pyogenes* infection. Decreased plasma factor V levels were associated with markedly increased mortality post infection, and this was correlated with diminished thrombin generation in these animals (Sun et al. 2009). Mice with a genetic deficiency in fibrinogen were used to demonstrate that fibrin generated in response to thrombin provides a crucial barrier against *S. pyogenes* dissemination (Sun et al. 2009). In 2010, another role for clot formation was described during the defence against dissemination of *S. pyogenes* (Shannon et al. 2010). Histidine-rich glycoprotein (HRG), which has been attributed multiple biological functions including a role in coagulation (Poon et al. 2011), was investigated in this study. Fibrin clots formed ex vivo in HRG deficient plasma failed to contain *S. pyogenes* at the clot surface and bacteria killing was diminished. Furthermore, HRG deficient mice failed to contain *S. pyogenes* infection at the local subcutaneous site and rapidly succumbed to invasive systemic infection (Shannon et al. 2010).

Recently, Massberg and colleagues presented another example of the close links between coagulation and innate immunity (Massberg et al. 2010). Neutrophil serine proteases, in concert with externalised nucleosomes, were shown to promote intravascular thrombus formation. The thrombi formed were shown to capture bacteria at the local site and reduce bacterial invasion (Massberg et al. 2010). Subsequent work has revealed that NETs provide a matrix for platelet activation and thrombus formation (Fuchs et al. 2010). As previously mentioned, the M1 protein from *S. pyogenes* is a potent activator of neutrophils (Herwald et al. 2004), resulting in degranulation of all four subsets of neutrophil granules (Soehnlein et al. 2008), and formation of NETs (Oehmcke et al. 2009a). It could be speculated that neutrophils activated by M proteins can also induce pathological thrombus formation and thereby corrupt this host defence strategy.

8 Plasminogen Activation

The importance of a fibrin network for host defence is implied by the fact that many pathogenic bacteria have evolved mechanisms to escape from the fibrin clot by modulating the host fibrinolytic system. Fibrinolysis is initiated when plasminogen is activated to plasmin, a host serine protease that degrades blood clots. Plasmin also plays a role in other physiological processes such as degradation of various components of the extracellular matrix and activation of metalloproteinases (Werb 1997). Binding and activation of host plasminogen is exploited by a number of pathogenic bacteria including *S. pyogenes* (Lähteenmäki et al. 2001).

As early as 1933, Tillett and Garner demonstrated that clinical isolates of haemolytic streptococci are capable of rapidly dissolving normal human fibrin clots (Tillett and Garner 1933). Streptokinase, a protein secreted by groups A, C and G streptococci, was found to bind and activate human plasminogen (Christensen and MacCleod 1945). Physiological activation of plasminogen by the mammalian proteases tissue-type plasminogen activator (tPA) or urokinase (uPA) involves a proteolytic cleavage of the Arg_{560}–Val_{561} peptide bond in the plasminogen molecule. Interestingly, plasminogen activation by streptokinase occurs without proteolytic cleavage. Streptokinase forms a tight 1:1 binding complex with plasminogen, whereby the active centre of plasminogen is exposed and becomes activated without hydrolysis of the Arg_{560}–Val_{561} peptide bond. This is followed by binding of another plasminogen molecule to the complex and plasmin is generated by proteolysis (Castellino 1979). Streptokinase also forms a complex with plasmin with at least 1000-fold higher affinity than the complex formed with plasminogen (Boxrud et al. 2000). Furthermore, streptokinase blocks inhibition of plasmin by its main physiological inhibitor α_2-antiplasmin, thereby acquiring an unregulated plasmin activity (Esmon and Mather 1998).

The interaction between streptokinase and plasminogen is highly species-specific; streptokinase produced by human isolates has a greater affinity for human plasminogen than mouse plasminogen (Marcum and Kline 1983; Yakovlev et al. 1995). Therefore, it has been proposed that the streptokinase–plasminogen interaction contributes to the fact that *S. pyogenes* is a strictly human pathogen. Consequently, a streptokinase-producing *S. pyogenes* strain and its streptokinase-deficient mutant show no difference in virulence in a mouse skin infection model (Khil et al. 2003), however if a source of human plasminogen is present at the infection site by co-injection of *S. pyogenes* with human plasminogen (Khil et al. 2003) or preincubation of *S. pyogenes* in human plasma (Li et al. 1999), the virulence of streptokinase-producing *S. pyogenes* is significantly enhanced. A humanised transgenic mouse, expressing human plasminogen, exhibits markedly increased mortality after *S. pyogenes* infection, which is largely abrogated by deletion of the streptokinase gene (*ska*) (Sun 2004). Taken together, streptokinase emerges as a key *S. pyogenes* virulence factor and inhibition of streptokinase gene expression has been proposed as an alternative treatment strategy (Sun et al. 2012).

9 Plasminogen Receptors

In addition to streptokinase several surface proteins of group A, C and G streptococci bind plasminogen directly [for a review see (Lähteenmäki et al. 2001; Walker et al. 2005)]. Bacterial plasminogen receptors immobilise plasminogen on the bacterial surface and enhance activation to plasmin, thus providing bacteria with proteolytic activity. Group C and G streptococci bind plasminogen via M like proteins and activate it to plasmin in the presence of streptokinase (Ben Nasr et al. 1994). To date, four plasminogen receptors have been described for *S. pyogenes;* the glycolytic enzymes glyceraldehyde-3-phosphate dehydrogenase (GAPDH) (Lottenberg et al. 1992; Pancholi and Fischetti 1992; Winram and Lottenberg 1996), streptococcal enolase (SEN) (Pancholi and Fischetti 1998), the plasminogen-binding group A streptococcal M protein (PAM) (Berge and Sjöbring 1993) and PAM-related protein (Prp) (Sanderson-Smith et al. 2007). Plasminogen bound to the streptococcal surface can be activated by bacterial streptokinase (Svensson et al. 2002) or the mammalian protease tPA (Kuusela et al. 1992; Berge and Sjöbring 1993), leading to bacterium bound plasmin activity.

GAPDH is a multifunctional enzyme involved in metabolic and non-metabolic processes. It was the first plasminogen-binding protein to be described for *S. pyogenes* (Broder et al. 1991). SEN is also a key glycolytic enzyme and contrary to GAPDH, SEN binds plasminogen with high affinity (Pancholi and Fischetti 1998). GAPDH and SEN can be located both in the cytoplasm and at the bacterial surface, but the mechanism involved in surface expression remains to be resolved (Walker et al. 2005). It has been proposed that surface M and M-related proteins are involved in anchoring GAPDH on the surface of GAS (D'Costa et al. 2000).

The plasminogen-binding region of PAM is located in the N-terminal variable region that contains 13–16 amino acid tandem repeats. Within the two internal repeat domains, interaction with plasminogen is mediated via a lysine, arginine or histidine residue (Wistedt et al. 1995; Sanderson-Smith et al. 2006) that interact with the lysine binding kringle 2 domain of plasminogen (Wistedt et al. 1998). *S. pyogenes* isolates that produce PAM have a strong tendency to cause impetigo (Svensson et al. 1999). McKay and colleagues have demonstrated a significant relationship between the acquisition of plasminogen via PAM and the propensity to cause invasive diseases (McKay et al. 2004). The PAM-related protein Prp is another high-affinity plasminogen-binding M protein, but it is phylogenetically distinct from PAM. Prp was identified in an *S. pyogenes* (*emm*98.1) isolated from severe invasive infection (Sanderson-Smith et al. 2007).

In addition to direct plasminogen binding, *S. pyogenes* also recruits plasminogen indirectly (Wang et al. 1995a, b). The bacteria can bind plasma fibrinogen via M and M-related proteins and this provides an anchoring site for the plasminogen–streptokinase complex (Christner et al. 1997; Hess and Boyle 2006). A trimolecular complex of streptokinase–plasminogen–fibrinogen (Takada and Takada 1989) is formed on the bacteria surface, and uncontrollable plasmin activity occurs.

10 Plasminogen Activation Facilitates *S. pyogenes* Dissemination

S. pyogenes isolates from invasive disease acquire more plasminogen by the trimolecular complex of fibrinogen–plasminogen–streptokinase or by PAM than the isolates from uncomplicated infections (McKay et al. 2004), indicating that plasminogen acquisition plays an essential role for bacterial dissemination. Importantly, Ikebe et al. (2005) showed that *ska* transcript levels were significantly higher in strains isolated from patients with severe invasive infection than those from noninvasive infection. This is supported by the findings of Rezcallah et al. (2004), who reported that bacteria isolated from the spleen of mice who succumbed to a fatal subcutaneous infection had an enhanced ability to bind and activate human plasminogen. The bacteria isolated were demonstrated to have an eightfold increase in *ska* transcription and also failed to produce SpeB. SpeB is a bacterial cysteine protease that inactivates numerous host proteins as well as multiple bacterial virulence factors including streptokinase [for a review see (Cole et al. 2011)]. In the absence of SpeB, there is a 75-fold increase in plasmin activity at the surface of *S. pyogenes* M1T1 (Cole 2006). The loss of SpeB activity at the local site of infection triggers the systemic dissemination of *S. pyogenes* M1T1 in a humanised plasminogen transgenic mouse model (Cole 2006). These data demonstrate that plasminogen acquisition and activation is an important virulence determinant during invasive *S. pyogenes* disease, and this in turn highlights the importance of fibrin formation in the host response to infection.

11 Therapeutic Targets Based on Coagulation

Modulation of the coagulation system is a potential target for the development of novel intervention strategies for severe infectious disease. As previously discussed, the genesis of coagulopathy is extremely complex and closely linked to the inflammatory response. The ideal drug candidate should limit the systemic coagulation activation, while maintaining local formation. Limited success has been achieved in clinical trials of patients treated with anticoagulation therapy aimed at the extrinsic pathway, such as administration of Antithrombin, Heparin or Tissue factor pathway inhibitor (Opal and Esmon 2003). The most recent clinical trials of one of the most promising drugs in this category, Xigris (activated protein C), have failed and the drug has been withdrawn from the market because of lack of efficiency combined with serious side effects (Williams 2012).

Platelet activation may also be a target for sepsis drug development and some success has been reported in animal models of gram-negative sepsis (Pu et al. 2001). Interestingly, in a systemic review of the literature it has been reported that a low-dose antiplatelet treatment during sepsis may reduce the risk of multi-organ failure (Winning et al. 2009).

The contact system is an interesting target for drug development, since contact system blockade might prevent an increased bleeding risk. Indeed, contact system inhibitors have been tested in some animal models. Stadnicki and co-workers reported in 1996 that a specific serine protease inhibitor, targeting one of the contact factors, prevents the induction of acute intestinal inflammatory reactions in rats challenged with streptococcal peptidoglycan polysaccharide (Stadnicki et al. 1996). More recently, an H-kininogen derived peptide was described that blocks the activation of the intrinsic pathway of coagulation and the formation of bradykinin in ex vivo experiments (Oehmcke et al. 2009b). This peptide was shown to protect mice from lung damage during invasive *S. pyogenes* infections, and to significantly prolong their survival when administered in combination with clindamycin (Oehmcke et al. 2009b). Only a few clinical trials have been conducted so far. Administration of a B2R antagonist to patients with severe systemic inflammatory response syndrome and sepsis revealed no significant effect on risk-adjusted 28-day survival, but a non-significant trend toward improvement on risk-adjusted 7-day survival was found. Moreover, a statistically significant improvement of patients with gram-negative infections was also observed (Fein et al. 1997). Unfortunately, a combination of a B1R and B2R antagonist was not considered in this study. A recent clinical trial was carried out for treatment of septic patients with the complement and contact system inhibitor, C1-esterase inhibitor. In a study with 61 patients, it was found that high-dose infusion of the protein

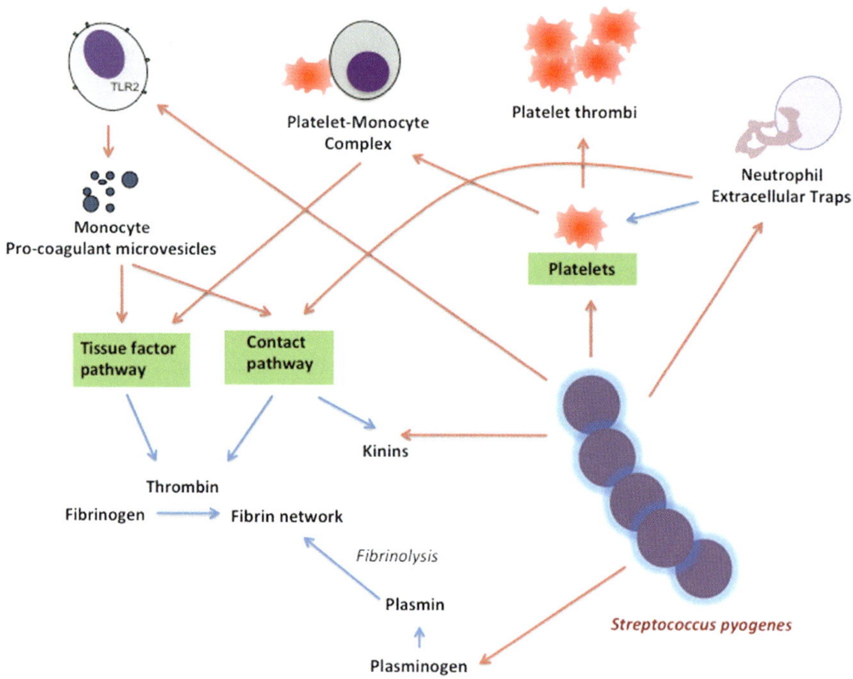

Fig. 1 Modulation of the coagulation system by *Streptococcus pyogenes*. The three central points of activation of the coagulation cascade are marked by *green boxes*; Platelet activation, Tissue factor pathway, Contact pathway. *Blue lines* depict pathways that contribute to normal haemostasis. *Red lines* depict components that are modulated by *Streptococcus pyogenes*

down-regulates the systemic inflammatory response and is associated with improved survival rates in sepsis patients (Igonin et al. 2012). These are very promising results and future work will hopefully show whether these findings can be validated using larger patient groups.

12 Conclusions

Coagulopathy is a central and significant finding in severe infection with *S. pyogenes*. These bacteria seem to be uniquely adapted to the human coagulation system and this is summarised in Fig. 1. The existence of multiple sites of intervention implies an important role for this interaction in the pathophysiological response to *S. pyogenes* infection. On one hand, coagulation suppresses bacterial dissemination, and *S. pyogenes* has multiple strategies to counteract this host defence locally. On the other hand, *S. pyogenes* can stimulate massive and uncontrolled local and systemic coagulation, which results in significant collateral damage to the host. These multiple and discriminatory effects on the coagulation system suggest a finely tuned host–parasite interaction. Further insights into these interactions may results in novel treatment strategies for these life-threatening infections.

References

Ashbaugh CD, Warren HB, Carey VJ, Wessels MR (1998) Molecular analysis of the role of the group A streptococcal cysteine protease, hyaluronic acid capsule, and M protein in a murine model of human invasive soft-tissue infection. J Clin Invest 102:550–560

Barker FG, Leppard BJ, Seal DV (1987) Streptococcal necrotising fasciitis: comparison between histological and clinical features. J Clin Pathol 40:335–341

Ben Nasr A, Wistedt A, Ringdahl U, Sjöbring U (1994) Streptokinase activates plasminogen bound to human group C and G streptococci through M-like proteins. Eur J Biochem 222:267–276

Ben Nasr AB, Herwald H, Müller-Esterl W, Björck L (1995) Human kininogens interact with M protein, a bacterial surface protein and virulence determinant. Biochem J 305:173–180

Ben Nasr A, Herwald H, Sjöbring U, Renné T, Müller-Esterl W, Björck L (1997) Absorption of kininogen from human plasma by *Streptococcus pyogenes* is followed by the release of bradykinin. Biochem J 326:657–660

Bengtson SH, Sandén C, Mörgelin M, Marx PF, Olin AI, Leeb-Lundberg LMF, Meijers JCM, Herwald H (2009) Activation of TAFI on the surface of *Streptococcus pyogenes* evokes inflammatory reactions by modulating the kallikrein/kinin system. J Innate Immun 1:18–28

Berge A, Björck L (1995) Streptococcal cysteine proteinase releases biologically active fragments of streptococcal surface proteins. J Biol Chem 270:9862–9867

Berge A, Sjöbring U (1993) PAM, a novel plasminogen-binding protein from *Streptococcus pyogenes*. J Biol Chem 268:25417–25424

Boxrud PD, Fay WP, Bock PE (2000) Streptokinase binds to human plasmin with high affinity, perturbs the plasmin active site, and induces expression of a substrate recognition exosite for plasminogen. J Biol Chem 275:14579–14589

Brinkmann V, Reichard U, Goosmann C, Fauler B, Uhlemann Y, Weiss DS, Weinrauch Y, Zychlinsky A (2004) Neutrophil extracellular traps kill bacteria. Science 303:1532–1535

Broder CC, Lottenberg R, von Mering GO, Johnston KH, Boyle MD (1991) Isolation of a prokaryotic plasmin receptor. Relationship to a plasminogen activator produced by the same micro-organism. J Biol Chem 266:4922–4928

Brosnahan AJ, Schlievert PM (2011) Gram-positive bacterial superantigen outside-in signaling causes toxic shock syndrome. FEBS J 278:4649–4667

Bryant AE, Hayes-Schroer SM, Stevens DL (2003) M type 1 and 3 group A streptococci stimulate tissue factor-mediated procoagulant activity in human monocytes and endothelial cells. Infect Immun 71:1903–1910

Bryant AE, Bayer CR, Chen RYZ, Guth PH, Wallace RJ, Stevens DL (2005) Vascular dysfunction and ischemic destruction of tissue in *Streptococcus pyogenes* infection: the role of streptolysin O-induced platelet/neutrophil complexes. J Infect Dis 192:1014–1022

Castellino F (1979) Unique enzyme-protein substrate modifier reaction—plasmin-streptokinase interaction. Trends Biochem Sci 4:1–5

Christensen L, MacCleod CM (1945) A proteolytic enzyme of serum—characterization, activation, and reaction with inhibitors. J Gen Physiol 28:559–583

Christner R, Li Z, Raeder R, Podbielski A, Boyle MD (1997) Identification of key gene products required for acquisition of plasmin-like enzymatic activity by group A streptococci. J Infect Dis 175:1115–1120

Clawsson CC, White JG (1971) Platelet interaction with bacteria I. Am J Pathol 65(2):1–14

Cole JN (2006) Trigger for group A streptococcal M1T1 invasive disease. FASEB J 20:1745–1747
Cole JN, Barnett TC, Nizet V, Walker MJ (2011) Molecular insight into invasive group A streptococcal disease. Nat Publ Group 9:724–736
Cox D, Kerrigan SW, Watson SP (2011) Platelets and the innate immune system: mechanisms of bacterial-induced platelet activation. J Thromb Haemost 9:1097–1107
D'Costa SS, Romer TG, Boyle MDP (2000) Analysis of expression of a cytosolic enzyme on the surface of Streptococcus pyogenes. Biochem Biophys Res Commun 278:826–832
de Jong HK, van der Poll T, Wiersinga WJ (2010) The systemic pro-inflammatory response in sepsis. J Innate Immun 2:422–430
de la Cadena RA, Laskin KJ, Pixley RA, Sartor RB, Schwab JH, Bedi GS, Colman RW (1991) Role of kallikrein-kinin system in pathogenesis of bacterial cell wall-induced inflammation. Am J Physiol 260:213–219
Esmon CT, Mather T (1998) Switching serine protease specificity. Nat Struct Biol 5:933–937
Fein AM, Bernard GR, Criner GJ, Fletcher EC, Good JT, Knaus WA, Levy H, Matuschak GM, Shanies HM, Taylor RW et al (1997) Treatment of severe systemic inflammatory response syndrome and sepsis with a novel bradykinin antagonist, deltibant (CP-0127). Results of a randomized, double-blind, placebo-controlled trial. CP-0127 SIRS and Sepsis Study Group. JAMA 277:482–487
Frick I-M, Akesson P, Herwald H, Mörgelin M, Malmsten M, Nägler DK, Björck L (2006) The contact system–a novel branch of innate immunity generating antibacterial peptides. EMBO J 25:5569–5578
Fuchs TA, Brill A, Duerschmied D, Schatzberg D, Monestier M, Myers DD, Wrobleski SK, Wakefield TW, Hartwig JH, Wagner DD (2010) Extracellular DNA traps promote thrombosis. Proc Natl Acad Sci USA 107:15880–15885
Furie B, Furie BC (1988) The molecular basis of blood coagulation. Cell 53:505–518
Gawaz M, Dickfeld T, Bogner C, Fateh-Moghadam S, Neumann FJ (1997) Platelet function in septic multiple organ dysfunction syndrome. Intensive Care Med 23:379–385
Herwald H, Cramer H, Mörgelin M, Russell W, Sollenberg U, Norrby-Teglund A, Flodgaard H, Lindbom L, Björck L (2004) M protein, a classical bacterial virulence determinant, forms complexes with fibrinogen that induce vascular leakage. Cell 116:367–379
Hess JL, Boyle MDP (2006) Fibrinogen fragment D is necessary and sufficient to anchor a surface plasminogen-activating complex in Streptococcus pyogenes. Proteomics 6:375–378
Hoffman M (2003) A cell-based model of coagulation and the role of factor VIIa. Blood Rev 17(Suppl 1):S1–S5
Igonin AA, Protsenko DN, Galstyan GM, Vlasenko AV, Khachatryan NN, Nekhaev IV, Shlyapnikov SA, Lazareva NB, Herscu P (2012) C1-esterase inhibitor infusion increases survival rates for patients with sepsis*. Crit Care Med 40:770–777
Ikebe T, Endoh M, Watanabe H (2005) Increased expression of the ska gene in emm49-genotyped Streptococcus pyogenes strains isolated from patients with severe invasive streptococcal infections. Japan J Infect Dis 58:272–275
Johansson D, Shannon O, Rasmussen M (2011) Platelet and neutrophil responses to gram positive pathogens in patients with bacteremic infection. PLoS One 6:e26928
Khil J, Im M, Heath A, Ringdahl U, Mundada L, Cary Engleberg N, Fay WP (2003) Plasminogen enhances virulence of group A streptococci by streptokinase-dependent and streptokinase-independent mechanisms. J Infect Dis 188:497–505
Kumar A, Roberts D, Wood KE, Light B, Parrillo JE, Sharma S, Suppes R, Feinstein D, Zanotti S, Taiberg L et al (2006) Duration of hypotension before initiation of effective antimicrobial therapy is the critical determinant of survival in human septic shock. Crit Care Med 34:1589–1596
Kurpiewski GE, Forrester LJ, Campbell BJ, Barrett JT (1983) Platelet aggregation by Streptococcus pyogenes. Infect Immun 39:704–708
Kuusela P, Ullberg M, Saksela O, Kronvall G (1992) Tissue-type plasminogen activator-mediated activation of plasminogen on the surface of group-A, group-C, and group-G streptococci. Infect Immun 60:196–201

Lähteenmäki K, Kuusela P, Korhonen TK (2001) Bacterial plasminogen activators and receptors. FEMS Microbiol Rev 25:531–552
Lamagni TL, Darenberg J, Luca-Harari B, Siljander T, Efstratiou A, Henriques-Normark B, Vuopio-Varkila J, Bouvet A, Creti R, Ekelund K et al (2008) Epidemiology of severe Streptococcus pyogenes disease in Europe. J Clin Microbiol 46:2359–2367
Leeb-Lundberg LMF, Marceau F, Müller-Esterl W, Pettibone DJ, Zuraw BL (2005) International union of pharmacology. XLV. Classification of the kinin receptor family: from molecular mechanisms to pathophysiological consequences. Pharmacol Rev 57:27–77
Li Z, Ploplis VA, French EL, Boyle MD (1999) Interaction between group A streptococci and the plasmin(ogen) system promotes virulence in a mouse skin infection model. J Infect Dis 179:907–914
Linder A, Johansson L, Thulin P, Hertzén E, Mörgelin M, Christensson B, Björck L, Norrby-Teglund A, Akesson P (2010) Erysipelas caused by group A streptococcus activates the contact system and induces the release of heparin-binding protein. J Invest Dermatol 130:1365–1372
Loof TG, Mörgelin M, Johansson L, Oehmcke S, Olin AI, Dickneite G, Norrby-Teglund A, Theopold U, Herwald H (2011a) Coagulation, an ancestral serine protease cascade, exerts a novel function in early immune defense. Blood 118:2589–2598
Loof TG, Schmidt O, Herwald H, Theopold U (2011b) Coagulation systems of invertebrates and vertebrates and their roles in innate immunity: the same side of two coins? J Innate Immun 3:34–40
Lorente JA, García-Frade LJ, Landín L, de Pablo R, Torrado C, Renes E, García-Avello A (1993) Time course of hemostatic abnormalities in sepsis and its relation to outcome. Chest 103:1536–1542
Lottenberg R, Broder CC, Boyle MD, Kain SJ, Schroeder BL, Curtiss R (1992) Cloning, sequence-analysis, and expression in Escherichia coli of a streptococcal plasmin receptor. J Bacteriol 174:5204–5210
Marcum JA, Kline DL (1983) Species specificity of streptokinase. Comp Biochem Physiol B Comp Biochem 75:389–394
Massberg S, Grahl L, von Bruehl M-L, Manukyan D, Pfeiler S, Goosmann C, Brinkmann V, Lorenz M, Bidzhekov K, Khandagale AB et al (2010) Reciprocal coupling of coagulation and innate immunity via neutrophil serine proteases. Nat Med 16:887–896
Mattsson E, Herwald H, Cramer H, Persson K, Sjöbring U, Björck L (2001) Staphylococcus aureus induces release of bradykinin in human plasma. Infect Immun 69:3877–3882
Mavrommatis AC, Theodoridis T, Orfanidou A, Roussos C, Christopoulou-Kokkinou V, Zakynthinos S (2000) Coagulation system and platelets are fully activated in uncomplicated sepsis. Crit Care Med 28:451–457
McKay FC, Mcarthur JD, Sanderson-Smith ML, Gardam S, Currie BJ, Sriprakash KS, Fagan PK, Towers RJ, Batzloff MR, Chhatwal GS et al (2004) Plasminogen binding by group A streptococcal isolates from a region of hyperendemicity for streptococcal skin infection and a high incidence of invasive infection. Infect Immun 72:364–370
Muszbek L, Bereczky Z, Bagoly Z, Komáromi I, Katona É (2011) Factor XIII: a coagulation factor with multiple plasmatic and cellular functions. Physiol Rev 91:931–972
Oehmcke S, Herwald H (2010) Contact system activation in severe infectious diseases. J Mol Med 88:121–126
Oehmcke SM, Ouml rgelin M, Herwald H (2009a) Activation of the human contact system on neutrophil extracellular traps. J Innate Immun 1:225–230
Oehmcke S, Shannon O, von Köckritz-Blickwede M, Mörgelin M, Linder A, Olin AI, Björck L, Herwald H (2009b) Treatment of invasive streptococcal infection with a peptide derived from human high-molecular weight kininogen. Blood 114:444–451
Oehmcke S, Shannon O, Mörgelin M, Herwald H (2010) Streptococcal M proteins and their role as virulence determinants. Clin Chim Acta 411:1172–1180
Oehmcke S, Mörgelin M, Malmström J, Linder A, Chew M, Thorlacius H, Herwald H (2012) Stimulation of blood mononuclear cells with bacterial virulence factors leads to the release of pro-coagulant and pro-inflammatory microparticles. Cell Microbiol 14:107–119

Opal SM, Esmon CT (2003) Bench-to-bedside review: functional relationships between coagulation and the innate immune response and their respective roles in the pathogenesis of sepsis. Crit Care 7:23–38

Påhlman LI, Mörgelin M, Eckert J, Johansson L, Russell W, Riesbeck K, Soehnlein O, Lindbom L, Norrby-Teglund A, Schumann RR et al (2006) Streptococcal M protein: a multipotent and powerful inducer of inflammation. J Immunol 177:1221–1228

Påhlman LI, Malmström E, Mörgelin M, Herwald H (2007) M protein from Streptococcus pyogenes induces tissue factor expression and pro-coagulant activity in human monocytes. Microbiology (Reading, Engl) 153:2458–2464

Pancholi V, Fischetti VA (1992) A major surface protein on group A streptococci is a glyceraldehyde-3-phosphate-dehydrogenase with multiple binding activity. J Exp Med 176:415–426

Pancholi V, Fischetti VA (1998) Alpha-enolase, a novel strong plasmin(ogen) binding protein on the surface of pathogenic streptococci. J Biol Chem 273:14503–14515

Poon IKH, Patel KK, Davis DS, Parish CR, Hulett MD (2011) Histidine-rich glycoprotein: the Swiss army knife of mammalian plasma. Blood 117:2093–2101

Pu Q, Wiel E, Corseaux D, Bordet R, Azrin MA, Ezekowitz MD, Lund N, Jude B, Vallet B (2001) Beneficial effect of glycoprotein IIb/IIIa inhibitor (AZ-1) on endothelium in Escherichia coli endotoxin-induced shock. Crit Care Med 29:1181–1188

Renné T (2012) The procoagulant and proinflammatory plasma contact system. Semin Immunopathol 34:31–41

Rezcallah MS, Boyle MDP, Sledjeski DD (2004) Mouse skin passage of Streptococcus pyogenes results in increased streptokinase expression and activity. Microbiology (Reading, Engl) 150:365–371

Sanderson-Smith ML, Walker MJ, Ranson M (2006) The maintenance of high affinity plasminogen binding by group A streptococcal plasminogen-binding M-like protein is mediated by arginine and histidine residues within the a1 and a2 repeat domains. J Biol Chem 281:25965–25971

Sanderson-Smith ML, Dowton M, Ranson M, Walker MJ (2007) The plasminogen-binding group A streptococcal M protein-related protein Prp binds plasminogen via arginine and histidine residues. J Bacteriol 189:1435–1440

Semeraro N, Ammollo CT, Semeraro F, Colucci M (2012) Sepsis, thrombosis and organ dysfunction. Thromb Res 129:290–295

Semple JW, Italiano JE, Freedman J (2011) Platelets and the immune continuum. Nat Rev Immunol 11(4):264–274

Shannon O, Hertzén E, Norrby-Teglund A, Mörgelin M, Sjöbring U, Björck L (2007) Severe streptococcal infection is associated with M protein-induced platelet activation and thrombus formation. Mol Microbiol 65:1147–1157

Shannon O, Rydengård V, Schmidtchen A, Mörgelin M, Alm P, Sørensen OE, Björck L (2010) Histidine-rich glycoprotein promotes bacterial entrapment in clots and decreases mortality in a mouse model of sepsis. Blood 116:2365–2372

Sjöbring U, Ringdahl U, Ruggeri ZM (2002) Induction of platelet thrombi by bacteria and antibodies. Blood 100:4470–4477

Smith SA (2009) The cell-based model of coagulation. J Vet Emerg Crit Care (San Antonio) 19:3–10

Smyth SS, McEver RP, Weyrich AS, Morrell CN, Hoffman MR, Arepally GM, French PA, Dauerman HL, Becker RC (2009) Platelet Colloquium Participants. Platelet functions beyond hemostasis. J Thromb Haemost 7:1759–1766

Soehnlein O, Oehmcke S, Ma X, Rothfuchs AG, Frithiof R, van Rooijen N, Mörgelin M, Herwald H, Lindbom L (2008) Neutrophil degranulation mediates severe lung damage triggered by streptococcal M1 protein. Eur Respir J 32:405–412

Sriskandan S, Kemball-Cook G, Moyes D, Canvin J, Tuddenham E, Cohen J (2000) Contact activation in shock caused by invasive group A Streptococcus pyogenes. Crit Care Med 28:3684–3691

Stadnicki A, de la Cadena RA, Sartor RB, Bender D, Kettner CA, Rath HC, Adam A, Colman RW (1996) Selective plasma kallikrein inhibitor attenuates acute intestinal inflammation in Lewis rat. Dig Dis Sci 41:912–920

Sun H (2004) Plasminogen is a critical host pathogenicity factor for group A streptococcal Infection. Science 305:1283–1286

Sun H, Wang X, Degen JL, Ginsburg D (2009) Reduced thrombin generation increases host susceptibility to group A streptococcal infection. Blood 113:1358–1364

Sun H, Xu Y, Sitkiewicz I, Ma Y, Wang X, Yestrepsky BD, Huang Y, Lapadatescu MC, Larsen MJ, Larsen SD et al (2012) Inhibitor of streptokinase gene expression improves survival after group A streptococcus infection in mice. Proc Natl Acad Sci USA 109:3469–3474

Svensson MD, Sjöbring U, Bessen DE (1999) Selective distribution of a high-affinity plasminogen-binding site among group A streptococci associated with impetigo. Infect Immun 67:3915–3920

Svensson MD, Sjöbring U, Luo F, Bessen DE (2002) Roles of the plasminogen activator streptokinase and the plasminogen-associated M protein in an experimental model for streptococcal impetigo. Microbiology (Reading, Engl) 148:3933–3945

Takada Y, Takada A (1989) Evidence for the formation of a trimolecular complex between streptokinase plasminogen and fibrinogen. Thromb Res 53:409–415

Taylor FB, Bryant AE, Blick KE, Hack E, Jansen PM, Kosanke SD, Stevens DL (1999) Staging of the baboon response to group A streptococci administered intramuscularly: a descriptive study of the clinical symptoms and clinical chemical response patterns. Clin Infect Dis 29:167–177

Tillett WS, Garner RLR (1933) The fibrinolytic activity of hemolytic streptococci. J Exp Med 58:485–502

van der Poll T, de Boer JD, Levi M (2011) The effect of inflammation on coagulation and vice versa. Curr Opin Infect Dis 24:273–278

Vandijck DM, Blot S, De Waele Jan J, Hoste EA, Vandewoude KH, Decruyenaere JM (2010) Thrombocytopenia and outcome in critically ill patients with bloodstream infection. Heart Lung J Acute Crit Care 39:21–26

Walker M, McArthur J, Mckay F, Ranson M (2005) Is plasminogen deployed as a virulence factor? Trends Microbiol 13:308–313

Wang H, Lottenberg R, Boyle MD (1995a) Analysis of the interaction of group A streptococci with fibrinogen, streptokinase and plasminogen. Microb Pathog 18:153–166

Wang H, Lottenberg R, Boyle MDP (1995b) A role for fibrinogen in the streptokinase-dependent acquisition of plasmin(ogen) by group-a streptococci. J Infect Dis 171:85–92

Werb Z (1997) ECM and cell surface proteolysis: regulating cellular ecology. Cell 91:439–442

Williams SCP (2012) After Xigris, researchers look to new targets to combat sepsis. Nat Publ Group 18:1001

Winning J, Reichel J, Eisenhut Y, Hamacher J, Kohl M, Deigner HP, Claus RA, Bauer M, Lösche W (2009) Anti-platelet drugs and outcome in severe infection: clinical impact and underlying mechanisms. Platelets 20:50–57

Winram SB, Lottenberg R (1996) The plasmin-binding protein Plr of group A streptococci is identified as glyceraldehyde-3-phosphate dehydrogenase. Microbiology (Reading, Engl) 142:2311–2320

Wistedt AC, Ringdahl U, Müller-Esterl W, Sjöbring U (1995) Identification of a plasminogen-binding motif in PAM, a bacterial surface protein. Mol Microbiol 18:569–578

Wistedt AC, Kotarsky H, Marti D, Ringdahl U, Castellino FJ, Schaller J, Sjöbring U (1998) Kringle 2 mediates high affinity binding of plasminogen to an internal sequence in streptococcal surface protein PAM. J Biol Chem 273:24420–24424

Yakovlev SA, Rublenko MV, Izdepsky VI, Makogonenko EM (1995) Activating effect of the plasminogen activators on plasminogens of different mammalia species. Thromb Res 79: 423–428

Group A Streptococcal Vaccine Candidates: Potential for the Development of a Human Vaccine

Anna Henningham, Christine M. Gillen and Mark J. Walker

Abstract Currently there is no commercial Group A *Streptococcus* (GAS; *S. pyogenes*) vaccine available. The development of safe GAS vaccines is challenging, researchers are confronted with obstacles such as the occurrence of many unique serotypes (there are greater than 150 M types), antigenic variation within the same serotype, large variations in the geographical distribution of serotypes, and the production of antibodies cross-reactive with human tissue which can lead to host auto-immune disease. Cell wall anchored, cell membrane associated, secreted and anchorless proteins have all been targeted as GAS vaccine candidates. As GAS is an exclusively human pathogen, the quest for an efficacious vaccine is further complicated by the lack of an animal model which mimics human disease and can be consistently and reproducibly colonized by multiple GAS strains.

Abbreviations

ADI	Arginine deiminase
ARF	Acute rheumatic fever
CFA	Complete Freund's adjuvant
CT	Cholera toxin
CTB	Cholera toxin B
ECM	Extracellular matrix
FbaA	Fibronectin-binding protein A
FBP54	Fibronectin-binding protein 54
GAS	Group A *Streptococcus*
GBS	Group B *Streptococcus*
IFA	Incomplete Freund's adjuvant

A. Henningham · C. M. Gillen · M. J. Walker (✉)
School of Chemistry and Molecular Biosciences
and Australian Infectious Disease Research Centre,
University of Queensland, St Lucia, QLD 4072, Australia
e-mail: mark.walker@uq.edu.au

Current Topics in Microbiology and Immunology (2013) 368: 207–242
DOI: 10.1007/82_2012_284

Published Online: 19 December 2012

Ig	Immunoglobulin
KLH	Keyhole limpet hemocyanin
MS	Mass spectrometry
SagP	Streptococcal acid glycoprotein
SfbI	Streptococcal fibronectin binding protein I
SfbII	Streptococcal fibronectin binding protein II
Shr	Streptococcal hemoprotein receptor
Sib35	Streptococcal immunoglobulin-binding protein 35
SLO	Streptolysin O
SOF	Serum opacity factor
Spa	Streptococcal protective antigen
SpeA	Streptococcal pyrogenic exotoxin A
SpeB	Streptococcal pyrogenic exotoxin B
SpyCEP	*Streptococcus pyogenes* cell envelope proteinase
Sse	Streptococcal secreted esterase
TF	Trigger factor

Contents

1 Introduction

Even though GAS is responsible for a high global disease burden in both developing and developed regions, there is no commercial vaccine available. Existing treatments for GAS infection include the administration of penicillin, macrolides or intravenous immunoglobulin (Ig). Penicillin, which has routinely been used for the treatment of GAS infection, is also administered prophylactically in attempts to prevent recurring GAS infection and subsequent immune sequelae. However, despite such measures, GAS auto-immune sequelae persist at endemic levels in developing countries and Indigenous populations (Carapetis et al. 2005). Whilst GAS has not been reported to possess resistance to penicillin, studies have measured a 20–40 % failure rate of penicillin in the treatment of GAS pharyngitis (Pichichero and Casey 2007). Possible factors contributing to the failure of penicillin during the treatment of GAS infections include in vivo co-colonization with other penicillin-resistant bacterial species, in vivo eradication of normal throat microflora, a carrier state, internalization of GAS into host cells, tolerance of GAS to penicillin, poor penetration of penicillin into host tissues, lack of patient compliance and reoccurring exposure to GAS (reviewed in (Pichichero and Casey 2007; Brook 2007). Macrolides are utilized as an alternative to penicillin, particularly in cases of penicillin allergy. There is concern that macrolide-resistant isolates may evolve and spread (Michos et al. 2009; Richter et al. 2005; Tse et al. 2012), and it has been suggested that macrolide-sensitive GAS isolates use biofilm formation as a mechanism to escape antibiotic treatment and persist within the host (Baldassarri et al. 2006). In cases of invasive GAS disease such as streptococcal toxic shock syndrome and necrotizing fasciitis (Young et al. 2005), intravenous Ig is recommended as an adjunctive therapy due to its capacity to neutralize superantigens and to promote opsonophagocytosis (Basma et al. 1998; Pandey et al. 2009). The timing of administration is critical, if administered too long after the initial diagnosis, intravenous Ig will not provide any benefits greater than if antibiotics alone were used for treatment (Pandey et al. 2009). Furthermore, intravenous Ig only offers short-term protection, as no immunological memory is generated. While in most cases penicillin or alternate antibiotics remain effective in the treatment of superficial GAS infection, the costs of medical care and time off work or school create a significant economic burden. The majority of persistent GAS infections and immune sequelae occur in geographic regions where individuals do not have ready access to high standard healthcare. A vaccine protective against all serotypes may be the only effective way to control and eliminate GAS disease worldwide.

In recent times, a new era of vaccine development has emerged, utilizing reverse vaccinology, aided by proteomics, whole genome sequencing, bioinformatics and DNA microarrays. Such technologies have allowed rapid discovery and identification of putative virulence factors, surface-associated proteins and possible vaccine candidates of many bacterial species including GAS. Despite the advent of such high throughput techniques, the development of GAS vaccines is challenging, facing obstacles such as the occurrence of many unique serotypes, antigenic variation within the same serotype, differences in geographical distribution of serotypes, and the production of antibodies cross-reactive with human tissue which can lead to host auto-immune disease. In the case of GAS vaccines, blocking the attachment of GAS to the host cell surface and its receptors (and hence preventing subsequent colonization) is likely to be an effective strategy in the prevention of infection (Wizemann et al. 1999). As attachment is usually mediated via bacterial adhesins, it follows that cell surface adhesins are among the key targets in GAS vaccine development. Prophylactic vaccination with adhesin-containing vaccines has successfully blocked GAS infection in mice. Proteins containing a cell wall anchor motif, signal peptide sequence or proteins experimentally determined to be located on, or associated with, the cell surface, may be exposed to the host immune system during infection potentially resulting in cellular and humoral immune responses. Since GAS is known to colonize mucosal tissues in the host, the upper respiratory tract and the skin, a vaccine promoting both systemic (serum IgG) and mucosal immunity (mucosal IgA) is likely to offer long lasting protection against a broad range of GAS disease etiology.

2 Animal Models for Evaluating Vaccine Candidates

GAS is an exclusively human pathogen that colonizes the upper respiratory tract and skin where it causes an array of diseases that vary extensively in clinical presentation. As such, there is no single reproducible animal model for studying GAS pathogenesis and evaluating vaccine efficacy. Numerous models have been devised for the different forms of GAS disease, although, in most cases these animal models do not adequately replicate human pathogenesis. However, even if the model is less than ideal a vaccine must first be tested in animals. Murine models are the most commonly employed for both pathogenesis studies and for vaccine development, with models for pharyngeal, cutaneous and systemic infections used extensively. In recent years other animal models have been developed including rheumatic heart disease in rats (Lymbury et al. 2003), invasive disease in rabbits (Piepmeier et al. 1995), streptococcal septic shock in pigs (Saetre et al. 2000), toxic shock in mice (Ulrich 2008), necrotizing fasciitis in mice (Matsui et al. 2009), pharyngitis in non-human primates (Skinner et al. 2011; Sumby et al. 2008), plus a zebrafish (Neely et al. 2002) and a wax worm model (Olsen et al. 2011) to study GAS virulence properties.

2.1 Shortcomings of Murine Models of GAS Infection

A major obstacle in establishing appropriate mouse models for GAS infection is the difficulty and variability in adapting the human pathogen to mice. Few GAS strains are naturally virulent in mice. In most cases prior to either mucosal or parenteral murine challenge, the virulence of GAS strains is increased by repeated intraperitoneal passage in mice (up to 12 times) (Batzloff et al. 2005; Bessen and Fischetti 1988, 1990; Brandt et al. 2000; Bronze et al. 1992; Fritzer et al. 2010; Olive et al. 2002; Sabharwal et al. 2006). However, in vivo passage of strains has been shown in several cases to lead to mutations that directly affect the virulence of strains (Cole et al. 2011; Eberhard et al. 2001; Fontaine et al. 2003), and infection with such strains may not be representative of the true pathogenetic mechanisms of GAS infection.

Another problem in the utilization of animal models to study human pathogenesis is that a virulence factor with a key role in human infection may be of little importance or completely negated in an animal model. For example, the GAS plasminogen activator streptokinase is a potent activator of human plasminogen but is ineffective against mouse plasminogen (Gladysheva et al. 2003), leading to low virulence of GAS strains particularly in skin infection models. To overcome this problem, a humanized plasminogen transgenic mouse model of skin infection has been developed for virulence studies (Sun et al. 2004). Other humanized transgenic mouse models used for virulence studies include a human CD46-expressing transgenic mouse model of subcutaneous infection (Matsui et al. 2009) and a HLA-DQR transgenic mouse model of toxic shock (Ulrich 2008). The latter model has also been employed in vaccine studies to ascertain the protective efficacy of the streptococcal pyrogenic exotoxin A (SpeA) to protect against toxic shock (Ulrich 2008).

A problem with using mouse models for vaccine studies is the variation observed between individual mice in an experiment and also the variable susceptibility of different mouse strains to infection by the same GAS strain. To reduce variation within individual experiments inbred mouse strains are commonly employed for vaccine studies. Strains utilized by streptococcal researchers include BALB/c (Bronze et al. 1992; Georgousakis et al. 2009; Gillen et al. 2008; Hall et al. 2004; Henningham et al. 2012; Schulze et al. 2003), B10.Br (Brandt et al. 2000; Olive et al. 2002), C3H/HeN (Okamoto et al. 2005; Stalhammar-Carlemalm et al. 1999), C57BL/6J (Henningham et al. 2012; Okamoto et al. 2005) and Swiss Webster (Severin et al. 2007). However, outbred mouse strains are a better representation of the genetic variation present in the human population and are potentially a more apt selection for GAS vaccinology studies. Outbred mouse strains employed in experimental models of GAS infection include Quackenbush (Batzloff et al. 2005, 2006; McMillan et al. 2004; Sanderson-Smith et al. 2006), CD1 (Bessen and Fischetti 1990; Cleary et al. 2004; Ji et al. 1997; Kapur et al. 1994), Swiss NIH mice (Courtney et al. 2003) and Crl:SKH1-hrBR (Liu et al. 2007).

Different mouse strains have varying susceptibilities to GAS infection, providing evidence that host genetic factors are important in determining susceptibility to GAS infection. In a study comparing the susceptibility of 5 different inbred mouse strains, BALB/c, C57BL/10, and DBA/2 mice were the most resistant to streptococcal infection, whereas C3H/HeN and CBA/J mice exhibited substantially higher bacterial growth and 100 % mortality (Goldmann et al. 2003; Medina et al. 2001). However, this innate difference in susceptibility to infection can be overcome by vaccination, with resistance to GAS infection promoted by vaccination. Innately resistant (BALB/c) and susceptible (C3H/HeN) mice vaccinated with Mprotein or heat killed GAS were equal in their capacity to control GAS infection (Siegert et al. 2006). Likewise, immunization with streptococcal immunolglobulin-binding protein 35 (Sib35) rendered both BALB/c and C3H/HeN mouse strains resistant to GAS infection (70 and 100 % survival, respectively), but the Streptococcal Immunoglobulin-Binding Protein 35 (Sib35) immunized outbred CD1 mouse strain was unable to control GAS infection (Okamoto et al. 2005).

2.2 Protective Efficacy Studies

Passive immunization experiments examine the capacity of antigen specific antibodies to protect against GAS infection. Most frequently these experiments have been performed using a model of systemic GAS infection. Antibodies are raised in rabbits via systemic immunization with complete Freund's adjuvant (CFA), this specific rabbit antiserum is then administered to an outbred mouse strain via the intraperitoneal route up to 24 h prior to an intraperitoneal GAS infection (Sabharwal et al. 2006; Stalhammar-Carlemalm et al. 1999; Kapur et al. 1994; Liu et al. 2007; Dale et al. 1999; Kawabata et al. 2001; Pandey et al. 2009). Passive immunization experiments have also been used to demonstrate passive protection at the mucosa. Mouse antiserum raised by intranasal immunization with C5a peptidase when administered to mice intranasally has been shown to protect against pharyngeal colonization by GAS (Park and Cleary 2005). Passive immunization demonstrates the capacity of antigen specific antibodies to protect against GAS infection, and may also provide an alternate treatment for invasive GAS diseases (Basma et al. 1998; Kaul et al. 1999). Additionally, passive immunization experiments demonstrate that protection observed in active immunization experiments is not due to non-specific stimulation of innate immune responses (Liu et al. 2007).

In the last ten years, two separate vaccine candidates have moved forward to human clinical trials (Kotloff et al. 2004; McNeil et al. 2005). In both cases advancement to clinical trials was based upon the success of the vaccines in eliciting systemic immunity in mouse models. Most murine models to evaluate systemic immunity in response to GAS antigens have utilized subcutaneous (Brandt et al. 2000; Sabharwal et al. 2006; Okamoto et al. 2005; McMillan et al. 2004; Kapur et al. 1994; Liu et al. 2007; Kawabata et al. 2001; Ma et al. 2009; Terao et al. 2005) and less frequently intraperitoneal (Henningham et al. 2012; Guzmán et al. 1999), intramuscular (Ulrich 2008; Turner et al. 2009), or intravenous (Gillen et al. 2008; Courtney et al. 2003) immunization in the presence of CFA/incomplete Freund's adjuvant (IFA). CFA is an efficient stimulator of cell-mediated immunity and augments the humoral immune response, however, CFA is banned for use in humans due to toxic side effects so it can be difficult to translate the results of these studies to humans [reviewed in Lindblad 2000]. Systemic immunization trials using Alum, the only parenterally delivered adjuvant approved for use in humans, have been performed. The protective efficacy of subcutaneous immunization with group A carbohydrate (CHO) and streptococcal secreted esterase (Sse) adsorbed to Alum was determined by survival of lethal challenge (Sabharwal et al. 2006; Liu et al. 2007). Most frequently, protective efficacy against systemic GAS infection is measured by survival against a lethal dose of GAS (Brandt et al. 2000; Gillen et al. 2008; Okamoto et al. 2005; McMillan et al. 2004; Sanderson-Smith et al. 2006; Kapur et al. 1994; Courtney et al. 2003; Ma et al. 2009; Terao et al. 2005; Schulze et al. 2006) or by measuring dissemination of GAS to organs (Turner et al. 2009; McArthur et al. 2004).

Parenteral immunization stimulates strong serum IgG titres, but generally fails to stimulate an immune response in mucosal secretions. Protection against GAS infection as a result of systemic immunization has primarily been attributed to bacterial clearance by opsonophagocytosis (McMillan et al. 2004; Sanderson-Smith et al. 2006; Kawabata et al. 2001; Olive et al. 2003). The primary route of GAS infection is the upper respiratory tract, therefore while systemic immunity has proved effective in preventing bacterial dissemination through opsonic serum IgG, it may not be the best approach for inducing immunity to the mucosal GAS pathogen. Most reports examining the efficacy of GAS antigens to elicit systemic immunity investigate only the efficacy of parenteral immunization to protect against a systemic challenge, and generally do not determine if parenteral immunization is sufficient to also protect against mucosal challenge. In only a few cases have researchers examined the efficacy of parenteral immunization to stimulate mucosal immunity. For instance, subcutaneous immunization with CHO adjuvanted with Alum was shown to protect against pharyngeal colonization in mice (Sabharwal et al. 2006). In another study, subcutaneous immunization with C5a peptidase adjuvanted with Alum and the mucosal adjuvant monophosphoryl lipid A (MPL) protected mice from pharyngeal colonization following intranasal challenge (Cleary et al. 2004). In contrast, intraperitoneal immunization of mice with the self adjuvanting streptococcal fibronectin binding protein I (SfbI) protein failed to protect against lethal intranasal challenge (Guzmán et al. 1999). This highlights the importance of utilizing multiple murine challenge models when assessing the efficacy of protective vaccine antigens.

Mucosal vaccination delivers vaccine antigens to the mucosal surfaces inducing a protective immune response at the site of colonization. Most murine models to evaluate mucosal immunity in response to GAS antigens have utilized intranasal or perioral immunization in the presence of the non-toxic B subunit of cholera toxin (CTB) (Bessen and Fischetti 1988, 1990; Bronze et al. 1992; Olive et al. 2002; Hall et al. 2004; Schulze et al. 2003, 2006; Batzloff et al. 2006; Kawabata et al. 2001; Park and Cleary 2005; Guzmán et al. 1999; Yokoyama and Harabuchi 2002). While CTB is not approved for human use it is one of the most commonly employed adjuvants for assessing vaccine potential at mucosal surfaces in animal models. The protective efficacy of GAS antigens when delivered with novel non-toxic mucosal adjuvants has also been investigated. Mucosal adjuvants include liposomes (Hall et al. 2004), MPL containing formulations (Ulrich 2008; Hall et al. 2004), proteosomes (Batzloff et al. 2005; Hall et al. 2004) and lipid entities (Batzloff et al. 2006; Olive et al. 2007; Zaman et al. 2011). Many of these formulations have proved to be effective stimulators of both mucosal and systemic immune responses (Batzloff et al. 2005, 2006; Hall et al. 2004; Olive et al. 2007; Zaman et al. 2011), and some have been shown to protect against mucosal GAS infection (Ulrich 2008; Batzloff et al. 2005, 2006; Hall et al. 2004; Olive et al. 2007). However, of these novel adjuvants only liposome vaccine formulations and MPL containing formulations have been approved for clinical use in humans for hepatitis A and influenza and for allergy vaccines, respectively (Ambrosch et al. 1997; Vajdy and Singh 2005).

Protection against mucosal GAS infection is most often measured by survival of lethal intranasal challenge (Hall et al. 2004; Schulze et al. 2001, 2003; Guzmán et al. 1999; Olive et al. 2007), and/or measuring pharyngeal colonization rates in a murine model of pharyngitis (Batzloff et al. 2005, 2006; Bessen and Fischetti 1990; Bronze et al. 1992; Olive et al. 2002; Ji et al. 1997). However, mice do not develop a true pharyngitis and only a relatively few animals become colonized even when high concentrations of inocula are used. This limits the relevance, significance and usefulness of the murine model of GAS pharyngitis. Pharyngeal colonization rates are measured by determining the colony forming units present in the pharynx from as early as 20 h post infection up to 10–15 days post intranasal or perioral challenge either by throat swab (Batzloff et al. 2005; Bessen and Fischetti 1988, 1990; Bronze et al. 1992, 2006; Ji et al. 1997) or by homogenizing mouse snout samples (Cleary et al. 2004). Alternatively, Park and Cleary (2005) visualized and quantified pharyngeal colonization by utilizing luciferase expressing GAS strains and a charge-coupled device camera system at both 24 and 48 h post infection.

Mucosal immunization stimulates both a humoral immune response in blood and mucosal secretions (Batzloff et al. 2005, 2006; Bessen and Fischetti 1988; Bronze et al. 1992; Hall et al. 2004; Ji et al. 1997; Kawabata et al. 2001; Park and Cleary 2005; Guzmán et al. 1999; Schulze et al. 2001, 2006; Yokoyama and Harabuchi 2002; Olive et al. 2007). Protection against GAS infection as a result of mucosal immunization is most likely due to prevention of colonization by secretory IgA and IgG at the site of infection. However, it has been shown that GAS can disseminate past the mucosal epithelium within 24 h after challenge (Park et al. 2003), thus the presence of opsonic serum IgG antibodies may provide additional protection by preventing bacterial dissemination from the mucosal site of infection. Batzloff et al. (2005) have shown that intranasal immunization with a minimal conformational epitope of the M protein (J14) formulated with proteosomes is capable of stimulating opsonic serum IgG. Only a handful of researchers have assessed the capacity of mucosal immunization to protect against systemic challenge. SfbI delivered intranasally failed to protect against bacterial dissemination following subcutaneous challenge (McArthur et al. 2004). On the other hand, intranasal or perioral immunization with Fibronectin-Binding Protein 54 (FBP54) stimulated strong serum IgG titres and both mucosal routes of vaccination protected against systemic subcutaneous challenge (Kawabata et al. 2001).

The protective efficacy of GAS vaccine antigens is predominantly tested in murine models. GAS vaccine preparations have been administered to humans and historically the protective efficacy of these vaccinations was tested by challenge with GAS (D'Alessandri et al. 1978; Fox et al. 1973; Polly et al. 1975; Waldman et al. 1975). However, due to the risk of contracting rapidly progressing invasive GAS diseases or post streptococcal sequelae as a result of challenge, this practice has not been continued. Contemporary vaccine experiments in human volunteers have consisted of vaccinating healthy human volunteers, however, the protective efficacy of the vaccines in these cases was not measured. The efficacy of vaccination was only measured by determining antibody titre and by using opsonophagocytic assays (Kotloff et al. 2004; McNeil et al. 2005). Following the argument that humans and non-human primates are closely related phylogenetically, a non-human primate model of experimental GAS pharyngitis has been developed (Sumby et al. 2008). This model has been successfully used to study transcriptional adaptation of GAS during upper respiratory tract infection (Virtaneva et al. 2003, 2005) and to determine the contribution of specific GAS virulence factors to infection (Gryllos et al. 2001, 2008; Sitkiewicz et al. 2006; Sumby et al. 2005) and would be easily adaptable for use in vaccine studies. Numerous historical studies and some present day studies have shown that non-human primates are successfully colonized by GAS and that infections mimic the humoral immune response characteristic for human disease (Skinner et al. 2011; Sumby et al. 2008; Watson et al. 1946; Zimmerman et al. 1970; Krause and Rammelkamp 1962).

3 Cell Wall Anchored GAS Vaccine Candidates

3.1 M Protein

M protein is an obvious target for GAS vaccine development as it is a major virulence factor expressed by all serotypes of GAS. M proteins are multifunctional, some function as adhesins, adhering to host epithelial cells via a variety of human extracellular matrix (ECM) and plasma proteins including fibrinogen (Carlsson et al. 2005), plasminogen (Berge and Sjöbring 1993; Ringdahl and Sjöbring 2000; Sanderson-Smith et al. 2008), CD46 (Okada et al. 1995), fibronectin (Cue et al. 2001), factor H (Perez-Casal et al. 1995) and collagen (Dinkla et al. 2003). M protein can also facilitate GAS evasion of phagocytosis by blocking the deposition of C3b onto the surface of GAS (Fischetti 1989). Historical studies involved the vaccination of human volunteers with crude M protein preparations followed by the administration of live GAS to the pharynx. This strategy effectively protected against GAS pharyngeal colonization (D'Alessandri et al. 1978; Fox et al. 1973; Polly et al. 1975; Massell et al. 1969), but an increased incidence of acute rheumatic fever (ARF) in some of the vaccinated children compared to the unvaccinated control children was observed (Massell et al. 1969), significantly hampering the development of whole M protein based vaccines. Since then, either the highly conserved C-repeat region of the M protein (Bessen and Fischetti 1990; Brandt et al. 2000; Medaglini et al. 1995) or the variable serotype specific N-terminal A-repeats of the M protein (Dale et al. 1999; Kotloff et al. 2004; Beachey et al. 1981, 1986; Dale et al. 1983, 1993, 1996) have been targeted. Type-specific antibodies against M protein have been found to neutralize the antiphagocytic effect of M protein, providing protection against GAS of the same serotype (Beachey et al. 1981). Exposure to GAS in childhood can lead to the development of antibodies raised against the conserved region of M protein (Fischetti 1991), which may offer protection against streptococcal pharyngitis in later life. If an M protein based vaccine can be produced acting against the C-repeat conserved regions shared by all GAS serotypes, the need to generate serotype-specific antibodies for protection would be eliminated.

3.1.1 Conserved C-Repeat-Based M Protein Vaccines

To date there have been several approaches for producing GAS vaccines based on the conserved, non-type specific C-repeats of the M protein. Whilst production of a vaccine based on this conserved region eliminates limitations associated with serotype specific antibody responses, antibodies directed against this region have been observed to be non-opsonic (Jones and Fischetti 1988). One approach produced four overlapping synthetic peptides encompassing the complete C-repeat region of the M6 protein. When Swiss CD1 mice were immunized intranasally with this construct conjugated to CTB followed by an intranasal challenge with either a homologous M6 or a heterologous M14 GAS isolate, a significant reduction in pharyngeal colonization was observed (Bessen and Fischetti 1988, 1990). A second approach synthesized two peptides comprising regions within the C-repeat region of the M5 protein (Bronze et al. 1992). In passive immunization studies, rabbit sera generated against M5 peptide-keyhole limpet hemocyanin (KLH) conjugates emulsified with CFA protected BALB/c mice from heterologous intranasal challenge with a M24 GAS isolate (Bronze et al. 1992). Likewise, in active immunization studies, BALB/c mice intranasally immunized with M5 peptide-CTB conjugates were significantly protected against heterologous intranasal challenge with M24 GAS (Bronze et al. 1992). Whilst an elevated level of serum IgG was detected against both the M5 peptide and CTB, similar to previous reports, the IgG antibodies raised against the C-repeat region peptides were unable to induce opsonophagocytosis in vitro (Bronze et al. 1992). A third approach identified B- and T cell epitopes within the C-repeat region of the M5 protein leading to the construction of a 55-amino acid protein, designated StreptInCor (Guilherme et al. 2006). Following subcutaneous immunization of BALB/c mice, StreptInCor co-administered with CFA/IFA was found to induce high titre serum IgG antibodies (Guilherme et al. 2009). When StreptInCor was co-administered with the mucosal adjuvant, Adjuvant Finlay Cochleate 1, via the intranasal route, a mucosal IgA and systemic IgG response resulted (Guilherme et al. 2009). A fourth approach utilized synthetic peptides based on minimal B- and T-cell epitopes. A-20 amino acid synthetic peptide designated p145 (Pruksakorn et al. 1994) and two derivative epitopes, designated J8 and J14 were designed (Hayman et al. 1997; Pruksakorn et al. 1992). Good et al. have investigated J8- and J14-mediated induction of murine serum-specific IgG (Brandt et al. 2000) and muscosal IgA (Batzloff et al. 2006; Brandt et al. 1999). The capacity of J8 and J14 to induce opsonic antibodies has also been examined (Batzloff et al. 2003, 2005; Brandt et al. 1996, 1997, 1999, 2000; Olive et al. 2002, 2003, 2004, 2005, 2006; Hayman et al. 2002). Formulations containing p145, J8 or J14 were observed to enhance the survival of immunized mice when co-administered with a number of different adjuvants including diphtheria toxin (Olive et al. 2002; Pandey et al. 2009; Batzloff et al. 2003), proteosomes (Batzloff et al. 2005), lipid core technology (Batzloff et al. 2006; Olive et al. 2002, 2003, 2005, 2006; Zaman et al. 2011; Hayman et al. 2002, Moyle et al. 2006a, b, c, Abdel-Aal et al. 2008, 2010; Zaman et al. 2010, 2012), liposaccharides (Simerska et al. 2008a, b; Fujita et al. 2008) and H12 (Georgousakis et al. 2009), a protective segment of Protein F1. Another approach utilized live vaccine delivery vectors for the delivery of C-repeat based M protein vaccine preparations. CD1 mice were intranasally immunized with a vaccinia virus expressing the C-repeat region of the M6 protein (Fischetti et al. 1989). Following homologous (M6) and heterologous (M14) intranasal challenge, a significant reduction in pharyngeal colonization by GAS was observed (Fischetti et al. 1989). *Lactococcus lactis*, an intestinal commensal flora, has also been used as a live vaccine vector harbouring the genes encoding the C-repeat region from M6 protein. CD1 mice were intranasally immunized with the live vaccine vector, eliciting a salivary IgA and serum IgG response (Mannam et al. 2004). Mice were also subcutaneously immunized with the *L. lactis* strain which elicited circulating serum IgG (Mannam et al. 2004). Following heterologous intranasal challenge with an M14 strain, mice immunized via intranasal and subcutaneous routes were significantly protected against pharyngeal infection (Mannam et al. 2004).

3.1.2 N-Terminal-Based M Protein Vaccines

The other major branch of M protein vaccinology focuses on the highly variable N-terminal serotype specific portion of the M protein. One example is the a1 peptide of the plasminogen-binding group A streptococcal M-like protein. Following subcutaneous immunization of outbred Quackenbush mice with a1-KLH/CFA, a systemic IgG antibody response capable of inducing opsonophagocytosis of GAS in vitro and protecting mice against intraperitoneal homologous GAS challenge in vivo ensued (Sanderson-Smith et al. 2006). Dale et al. have developed multivalent vaccine preparations containing protective N-terminal M protein fragments. Two of these preparations have reached human clinical trials. A hexavalent preparation contains fragments from six serotypes; M1, M3, M5, M6, M19 and M24; selected due to a frequent association with pharyngitis and ARF in the US (Kotloff et al. 2004; Dale 1999). Administration of this hybrid vaccine with either CFA or Alum generated high titre opsonizing antibodies in rabbits (Dale 1999) and protected BALB/c mice against mucosal challenge with M24 GAS when co-administered with either liposomes/MPL, CTB, CTB/cholera toxin (CT) or proteosomes/lipoplysaccharide (Hall et al. 2004). In 2004, this vaccine preparation reached phase I clinical trials and was found to result in a statistically significant increase in antibody titre for all six M protein-based fragments, with five of the six targeted GAS serotypes opsonized in vitro by the resultant antisera (Kotloff et al. 2004). Additionally, there was no evidence the antibodies were cross-reactive with human tissue (Kotloff et al. 2004). Although this vaccine was considered successful in phase I clinical trials, a major shortcoming of this preparation is that it only offers protection against six of at least 150 GAS M types. In an attempt to broaden the protection, a multivalent vaccine based on the variable amino terminal of 26 different M proteins was produced (Hu et al. 2002). Following immunization of rabbits with the multivalent preparation adsorbed to Alum, type-specific serum IgG antibodies were generated. These antibodies recognized 25 of the 26 M protein fragments in the vaccine and did not cross-react with host tissue (Hu et al. 2002). In vitro opsonophagocytosis assays were performed and 18 of the 26 targeted vaccine strains had rates of $\geq$30 % opsonization (Hu et al. 2002). Additionally, this vaccine preparation was observed to be safe and immunogenic in phase I human clinical trials (McNeil et al. 2005). In a study of 5,400 invasive GAS cases in the US from 2000 to 2004, the *emm* types in the 26-valent vaccine covered 79 % of serotypes isolated from patients (O'Loughlin et al. 2007). A separate epidemiological study conducted from 2001 to 2005 in the US predicted coverage of the vaccine serotypes to be approximately 60 % (Nir-Paz et al. 2010). Vaccine serotype coverage in other countries has been reported as follows: 81 % for invasive isolates in Japan (Ikebe et al. 2007), 54 % for isolates recovered from healthy school children in Ethiopia (Abdissa et al. 2006), and serotype coverage within Africa, Asia, Latin America, the Middle East and the Pacific region estimated to be 39, 61, 72, 63 and 34 % respectively (Steer et al. 2009). Recently, Dale et al. (2011) have formulated a 30-valent GAS vaccine preparation based on the serotypes most prevalent in North America and Europe. This preparation was immunogenic in rabbits and the resultant antisera opsonized all 30 vaccine containing serotypes. Furthermore, the antisera contained bactericidal antibodies against 24 (of 40) non-vaccine serotypes tested (Dale et al. 2011). These findings suggest that multivalent M protein amino terminal based preparations may provide a greater coverage than estimated based solely on the type-specific fragments present. Nonetheless, the serotypes represented in the 30-valent preparation account for only 40 and 59 % of pharyngeal infections in Mali and South Africa, respectively (Dale et al. 2011). Strategies that ideally cover all GAS serotypes are required for the production of a vaccine that will be efficacious in all geographical regions.

3.2 Fibronectin-Binding Protein A

Fibronectin-Binding Protein A (FbaA) is a GAS cell wall anchored protein reported to bind fibronectin (Terao et al. 2001) and host regulators of complement, Factor H and Factor H-like protein 1, thereby promoting resistance to phagocytosis (Pandiripally et al. 2002). The *fba* gene was reported in M1, M2, M4, M22, M28 and M49 GAS, but was not detected in other serotypes (Terao et al. 2001). Intraperitoneal immunization of BALB/c mice with FbaA/CFA produced an elevated humoral IgG response similar to that of M protein and conferred a level of protection on par with M1 protein following homologous (M1) intraperitoneal challenge (Ma et al. 2009).

3.3 R28 Protein

R28 protein of GAS is a cell wall anchored protein with sequence similarity to the α, β and Rib proteins of group B *Streptococcus* (GBS). R28 protein was initially identified in M28 GAS (Lancefield and Perlmann 1952), but has also been reported in M13 and M48 GAS (Lancefield 1957). C3H/NeH mice administered R28 antisera (generated in rabbits administered R28/CFA) via passive intraperitoneal immunization were significantly protected against lethal homologous (M28) intraperitoneal challenge (Stalhammar-Carlemalm et al. 1999). In addition, both passive intraperitoneal and active subcutaneous immunization of C3H/NeH mice with R28 or Rib (co-administered with CFA/IFA) resulted in protection against intraperitoneally administered GBS or GAS respectively (Stalhammar-Carlemalm et al. 2000). Thus, R28 protein shows promise as a cross-species streptococcal vaccine antigen.

3.4 Protein F1/Streptococcal Fibronectin Binding Protein I

Protein F1 (also designated SfbI) is a GAS cell wall anchored protein which binds host fibronectin and fibrinogen (Hanski and Caparon 1992; Okada et al. 1994; Talay et al. 1992; Katerov et al. 1998), enabling GAS adherence to and internalization into host epithelial cells (Hanski and Caparon 1992; Okada et al. 1994; Molinari et al. 1997, 1999; Jadoun et al. 1998). In addition, protein F1 confers antiphagocytic effects (Hyland et al. 2007). The *sfbI* gene encoding protein F1 is not present in all serotypes of GAS. In a study of 69 clinical isolates from the Northern Territory of Australia, the *sfbI* gene was present in 64 % of isolates (Goodfellow et al. 2000). CD1 mice intranasally immunized with either protein F1 alone or protein F1 coupled to CTB, produced specific humoral IgG and mucosal IgA immune responses (Guzmán et al. 1999). Intranasal challenge of BALB/c mice intranasally immunized with protein F1/CTB with homologous (M23) or heterologous (NS239; Vir type 17.2) GAS isolates, produced 80 or 90 % survival (5 days post infection) respectively (Guzmán et al. 1999). However, intranasal immunization of BALB/c mice with protein F1/CTB did not protect mice against subcutaneous GAS challenge with a homologous (M23) strain (McArthur et al. 2004). McArthur et al. (2004) found that sera from mice immunized with protein F1/CTB did not result in in vitro opsonophagocytosis of homologous GAS. Protein F1, specifically the H12 region responsible for binding fibronectin, has also being investigated for use as a mucosal adjuvant as it has self-adjuvanting properties (Schulze et al. 2003; Schulze and Guzman 2003). The co-administration of H12 enhanced the serum IgG immune response against the M protein-based peptide J14 (Georgousakis et al. 2009).

3.5 Serum Opacity Factor/Streptococcal Fibronectin Binding Protein II

Serum Opacity Factor (SOF), also known as Streptococcal Fibronectin Binding Protein II (SfbII) is a cell wall anchored fibronectin, fibrinogen and fibulin-1 binding protein of GAS possessing the capacity to opacify mammalian serum (Rakonjac et al. 1995; Courtney et al. 2002, 2009). SOF has also been reported to contribute to GAS adherence to and internalization into host cells (Gillen et al. 2008; Oehmcke et al. 2004; Timmer et al. 2006). SOF is capable of generating opsonic antibodies in humans, rabbits and mice (Courtney et al. 2003). SOF is expressed by approximately 32–60 % of circulating strains depending on geographical location (Goodfellow et al. 2000; Kreikemeyer et al. 2002; Prakash and Dutta 1991; Beall et al. 2000). Swiss NIH mice parenterally immunized with SOF/CFA were protected against intraperitoneal challenge with homologous M2 GAS (Courtney et al. 2003). Likewise, BALB/c mice immunized subcutaneously and boosted intramuscularly with a SOF protein lacking the signal sequence and fibronectin-binding repeat region (designated SOFΔFn) adjuvanted with IFA were protected against lethal intraperitoneal challenge with heterologous M49 GAS (Gillen et al. 2008). The administration of SOF coupled to CTB via the intranasal route resulted in elevated mucosal IgA and IgG antibody responses (Schulze et al. 2006). However, BALB/c mice were not protected against lethal intranasal challenge with a heterologous (M100) GAS isolate in this model (Schulze et al. 2006). Given the ability of SOF to opacify serum via the disruption of high density lipoprotein, SOF, at least administered in an active form, would presumably not be safe as a vaccine intended for human use. In preliminary studies, immunization with two inactive SOF mutant proteins failed to protect against lethal heterologous (M49) intraperitoneal GAS challenge (Gillen et al. 2008).

3.6 Streptococcal Protective Antigen

Streptococcal Protective Antigen (Spa) is a GAS vaccine antigen originally identified in a M protein-negative mutant GAS strain (M18), capable of inducing bactericidal antibodies, promoting resistance to phagocytosis and conferring virulence in mice (Dale et al. 1999; McLellan et al. 2001). Spa has also been described in M36 GAS and Spa or Spa-like proteins were detected on the surface of 35.7 % of GAS serotypes tested (Ahmed et al. 2010). Intraperitoneal administration of Spa antisera to Swiss mice mediated passive protection against subsequent homologous (M18) intraperitoneal GAS challenge (Dale et al. 1999). In addition, anti-Spa antibodies were found to opsonize M3 and M28 GAS in vitro (Dale et al. 1999).

3.7 Streptococcus pyogenes Cell Envelope Proteinase/ Spy0416

Spy0416 was initially identified as a GAS surface protein in a bioinformatic and proteomic study. When co-administered with CFA/IFA, Spy0416 elicited a protective immune response against homologous (M23) mucosal infection in CD1 mice (Rodreiguez-Ortega et al. 2006). Spy0416 was also one of three protective antigens conferring protection against four different GAS serotypes in a recent high throughput GAS vaccine study (Bensi et al. 2012). Spy0416 has been identified as *Streptococcus pyogenes* Cell Envelope Proteinase (SpyCEP), a protein known to cleave and inactivate interleukin-8 (Edwards et al. 2005) and play a key role in systemic bacterial dissemination (Turner and Nohadani 2010). SpyCEP is a highly conserved, cell wall anchored protein that also exists in a secreted form (Turner et al. 2009). A recent study assessed the protective efficacy of SpyCEP (co-administered with CFA) via intramuscular immunization of BALB/c mice. Upon subsequent intramuscular and intranasal infection with homologous M81 GAS, a decrease in bacterial dissemination to both the liver and the spleen was observed in both cases (Turner et al. 2009). Furthermore, immunization of BALB/c mice with GAS SpyCEP also reduced organ dissemination of the equine pathogen *Streptococcus equi* (Turner et al. 2009). As such, SpyCEP shows potential as a cross-species streptococcal vaccine candidate.

4 Cell Membrane Associated and/or Secreted GAS Vaccine Candidates

4.1 C5a Peptidase

Streptococcal C5a peptidase is a surface expressed endopeptidase which cleaves the C5a component of complement, inactivating C5a so it can no longer act as a chemo-attractant (Koroleva et al. 2002). This delays the infiltration of phagocytes to sites of GAS infection, slowing the clearance of GAS from host mucosal surfaces (Ji et al. 1997), allowing GAS to further colonize the host (Stafslien and Cleary 2000). C5a peptidase has also been reported to mediate invasion into HEp-2 cells (Purushothaman et al. 2004). C5a peptidase is highly conserved and is ubiquitously expressed by GAS (Ji et al. 1997). In CD1 mice, following intranasal immunization with truncated, enzymatically inactive recombinant C5a peptidase (no adjuvant), significant levels of IgA salivary antibodies and serum IgG were detected (Ji et al. 1997). Anti-C5a peptidase antibodies caused a reduction in pharyngeal colonization mediated by homologous (M49) and heterologous (M1, 2, 6, 11) GAS serotypes (Ji et al. 1997; Park and Cleary 2005). The same inactive C5a peptidase protein was administered with Alum and MPL to CD1 mice via the subcutaneous route. This preparation was highly immunogenic, resulting in significantly elevated levels of specific serum IgG compared to control mice administered tetanus toxoid only (Cleary et al. 2004). Following homologous (M49) and heterologous (M1) mucosal challenge, GAS were cleared more rapidly from the oral/nasal mucosa of immunized mice (Cleary et al. 2004). Antibodies reactive against C5a peptidase have also been detected in the sera of children with GAS-associated pharyngitis, indicating that C5a peptidase is expressed in vivo during the natural course of infection and is immunogenic in humans (Shet et al. 2003, 2004). GBS also possess C5a peptidase and the *scpB* gene shares 97–98 % identity with GAS *scpA* (Cleary et al. 1992; Chmouryguina et al. 1996). Subcutaneous immunization of CD1 mice with the GBS inactive C5a peptidase co-administered with Alum and MPL resulted in enhanced clearance of GAS and GBS from the lung and nasal mucosa (Cleary et al. 2004; Cheng et al. 2002). Thus, C5a peptidase shows promise as a broad spectrum streptococcal vaccine candidate, eliciting protection against both GAS and GBS.

4.2 Streptococcal Hemoprotein Receptor

Streptococcal Hemoprotein Receptor (Shr) is a secreted GAS protein which binds hemoglobin, myoglobin, the hemoglobin-haptoglobin complex, fibronectin and laminin (Fisher et al. 2008). The *shr* gene is part of the highly conserved *sia* operon (streptococcal iron acquisition) (Bates et al. 2003), although, in one report Shr antiserum only reacted with 82.4 % of GAS isolates tested (Fisher et al. 2008), indicating Shr is not present on the surface of all serotypes. Intraperitoneal immunization of CD1 mice with Shr (CFA/IFA) resulted in a robust serum IgG response which protected mice against homologous (M1) intraperitoneal GAS challenge (Huang et al. 2011). Mucosal immunization of CD1 mice with Shr-expressing *L. lactis* resulted in specific serum IgG and mucosal IgA responses which elicited protection against homologous (M1) intraperitoneal challenge (Huang et al. 2011). In addition, rabbit Shr antiserum opsonized whole GAS in vitro and mediated passive protection against homologous (M1) and heterologous (M3) intraperitoneal GAS challenge (Huang et al. 2011).

4.3 Streptococcal Pyrogenic Exotoxin B

Streptococcal Pyrogenic Exotoxin B (SpeB) is a highly conserved ubiquitously expressed broad spectrum cysteine proteinase located on the surface of GAS which in vitro is also secreted into the culture supernatant. SpeB degrades human ECM components such as fibronectin and vitronectin (Kapur et al. 1993), the antibacterial peptide LL-37 (Schmidtchen et al. 2002), properdin (a positive complement activation regulator found in serum) (Tsao et al. 2006), complement component C3 (Hsu et al. 2008; Kuo et al. 2008), in addition to IgG, IgA, IgD, IgE and IgM (Von Pawel-Rammingen and Björck 2003). SpeB also binds laminin (Hytönen et al. 2001) and integrins (Stockbauer et al. 1999), which may facilitate the invasion of host cells by GAS (Tsai et al. 1998). The actions of SpeB promote GAS resistance to phagocytosis at the infection site (Chiang-Ni et al. 2006; Terao et al. 2008). The use of SpeB as a vaccine candidate in its native form raises safety concerns given its proteolytic activity. One group tested the ability of SpeB to elicit a protective immune response following intraperitoneal challenge of CD1 mice. Active subcutaneous immunization with SpeB (no adjuvant) using a combination of subcutaneous and intraperitoneal routes and also passive intraperitoneal immunization with SpeB antibodies enhanced the survival of mice following intraperitoneal heterologous (M3) GAS challenge (Kapur et al. 1994). More recently, a chimeric protein comprised of an inactive mutant form of SpeB and the receptor binding surface of SpeA, was intramuscularly administered (with MPL) to HLA-DQ8/human $CD4^{+}$transgenic mice (Ulrich 2008). This construct was observed to elicit a protective response following homologous intravenous GAS challenge (Ulrich 2008). The use of SpeB as a vaccine candidate, at least in the active form, is questionable given the inherent proteolytic activity of SpeB. In addition, the observation that SpeB may be associated with the induction of acute post streptococcal glomerulonephritis (Luo et al. 2007) also raises questions to the suitability of this secreted GAS protein as a vaccine candidate.

4.4 Streptococcal Secreted Esterase

Streptococcal Secreted Esterase (Sse) is a secreted GAS protein which hydrolyzes platelet-activating factor, thereby impeding neutrophil recruitment to the site of infection (Liu et al. 2012). Sse is also documented to play an important role in invasive skin infections and GAS dissemination in mice (Zhu et al. 2009). Two alternate variants of Sse were identified within 10 serotypes of GAS (Liu et al. 2007). However, it is not yet known if these two Sse variants are expressed by all serotypes of GAS, evoking questions about the capability of Sse to provide broad serotype protection. Following subcutaneous immunization of Sse utilizing Alum as an adjuvant, CD1 mice were protected against lethal subcutaneous challenge by homologous (M1) and heterologous (M3) GAS strains (Liu et al. 2007). Furthermore, passive immunization of CD1 mice with anti-Sse serum also resulted in protection against homologous (M1) GAS challenge (Liu et al. 2007). Anti-Sse antibodies were detected in convalescent serum from pharyngitis patients, however, experimentally derived anti-Sse serum was not opsonic in an in vitro assay (Liu et al. 2007).

5 Anchorless GAS Vaccine Candidates

5.1 Fibronectin-Binding Protein 54

FBP54 is an anchorless GAS surface protein known to bind both fibronectin and fibrinogen (Courtney et al. 1994, 1996). In a study of invasive disease isolates from the Northern Territory of Australia, the *fbp54* gene was present in all 75 isolates tested (Delvecchio et al. 2002). Likewise, the *fbp54* gene was present in all nine serotypes tested by Southern blot in another study (Kawabata et al. 2001). FBP54 adjuvanted with CFA or CT was used to immunize BALB/c mice via subcutaneous, oral or intranasal routes respectively. An elevated FBP54-specific IgG response ensued for all three routes, with an increase in salivary IgA following oral and intranasal immunization. Immunization via all three routes protected mice against homologous (M3) intraperitoneal challenge (Kawabata et al. 2001). Mice subcutaneously immunized with FBP54 had significantly enhanced survival following heterologous (M1 or M12) challenge with GAS. Furthermore, anti-FBP54 antisera was opsonic in in vitro assays (Kawabata et al. 2001).

5.2 Streptococcal Immunoglobulin-Binding Protein 35

Sib35 binds IgG, IgA and IgM (Kawabata et al. 2002). It is thought Sib35 is secreted into the culture supernatant and it is postulated to be surface-localized, although, it does not contain the typical Gram-positive LPXTG cell-wall anchor motif (Kawabata et al. 2002). The *sib35* gene was detected by Southern blot in all nine isolates tested in one report (Kawabata et al. 2002). Subcutaneous immunization of BALB/c, C3H/HeN, C57BL/6J and CD1 mice with Sib35 (adjuvanted with CFA/IFA) resulted in significant protection of BALB/c, C3H/HeN and C57/BL6 mice strains (but not outbred CD1 mice) against homologous (M3) challenge (Okamoto et al. 2005). In addition, polyclonal rabbit antiserum raised against Sib35 was bactericidal in vitro (Okamoto et al. 2005).

5.3 Arginine Deiminase/Streptococcal Acid Glycoprotein

Arginine Deiminase (ADI) also known as Streptococcal Acid glycoprotein (SagP) is an anchorless virulence factor of GAS capable of inhibiting the proliferation of human T cells in vitro (Degnan et al. 1998). ADI forms part of the arginine deiminase system, converting arginine to citrulline with concomitant production of ammonia. It is thought that ADI may play an indirect regulatory role in GAS, influencing the expression of GAS virulence factors and the consequent internalization of GAS by host cells (Marouni et al. 2003). ADI is expressed across multiple GAS serotypes exhibiting $\geq$99 % amino acid sequence identity amongst sequenced GAS genomes (Henningham et al. 2012). Subcutaneous immunization with ADI adjuvanted with CFA protected BALB/c mice against intraperitoneal homologous (M1) challenge. Intraperitoneal immunization of CD1 mice with ADI (adjuvanted with Alum) conferred protection against lethal heterologous (M12) challenge. Co-administration of ADI and trigger factor (TF; described below) with CFA via the intraperitoneal route protected C57BL/6J mice against intraperitoneal homologous (M1) challenge with the hypervirulent 5448AP GAS isolate (Henningham et al. 2012). Vaccine safety concerns are addressed by the observation that ADI lacks a human homolog and sera from human populations suffering repeated GAS infections and high levels of autoimmune complications do not recognize ADI (Henningham et al. 2012).

5.4 Trigger Factor

TF is an anchorless protein of GAS indentified following proteomic analysis of cell wall extracts (Henningham et al. 2012; Cole et al. 2005). TF is a ribosome-associated chaperone and peptidyl–prolyl cis–trans isomerase implicated in SpeB protease maturation in GAS (Lyon and Caparon 2003; Lyon et al. 1998). Subcutaneous immunization with TF adjuvanted with CFA protected BALB/c mice against homologous (M1) intraperitoneal GAS challenge. Intraperitoneal immunization of CD1 mice with TF co-administered with Alum conferred protection against lethal intraperitoneal heterologous (M12) challenge (Henningham et al. 2012).

6 Elucidation of Novel GAS Vaccine Candidates

Modern vaccinology utilizes techniques such as bioinformatics and data mining to readily select cell wall, membrane and surface associated proteins (Pizza et al. 2000). Reverse vaccinology utilizes other high throughput techniques including immunoblotting and protein arrays are used to identify immunogenic antigens (Bombaci et al. 2009; Meinke et al. 2005) and proteomic-based studies (Henningham et al. 2012; Rodreiguez-Ortega et al. 2006; Cole et al. 2005) are used to elucidate surface exposed antigens. In the case of GAS, one such approach used bioinformatics to identify 30 open reading frames in M1 GAS which were thought to encode putative extracellular lipoproteins (Lei et al. 2004). Antisera obtained from mice infected with whole GAS and from patients suffering GAS pharyngitis and invasive infection was used to probe recombinant forms of the lipoproteins in Western blots. Six of the recombinant proteins were observed to react with both the mouse antisera and the human sera (Lei et al. 2004), suggesting they are expressed during natural infection. Recombinant forms of 16 of the 30 proteins (adjuvanted with MPL-synthetic trehalose dicorymycolate) were used to individually immunize CD1 mice via the subcutaneous route. Following immunization, the resultant antisera raised against five of the proteins (Spy0385, Spy1245, Spy1274, Spy1390 and Spy1558) was observed to inhibit the growth of GAS in an in vitro bactericidal assay (Lei et al. 2004).

An alternative approach utilized a combination of biochemical, bioinformatic and proteomic techniques to identify surface exposed proteins of a M1 GAS isolate, SF370. Whole GAS cells were treated with trypsin or proteinase K, or a combination of both, in order to 'shave' the surface exposed proteins (Rodreiguez-Ortega et al. 2006). The resultant peptide fragments were identified using nano-liquid chromatography-mass spectrometry (MS)/MS followed by interrogation of the SF370 genomic database. A total of 68 proteins were identified, 12 contained a LPXTG cell-wall anchor motif, 11 lipoproteins, 37 transmembrane proteins and eight secreted proteins (Rodreiguez-Ortega et al. 2006). Only four of the proteins identified were predicted to be cytosolic in nature (Rodreiguez-Ortega et al. 2006). Fourteen proteins were recombinantly expressed and used to immunize CD1 mice intraperitoneally (with CFA/IFA). Only one protein, SpyCEP, elicited protection against heterologous (M23) mucosal infection with GAS (Rodreiguez-Ortega et al. 2006).

Another research group also utilized proteomics to identify SF370-surface exposed proteins (Severin et al. 2007). Severin et al. (2007) found a total of 79 proteins, 33 of which had not previously been localized to the cell surface of GAS. Amongst the proteins identified were four proteins containing a LPXTG cell-wall anchor motif, 12 lipoproteins, nine secreted proteins, 22 membrane-associated proteins, one bacteriophage-associated protein and 21 proteins traditionally known to exist in the cytoplasm (Severin et al. 2007). Sixteen of the proteins were selected to individually immunize Swiss Webster mice via the subcutaneous route with Alum and MPL. Whilst the surface localization of the 16 proteins was confirmed using the resultant polyclonal antiserum in a whole cell enzyme-linked immunoabsorbant assay (Severin et al. 2007), the ability of these putative surface antigens to protect against lethal GAS challenge is yet to be examined.

Another strategy adopted by researchers to identify novel GAS surface proteins is searching genomic surface display libraries. In short, a GAS genomic library was produced in which sheared genomic DNA (M1; SF370) was ligated into appropriate plasmids for bacterial surface display screens using human serum (Fritzer et al. 2010). Biotinylated human serum was incubated with the *Escherichia coli* cells expressing the genomic library and surface exposed antigenic fragments that bound the serum were detected using streptavidin-conjugated microbeads followed by identification with MS (Fritzer et al. 2010). This resulted in the elucidation of 95 putative antigens, some of which conferred protection against GAS challenge in mouse models. Six antigens (Spy0269, Spy0292, SpyCEP, Spy0872, Spy0895 and Spy1666) were observed to protect against heterologous (M23) intranasal challenge; whilst Spy0292, SpyCEP and Spy0872 also elicited protective immunity against homologous (M1) intravenous GAS challenge (Fritzer et al. 2010). Each of the protective antigens identified in this study were highly conserved, each gene was present in 50 isolates analysed using PCR, with greater than 97 % amino acid sequence identity amongst sequenced GAS genomes.

The observation that efficacious vaccine candidates are often conserved surface associated and/or secreted proteins inspired researchers to adopt a tripartite approach consisting of mass spectrometry based proteomics, protein array and flow cytometry to identify novel GAS vaccine antigens (Bensi et al. 2012). Using this approach, six of the 40 antigens tested in murine challenge models were identified by each of the three technologies; SpyCEP, streptolysin O (SLO), SPy0269 (putative surface exclusion protein), SPy0019 (putative secreted protein), SPy1361 (belonging to internalin A protein family) and C5a peptidase (Bensi et al. 2012). A combination of SpyCEP, SPy0269 and SLO was administered with Alum to CD1 mice via the intraperitoneal route, resulting in protection against intranasal or intraperitoneal challenge with four GAS serotypes, M1, M6, M12 and M23 (Bensi et al. 2012). Such formulations, combining highly conserved surface antigens, show promise for the global eradication of GAS infection and disease. Furthermore, high throughput multifactorial approaches such as this may be applied to other pathogenic organisms for which no vaccine exists.

7 Human Trials of GAS Vaccine Candidates

Although many proteins and peptides have been tested as GAS vaccine candidates in pre-clinical animal models, only a select number of preparations have been tested in human volunteers and clinical trials. Early studies documented the immunization of individuals and subsequently monitored the incidence of natural GAS infection in immunized and unimmunized/placebo groups. Studies conducted in the period 1923–1949 utilized heat- or UV-killed GAS or acid precipitated toxins (Bloomfield and Felty 1923; Rantz et al. 1949; Wasson and Brown 1943;

Wilson and Swift 1931; Young et al. 1946). For the most part, localized responses to these preparations were quite variable with some preparations inducing systemic reactions with chills, malaise and aching, nodule formation and swelling and soreness at the injection site (Rantz et al. 1949). Typically, these early formulations had a low efficacy, with minimal or no reduction in the frequency of GAS infection and disease in vaccinated individuals. Subsequently, whole GAS cells were no longer incorporated into preparations used in human GAS vaccine trials. Instead, later vaccine studies conducted in the 1960s focused on the cell-surface of GAS, administering either cell wall extracts or purified M protein. On the whole, the administration of M protein based preparations resulted in type-specific secondary bactericidal responses (Fox et al. 1966; Potter et al. 1962). A subsequent series of human trials conducted in the 1970s tested the protective efficacy of purified M proteins. Individuals were immunized with either M1 (Fox et al. 1973; Polly et al. 1975), M3 or M12 protein (D'Alessandri et al. 1978), followed by homologous pharyngeal challenge. In each of these studies, the majority of immunized individuals had a significant increase in type-specific M protein antibodies and overall there was a significant decrease in GAS colonization rate in immunized individuals compared to individuals given a placebo. However, one study noted that the type-specific antibodies were not bactericidal against homologous strains (D'Alessandri et al. 1978).

Following reports that purified M protein preparations were well tolerated by adults, these preparations were subsequently tested in children. The administration of partially purified M3 protein to healthy children who were siblings of ARF patients produced controversial results. In 95 % of the immunized children, the M3-based vaccine elicited a type-specific bactericidal response (Massell et al. 1968). During and following vaccination there were 18 reported natural GAS infections (heterologous serotypes), evident due to a significant elevation in anti-SLO titre (Massell et al. 1969). Two of these GAS infections were followed by definite cases of ARF, and one by a probable case of ARF (Massell et al. 1969). The two definite ARF cases comprise 11.1 % of the GAS infections recorded in the vaccinated children, which is in stark contrast to the incidence of ARF amongst 447 natural GAS infections in unvaccinated siblings (1.1 %); or the incidence reported in patients with a significant rise in anti-SLO titre in a preceding study (0.9 %) (Siegel et al. 1961). There are a number of factors (other than the administration of M3 protein) that may have contributed to the onset of ARF in these children. Such factors include the number and frequency of GAS infections the children had prior to the vaccination regime and the health of each child. Nonetheless, the unprecedented high frequency of ARF coinciding with the immunization of children with M protein-based GAS vaccines was of great concern and consequently there were very few human GAS vaccine trials from the 1980s to the early 2000s.

The capacity of certain M protein epitopes to elicit an auto-immune response has been well documented elsewhere (Kaplan and Meyeserian 1962; Dale and Beachey 1985; Cunningham 2003). The aim of developing a GAS vaccine which does not generate auto-antibodies motivated researchers to move from whole M protein based vaccines to preparations containing M protein fragments which do not react with host tissue. In recent times, two M protein based multivalent vaccine preparations have reached phase I human clinical trials. In the initial study, healthy adult volunteers ($n = 28$; 18–50 years old) were administered either 50, 100 or 200 μg of hexavalent preparation formulated with Alum via intramuscular immunization. The immunizations were spaced 0, 28 and 56 or 112 days apart in a single-center, open-label trial (Kotloff et al. 2004). The preparation was well tolerated in all but one volunteer and reactions to the vaccine were regarded as mild. In this preliminary study, there was no evidence of antibodies cross-reactive with human tissue or immunological complications. For the highest dose tested (200 μg), there was a significant increase in serum IgG antibody titre to all six vaccine antigens. The resultant sera killed 5/6 M types (not M6) in an opsonophagocytosis assay (Kotloff et al. 2004). Given the positive results following the administration of the hexavalent preparation, in an attempt to achieve greater serotype coverage, a second clinical trial was undertaken administering Streptavax, a 26-valent preparation (McNeil et al. 2005). Strain selection was based on surveillance programs reporting the predominant M types associated with invasive diseases and pediatric pharyngitis in the US. M types historically associated with ARF were also included, in addition to a fragment derived from Spa. A 400 μg dose of the vaccine was administered intramuscularly with Alum on days 0, 30 and 120 (McNeil et al. 2005). There was no evidence of rheumatogenicity or nephritogenicity, nor induction of antibodies cross-reactive with human tissue. Reported side effects to the vaccine were all self-limited. For 26/27 peptides, a four-fold or greater increase in antibody titre above baseline was recorded (McNeil et al. 2005). There was a significant reduction in bacterial counts in opsonization assays for all included M serotypes (McNeil et al. 2005). In January 2005, a larger phase II clinical trial commenced with 70 individuals administered Streptavax and 20 individuals receiving a hepatitis A vaccine (control group) (McMillan 2006). An average 11.3-fold increase in serum antibody titre was measured for each of the 27 peptides. The US Food and Drug Administration requested more safety information from the adult trials before the commencement of pediatric studies, are yet to commence (McMillan 2006). There have been no further reports regarding the safety of these promising multivalent N-terminal M protein based preparations.

8 Conclusion

Even though the GAS surface proteins reviewed herein elicit protective immunity in animal models and show promise as effective GAS vaccines, there is still no commercial human GAS vaccine available. Researchers should carefully consider a number of factors when developing GAS vaccines: (i) the conservation and serotype coverage of antigens, (ii) differences in the geographical distribution of serotypes, (iii) the possibility that antigens may contain auto-immune epitopes, (iv) the selection of human approved adjuvants, and v) the design and validity of experimental animal models. Coverage of all GAS serotypes and global elimination of GAS disease can only be achieved by preparations which optimally address each of these factors.

Acknowledgments The authors would like to acknowledge the National Health and Medical Research Council for funding.

References

Abdel-Aal AB, Batzloff MR, Fujita Y, Barozzi N, Faria A, Simerska P, Moyle PM, Good MF, Toth I (2008) Structure-activity relationship of a series of synthetic lipopeptide self-adjuvanting group a streptococcal vaccine candidates. J Med Chem 51:167–172

Abdel-Aal A-BM, Zaman M, Fujita Y, Batzloff MR, Good MF, Toth I (2010) Design of three-component vaccines against group A streptococcal infections: importance of spatial arrangement of vaccine components. J Med Chem 53:8041–8046

Abdissa A, Asrat D, Kronvall G, Shittu B, Achiko D, Zeidan M, Yamuah LK, Aseffa A (2006) High diversity of group A streptococcal *emm* types among healthy schoolchildren in Ethiopia. Clin Infect Dis 42:1362–1367

Ahmed EA, Penfound TA, Brewer SC, Tennant PA, Chiang EY, Dale JB (2010) Streptococcal protective antigens (Spa): a new family of type-specific proteins of group A streptococci. Eur J Clin Microbiol Infect Dis 29:51–57

Ambrosch F, Wiedermann G, Jonas S, Althaus B, Finkel B, Gluck R, Herzog C (1997) Immunogenicity and protectivity of a new liposomal hepatitis A vaccine. Vaccine 15:1209–1213

Baldassarri L, Creti R, Recchia S, Imperi M, Facinelli B, Giovanetti E, Pataracchia M, Alfarone G, Orefici G (2006) Therapeutic failures of antibiotics used to treat macrolide-susceptible *Streptococcus pyogenes* infections may be due to biofilm formation. J Clin Microbiol 44:2721–2727

Basma H, Norrby-Teglund A, McGeer A, Low DE, El-Ahmedy O, Dale JB, Schwartz B, Kotb M (1998) Opsonic antibodies to the surface M protein of group A streptococci in pooled normal immunoglobulins (IVIG): potential impact on the clinical efficacy of IVIG therapy for severe invasive group A streptococcal infections. Infect Immun 66:2279–2283

Bates CS, Montanez GE, Woods CR, Vincent RM, Eichenbaum Z (2003) Identification and characterization of a *Streptococcus pyogenes* operon involved in binding of hemoproteins and acquisition of iron. Infect Immun 71:1042–1055

Batzloff MR, Hayman WA, Davies MR, Zeng M, Pruksakorn S, Brandt ER, Good MF (2003) Protection against group A streptococcus by immunisation with J8-Diphtheria toxoid: contribution of J8- and Diphtheria toxoid-specific antibodies to protection. J Infect Dis 187:1598–1608

Batzloff MR, Yan H, Davies MR, Hartas J, Lowell GH, White G, Burt DS, Leanderson T, Good MF (2005) Toward the development of an anti-disease, transmission-blocking intranasal vaccine for group A streptococcus. J Infect Dis 192:1450–1455

Batzloff MR, Hartas J, Zeng W, Jackson DC, Good MF (2006) Intranasal vaccination with a lipopeptide containing a conformationally constrained conserved minimal peptide, a universal T cell epitope, and a self-adjuvanting lipid protects mice from group a streptococcus challenge and reduces throat colonization. J Infect Dis 194:325–330

Beachey EH, Seyer JM, Dale JB, Simpson WA, Kang AH (1981) Type-specific protective immunity evoked by synthetic peptide of *Streptococcus pyogenes* M protein. Nature 292: 457–459

Beachey EH, Gras-Masse H, Tarter A, Jolivet M, Audibert F, Chedid L, Seyer JM (1986) Opsonic antibodies evoked by hybrid peptide copies of types 5 and 24 streptococcal M proteins synthesized in tandem. J Exp Med 163:1451–1458

Beall B, Gherardi G, Lovgren M, Facklam RR, Forwick BA, Tyrrell GJ (2000) *emm* and *sof* gene sequence variation in relation to serological typing of opacity-factor-positive group A streptococci. Microbiology 146:1195–1209

Bensi G, Mora M, Tuscano G, Biagini M, Chiarot E, Bombaci M, Capo S, Falugi F, Manetti AG, Donato P, Swennen E, Gallotta M, Garibaldi M, Pinto V, Chiappini N, Musser JM, Janulczyk R, Mariani M, Scarselli M, Telford JL, Grifantini R, Norais N, Margarit I, Grandi G (2012) Multi high-throughput approach for highly selective identification of vaccine candidates: the Group A Streptococcus case. Mol Cell Proteomics 11:M111.015693

Berge A, Sjöbring U (1993) PAM, a novel plasminogen-binding protein from *Streptococcus pyogenes*. J Biol Chem 268:25417–25424

Bessen D, Fischetti VA (1988) Influence of intranasal immunization with synthetic peptides corresponding to conserved epitopes of M protein on mucosal colonization by group A streptococci. Infect Immun 56:2666–2672

Bessen D, Fischetti VA (1990) Synthetic peptide vaccine against mucosal colonization by group A streptococci. I. Protection against a heterologous M serotype with shared C repeat region epitopes. J Immunol 145:1251–1256

Bloomfield AL, Felty AR (1923) Prophylactic vaccination against acute tonsillitis. Johns Hopkins Hosp Bull 34:251–253

Bombaci M, Grifantini R, Mora M, Reguzzi V, Petracca R, Meoni E, Balloni S, Zingaretti C, Falugi F, Manetti AGO, Margarit I, Musser JM, Cardona F, Orefici G, Grandi G, Bensi G (2009) Protein array profiling of tic patient sera reveals a broad range and enhanced immune response against Group A Streptococcus antigens. PLoS ONE 4:e6332

Brandt ER, Hayman WA, Currie B, Carapetis JR, Wood Y, Jackson DC, Cooper J, Melrose WD, Saul AJ, Good MF (1996) Opsonic human antibodies from an endemic population specific for a conserved epitope on the M protein of group A streptococci. Immunology 89:331–337

Brandt ER, Hayman WA, Currie B, Pruksakorn S, Good MF (1997) Human antibodies to the conserved region of the M protein: opsonization of heterologous strains of group A streptococci. Vaccine 15:1805–1812

Brandt ER, Hayman WA, Currie B, Carapetis J, Jackson DC, Do KA, Good MF (1999) Functional analysis of IgA antibodies specific for a conserved epitope within the M protein of group A streptococci from Australian Aboriginal endemic communities. Int Immunol 11: 569–576

Brandt ER, Sriprakash KS, Hobb RI, Hayman WA, Zeng W, Batzloff MR, Jackson DC, Good MF (2000) New multi-determinant strategy for a group A streptococcal vaccine designed for the Australian Aboriginal population. Nat Med 6:455–459

Bronze MS, Courtney HS, Dale JB (1992) Epitopes of group A streptococcal M protein that evoke cross-protective local immune responses. J Immunol 148:888–893

Brook I (2007) Overcoming penicillin failures in the treatment of group A streptococcal pharyngotonsillitis. Int J Pediatr Otorhinolaryngol 71:1501–1508

Carapetis JR, Steer AC, Mulholland EK, Weber M (2005) The global burden of group A streptococcal diseases. Lancet Infect Dis 5:685–694

Carlsson F, Sandin C, Lindahl G (2005) Human fibrinogen bound to *Streptococcus pyogenes* M protein inhibits complement deposition via the classical pathway. Mol Microbiol 56:28–39

Cheng Q, Debol S, Lam H, Eby R, Edwards L, Matsuka Y, Olmsted SB, Cleary PP (2002) Immunization with C5a peptidase or peptidase-type III polysaccharide conjugate vaccines enhances clearance of group B Streptococci from lungs of infected mice. Infect Immun 70:6409–6415

Chiang-Ni C, Wang CH, Tsai PJ, Chuang WJ, Lin YS, Lin MT, Liu CC, Wu JJ (2006) Streptococcal pyrogenic exotoxin B causes mitochondria damage to polymorphonuclear cells preventing phagocytosis of group A streptococcus. Med Microbiol Immun 195:55–63

Chmouryguina I, Suvorov A, Ferrieri P, Cleary PP (1996) Conservation of the C5a peptidase genes in group A and B streptococci. Infect Immun 64:2387–2390

Cleary PP, Handley J, Suvorov AN, Podbielski A, Ferrieri P (1992) Similarity between the group B and A streptococcal C5a peptidase genes. Infect Immun 60:4239–4244

Cleary PP, Matsuka YV, Hunyh T, Lam H, Olmsted SB (2004) Immunisation with C5a peptidase from either group A or B streptococci enhances clearance of group A streptococci from intranasally infected mice. Vaccine 22:4332–4341

Cole JN, Ramirez RD, Currie BJ, Cordwell SJ, Djordjevic SP, Walker MJ (2005) Surface analysis and immune reactivity of major cell-wall associated proteins of group A streptococcus. Infect Immun 73:3137–3146

Cole JN, Barnett TC, Nizet V, Walker MJ (2011) Molecular insight into invasive group A streptococcal disease. Nat Rev Microbiol 9:724–736

Courtney HS, Li Y, Dale JB, Hasty DL (1994) Cloning, sequencing, and expression of a fibronectin/fibrinogen-binding protein from group A streptococci. Infect Immun 62:3937–3946

Courtney HS, Dale JB, Hasty DL (1996) Differential effects of the streptococcal fibronectin-binding protein, FPB54, on adhesion of group A streptococci to human buccal cells and HEp-2 tissue culture cells. Infect Immun 64:2415–2419

Courtney HS, Dale JB, Hasty DL (2002) Mapping the fibrinogen-binding domain of serum opacity factor of group A streptococci. Curr Microbiol 44:236–240

Courtney HS, Hasty DL, Dale JB (2003) Serum opacity factor (SOF) of *Streptococcus pyogenes* evokes antibodies that opsonize homologous and heterologous SOF-positive serotypes of group A streptococci. Infect Immun 71:5097–5103

Courtney HS, Li Y, Twal WO, Argraves WS (2009) Serum opacity factor is a streptococcal receptor for the extracellular matrix protein fibulin-1. J Biol Chem 284:12966–12971

Cue D, Lam H, Cleary PP (2001) Genetic dissection of the *Streptococcus pyogenes* M1 protein: regions involved in fibronectin binding and intracellular invasion. Microb Pathog 31:231–242

Cunningham MW (2003) Autoimmunity and molecular mimicry in the pathogenesis of post-streptococcal heart disease. Front Biosci 8:s533–543

Dale JB (1999) Multivalent group A streptococcal vaccine designed to optomize the immunogenicity of six tandem M protein fragments. Vaccine 17:193–200

Dale JB, Beachey EH (1985) Epitopes of streptococcal M proteins shared with cardiac myosin. J Exp Med 162:583–591

Dale JB, Seyer JM, Beachey EH (1983) Type-specific immunogenicity of a chemically synthesized peptide fragment of type 5 streptococcal M protein. J Exp Med 158:1727–1732

Dale JB, Chiang EY, Lederer JW (1993) Recombinant tetravalent group A streptococcal M protein vaccine. J Immunol 151:2188–2194

Dale JB, Simmons M, Chiang EC, Chiang EY (1996) Recombinant, octavalent group A streptococcal M protein vaccine. Vaccine 14:944–948

Dale JB, Chiang EY, Liu S, Courtney HS, Hasty DL (1999) New protective antigen of group A streptococci. J Clin Invest 103:1261–1268

Dale JB, Penfound TA, Chiang EY, Walton WJ (2011) New 30-valent M protein-based vaccine evokes cross-opsonic antibodies against non-vaccine serotypes of group A streptococci. Vaccine 29:8175–8178

D'Alessandri R, Plotkin G, Kluge RM, Wittner MK, Fox EN, Dorfman A, Waldman RH (1978) Protective studies with group A streptococcal M protein vaccine. III. Challenge of volunteers after systemic or intranasal immunization with Type 3 or Type 12 group A *Streptococcus*. J Infect Dis 138:712–718

Degnan BA, Palmer JM, Robson T, Jones CED, Fischer M, Glanville M, Mellor GD, Diamond AG, Kehoe MA, Goodacre JA (1998) Inhibition of human peripheral blood mononuclear cell proliferation by *Streptococcus pyogenes* cell extract is associated with arginine deiminase activity. Infect Immun 66:3050–3058

Delvecchio A, Currie BJ, McArthur JD, Walker MJ, Sriprakash KS (2002) *Streptococcus pyogenes prtFII*, but not *sfbI*, *sfbII*, or *fbp54*, is represented more frequently among invasive-disease isolates of tropical Australia. Epidemiol Infect 128:391–396

Dinkla K, Rohde M, Jansen WTM, Kaplan EL, Chhatwal GS, Talay SR (2003) Rheumatic fever-associated *Streptococcus pyogenes* isolates aggregate collagen. J Clin Invest 11:1905–1912

Eberhard TH, Sledjeski DD, Boyle MD (2001) Mouse skin passage of a *Streptococcus pyogenes* Tn917 mutant of sagA/pel restores virulence, beta-hemolysis and sagA/pel expression without altering the position or sequence of the transposon. BMC Microbiol 1:33

Edwards RJ, Taylor GW, Ferguson M, Murray S, Rendell N, Wrigley A, Bai Z, Boyle J, Finney SJ, Jones A, Russell HH, Turner C, Cohen J, Faulkner L, Sriskandan S (2005) Specific C-terminal cleavage and inactivation of interleukin-8 by invasive disease isolates of *Streptococcus pyogenes*. J Infect Dis 192:783–790

Fischetti V (1989) Streptococcal M protein: molecular design and biological behaviour. Clin Microbiol Rev 2:285–314

Fischetti VA (1991) Streptococcal M protein. Sci Am 264:32–39

Fischetti VA, Hodges WM, Hruby DE (1989) Protection against streptococcal pharyngeal colonization with a vaccinia: M protein recombinant. Science 244:1487–1490

Fisher M, Huang YS, Li X, McIver KS, Toukoki C, Eichenbaum Z (2008) Shr is a broad-spectrum surface receptor that contributes to adherence and virulence in group A streptococcus. Infect Immun 76:5006–5015

Fontaine MC, Lee JJ, Kehoe MA (2003) Combined contributions of streptolysin O and streptolysin S to virulence of serotype M5 *Streptococcus pyogenes* strain Manfredo. Infect Immun 71:3857–3865

Fox EN, Wittner MK, Dorfman A (1966) Antigenicity of the M proteins of group A hemolytic streptococci. 3. Antibody responses and cutaneous hypersensitivity in humans. J Exp Med 124:1135–1151

Fox EN, Waldman RH, Wittner MK, Mauceri AA, Dorfman A (1973) Protective study with a group A streptococcal M protein vaccine. Infectivity challenge of human volunteers. J Clin Invest 52:1885–1892

Fritzer A, Senn BM, Minh DB, Hanner M, Gelbmann D, Noiges B, Henics T, Schulze K, Guzman CA, Goodacre J, von Gabain A, Nagy E, Meinke AL (2010) Novel conserved group A streptococcal proteins identified by the antigenome technology as vaccine candidates for a non-M protein-based vaccine. Infect Immun 78:4051–4067

Fujita Y, Abdel-Aal AB, Wimmer N, Batzloff MR, Good MF, Toth I (2008) Synthesis and immunological evaluation of self-adjuvanting glycolipopeptide vaccine candidates. Bioorg Med Chem 16:8907–8913

Georgousakis MM, Hofmann A, Batzloff MR, McMillan DJ, Sriprakash KS (2009) Structural optimisation of a conformational epitope improves antigenicity when expressed as a recombinant fusion protein. Vaccine 27:6799–6806

Gillen CM, Courtney HS, Schulze K, Rohde M, Wilson MR, Timmer AM, Guzman CA, Nizet V, Chhatwal GS, Walker MJ (2008) Opacity factor activity and epithelial cell binding by the serum opacity factor protein of *Streptococcus pyogenes* are functionally discrete. J Biol Chem 283:6359–6366

Gladysheva IP, Turner RB, Sazonova IY, Liu L, Reed GL (2003) Coevolutionary patterns in plasminogen activation. Proc Natl Acad Sci USA 100:9168–9172

Goldmann O, Chhatwal GS, Medina E (2003) Immune mechanisms underlying host susceptibility to infection with group A streptococci. J Infect Dis 187:854–861

Goodfellow AM, Hibble M, Talay SR, Kreikemeyer B, Currie BJ, Sriprakash KS, Chhatwal GS (2000) Distribution and antigenicity of fibronectin binding proteins (SfbI and SfbII) of *Streptococcus pyogenes* clinical isolates from the Northern Territory, Australia. J Clin Microbiol 38:389–392

Gryllos I, Cywes C, Shearer MH, Cary M, Kennedy RC, Wessels MR (2001) Regulation of capsule gene expression by group A Streptococcus during pharyngeal colonization and invasive infection. Mol Microbiol 42:61–74

Gryllos I, Grifantini R, Colaprico A, Cary ME, Hakansson A, Carey DW, Suarez-Chavez M, Kalish LA, Mitchell PD, White GL, Wessels MR (2008) PerR confers phagocytic killing resistance and allows pharyngeal colonization by group A Streptococcus. PLoS Pathog 4:e1000145

Guilherme L, Faé KC, Higa F, Chaves L, Oshiro SE, Freschi de Barros S, Puschel C, Juliano MA, Tanaka AC, Spina G, Kalil J (2006) Towards a vaccine against rheumatic fever. Clin Dev Immunol 13:125–132

Guilherme L, Postol E, Freschi de Barros S, Higa F, Alencar R, Lastre M, Zayas C, Puschel CR, Silva WR, Sa-Rocha LC, Sa-Rocha VM, Pérez O, Kalil J (2009) A vaccine against *S. pyogenes*: design and experimental immune response. Methods 49:316–321

Guzmán CA, Talay SR, Molinari G, Medina E, Chhatwal GS (1999) Protective immune response against *Streptococcus pyogenes* in mice after intranasal vaccination with fibronectin-binding protein SfbI. J Infect Dis 179:901–906

Hall MA, Stroop SD, Hu MC, Walls MA, Reddish MA, Burt DS, Lowell GH, Dale JB (2004) Intranasal immunization with multivalent group A streptococcal vaccines protects mice against intranasal challenge infections. Infect Immun 72:2507–2512

Hanski E, Caparon MG (1992) Protein F, a fibronectin-binding protein is an adhesin of the group A streptococcus *Streptococcus pyogenes*. Proc Natl Acad Sci USA 89:6172–6176

Hayman WA, Brandt ER, Relf WA, Cooper J, Saul A, Good MF (1997) Mapping the minimal murine T cell and B cell epitopes within a peptide vaccine candidate from the conserved region of the M protein of group A streptococcus. Int Immunol 9:1723–1733

Hayman WA, Toth I, Flinn N, Scanlin M, Good MF (2002) Enhancing the immunogenicity and modulating the fine epitope recognition of antisera to a helical group A streptococcal peptide vaccine candidate from the M protein using lipid-core peptide technology. Immunol Cell Biol 80:178–187

Henningham A, Chiarot E, Gillen CM, Cole JN, Rohde M, Fulde M, Ramachandran V, Cork AJ, Hartas J, Magor G, Djordjevic SP, Cordwell SJ, Kobe B, Sriprakash KS, Nizet V, Chhatwal GS, Margarit IYR, Batzloff MR, Walker MJ (2012) Conserved anchorless surface proteins as group A streptococcal vaccine candidates. J Mol Med. 90:1197–1207. doi:10.1007/s00109-012-0897-9

Hsu JF, Chuang WJ, Shiesh SC, Lin YS, Liu CC, Wang CC, Fu TF, Tsai JH, Tsai WL, Huang YJ, Hsieh YH, Wu JJ, Lin MT, Huang W (2008) Streptococcal pyrogenic exotoxin B cleaves human S-adenosylhomocysteine hydrolase and induces hypermethioninemia. J Infect Dis 198:367–374

Hu MC, Walls MA, Stroop SD, Reddish MA, Beall B, Dale JB (2002) Immunogenicity of a 26-valent group A streptococcal vaccine. Infect Immun 70:2171–2177

Huang YS, Fisher M, Nasrawi Z, Eichenbaum Z (2011) Defense from the group a streptococcus by active and passive vaccination with the streptococcal hemoprotein receptor. J Infect Dis 203:1595–1601

Hyland KA, Wang B, Cleary PP (2007) Protein F1 and *Streptococcus pyogenes* resistance to phagocytosis. Infect Immun 75:3188–3191

Hytönen J, Haataja S, Gerlach D, Podbielski A, Finne J (2001) The SpeB virulence factor of *Streptococcus pyogenes*, a multifunctional secreted and cell surface molecule with strepadhesin, laminin-binding and cysteine protease activity. Mol Microbiol 39:512–519

Ikebe T, Hirasawa K, Suzuki R, Ohya H, Isobe J, Tanaka D, Katsukawa C, Kawahara R, Tomita M, Ogata K, Endoh M, Okuno R, Tada Y, Okabe N, Watanabe H, Japan WGfB-hSi (2007) Distribution of *emm* genotypes among group A streptococcus isolates from patients with severe invasive streptococcal infections in Japan, 2001–2005. Epidemiol Infect 135:1227–1229

Jadoun J, Ozeri V, Burstein E, Skutelsky E, Hanski E, Sela S (1998) Protein Fl is required for efficient entry of *Streptococcus pyogenes* into epithelial cells. J Infect Dis 178:147–158

Ji Y, Carlson B, Kondagunta A, Cleary PP (1997) Intranasal immunisation with C5a peptidase prevents nasopharyngeal colonisation of mice by the group A *Streptococcus*. Infect Immun 65:2080–2087

Jones KF, Fischetti VA (1988) The importance of the location of antibody binding on the M6 protein for opsonisation and phagocytosis of group A M6 streptococci. J Exp Med 167:1114–1123

Kaplan MH, Meyeserian M (1962) An immunological cross-reaction between group A streptococcal cells and human heart tissue. Lancet 1:706–710

Kapur V, Topouzis S, Majesky MW, Li LL, Hamrick MR, Hamill RJ, Patti JM, Musser JM (1993) A conserved *Streptococcus pyogenes* extracellular cysteine protease cleaves human fibronectin and degrades vitronectin. Microb Pathog 15:327–346

Kapur V, Maffei JT, Greer RS, Li LL, Adams GJ, Musser JM (1994) Vaccination with streptococcal extracellular cysteine protease (interleukin-1 beta convertase) protects mice against challenge with heterologous group A streptococci. Microb Pathog 16:443–450

Katerov V, Andreev A, Schalén C, Totolian AA (1998) Protein F, a fibronectin-binding protein of *Streptococcus pyogenes*, also binds human fibrinogen: isolation of the protein and mapping of the binding region. Microbiology 144:119–126

Kaul R, McGeer A, Norrby-Teglund A, Kotb M, Schwartz B, O'Rourke K, Talbot J, Low DE (1999) Intravenous immunoglobulin therapy for streptococcal toxic shock syndrome—a comparative observational study. The Canadian Streptococcal Study Group. Clin Infect Dis 28:800–807

Kawabata S, Kunitomo E, Terao Y, Nakagawa I, Kikuchi H, Totsuka K-I, Hamada S (2001) Systemic and mucosal immunisations with fibronectin-binding protein FBP54 induce protective immune response against *Streptococcus pyogenes* challenge in mice. Infect Immun 69:924–930

Kawabata S, Tamura Y, Murakami J, Terao Y, Nakagawa I, Hamada S (2002) A novel, anchorless streptococcal surface protein that binds to human immunoglobulins. Biochem Biophys Res Commun 296:1329–1333

Koroleva IV, Efstratiou A, Suvorov AN (2002) Structural heterogeneity of the streptococcal C5a peptidase gene in *Streptococcus pyogenes*. J Bacteriol 184:6384–6386

Kotloff KL, Corretti M, Palmer K, Campbell JD, Reddish MA, Hu MC, Wasserman SS, Dale JB (2004) Safety and immunogenicity of a recombinant multivalent group A streptococcal vaccine in healthy adults: phase I trial. JAMA 292:709–715

Krause RM, Rammelkamp CH Jr (1962) Studies of the carrier state following infection with group A streptococci. II. Infectivity of streptococci isolated during acute pharyngitis and during the carrier state. J Clin Invest 41:575–578

Kreikemeyer B, Beckert S, Braun-Kiewnick A, Podbielski A (2002) Group A streptococcal RofA-type global regulators exhibit a strain-specific genomic presence and regulation pattern. Microbiology 148:1501–1511

Kuo CF, Lin YS, Chuang WJ, Wu JJ, Tsao N (2008) Degradation of complement 3 by streptococcal pyrogenic exotoxin B inhibits complement activation and neutrophil opsonophagocytosis. Infect Immun 76:1163–1169

Kurupati P, Turner CE, Tziona I, Lawrenson RA, Alam FM, Nohadani M, Stamp GW, Zinkernagel AS, Nizet V, Edwards RJ, Sriskandan S (2010) Chemokine-cleaving *Streptococcus pyogenes* protease SpyCEP is necessary and sufficient for bacterial dissemination within soft tissues and the respiratory tract. Mol Microbiol 76:1197–1387

Lancefield RC (1957) Differentiation of group A streptococci with a common R antigen into three serological types, with special reference to the bactericidal test. J Exp Med 106:525–544

Lancefield RC, Perlmann GE (1952) Preparation and properties of a protein (R antigen) occurring in streptococci of group A, type 28 and in certain streptococci of other serological groups. J Exp Med 96:83–97

Lei B, Liu M-Y, Chesney GL, Musser JM (2004) Identification of new candidate vaccine antigens made by *Streptococcus pyogenes*: purification and characterisation of 16 putative extracellular lipoproteins. J Infect Dis 189:79–89

Lindblad EB (2000) Freund's adjuvants. In: O'Hagan DT (ed) Vaccine adjuvants: preparation methods and research protocols, 1st edn. Human Press, Totowa

Liu M, Zhu H, Zhang J, Lei B (2007) Active and passive immunizations with the streptococcal esterase Sse protect mice against subcutaneous infection with group A streptococci. Infect Immun 75:3651–3657

Liu M, Zhu H, Li J, Garcia CC, Feng W, Kirpotina LN, Hilmer J, Tavares LP, Layton AW, Quinn MT, Bothner B, Teixeira MM, Lei B (2012) Group A Streptococcus secreted esterase hydrolyzes platelet-activating factor to impede neutrophil recruitment and facilitate innate immune evasion. PLoS Pathog 8:e1002624

Luo YH, Kuo CF, Huang KJ, Wu JJ, Lei HY, Lin MT, Chuang WJ, Liu CC, Lin CF, Lin YS (2007) Streptococcal pyrogenic exotoxin B antibodies in a mouse model of glomerulonephritis. Kidney Int 72:716–724

Lymbury RS, Olive C, Powell KA, Good MF, Hirst RG, LaBrooy JT, Ketheesan N (2003) Induction of autoimmune valvulitis in Lewis rats following immunization with peptides from the conserved region of group A streptococcal M protein. J Autoimmun 20:211–217

Lyon WR, Caparon MG (2003) Trigger Factor-mediated prolyl isomerization influences maturation of the *Streptococcus pyogenes* cysteine protease. J Bacteriol 185:1667–3661

Lyon WR, Gibson CM, Caparon MG (1998) A role for trigger factor and an Rgg-like regulator in the transcription, secretion and processing of the cysteine proteinase of *Streptococcus pyogenes*. EMBO J 17:6263–6275

Ma CQ, Li CH, Wang XR, Zeng RH, Yin XL, Feng HD, Wei L (2009) Similar ability of FbaA with M protein to elicit protective immunity against group A *Streptococcus* challenge in mice. Cell Mol Immunol 6:73–77

Mannam P, Jones KF, Geller BL (2004) Mucosal vaccine made from live, recombinant *Lactococcus lactis* protects mice against pharyngeal infection with *Streptococcus pyogenes*. Infect Immun 72:3444–3450

Marouni MJ, Ziomek E, Sela S (2003) Influence of group A streptococcal acid glycoprotein on expression of major virulence factors and internalisation by epithelial cells. Microb Pathog 35:63–72

Massell BF, Michael JG, Amezcua J, Siner M (1968) Secondary and apparent primary antibody responses after group A streptococcal vaccination of 21 children. Appl Microbiol 16:509–518

Massell BF, Honikman LH, Amezcua J (1969) Rheumatic fever following streptococcal vaccination. Report of three cases. JAMA 207:1115–1119

Matsui H, Sekiya Y, Nakamura M, Murayama SY, Yoshida H, Takahashi T, Imanishi K, Tsuchimoto K, Uchiyama T, Sunakawa K, Ubukata K (2009) CD46 transgenic mouse model of necrotizing fasciitis caused by *Streptococcus pyogenes* infection. Infect Immun 77:4806–4814

McArthur J, Medina E, Mueller A, Chin J, Currie BJ, Sriprakash KS, Talay SR, Chhatwal GS, Walker MJ (2004) Intranasal vaccination with streptococcal fibronectin binding protein Sfb1 fails to prevent growth and dissemination of *Streptococcus pyogenes* in a murine skin infection model. Infect Immun 72:7342–7345

McLellan DG, Chiang EY, Courtney HS, Hasty DL, Wei SC, Hu MC, Walls MA, Bloom JJ, Dale JB (2001) Spa contributes to the virulence of type 18 group A streptococci. Infect Immun 69:2943–2949

McMillan D (2006) StreptAvax (ID Biomedical). Curr Opin Investig Drugs 7:186–190

McMillan DJ, Davies MR, Good MF, Sriprakash KS (2004) Immune response to superoxide dismutase in group A streptococcal infection. FEMS Immunol Med Microbiol 40:249–256

McNeil SA, Halperin SA, Langley JM, Smith B, Warren A, Sharratt GP, Baxendale DM, Reddish MA, Hu MC, Stroop SD, Linden J, Fries LF, Vink PE, Dale JB (2005) Safety and immunogenicity of 26-valent group a streptococcus vaccine in healthy adult volunteers. Clin Infect Dis 41:1114–1122

Medaglini D, Pozzi G, King TP, Fischetti VA (1995) Mucosal and systemic immune responses to a recombinant protein expressed on the surface of the oral commensal bacterium *Streptococcus gordonii* after oral colonization. Proc Natl Acad Sci USA 92:6868–6872

Medina E, Goldmann O, Rohde M, Lengeling A, Chhatwal GS (2001) Genetic control of susceptibility to group A streptococcal infection in mice. J Infect Dis 184:846–852

Meinke A, Henics T, Hanner M, Minh DB, Nagy E (2005) Antigenome technology: a novel approach for the selection of bacterial vaccine candidate antigens. Vaccine 23:2035–2041
Michos AG, Bakoula CG, Braoudaki M, Koutouzi FI, Roma ES, Pangalis A, Nikolopoulou G, Kirikou E, Syriopoulou VP (2009) Macrolide resistance in *Streptococcus pyogenes*: prevalence, resistance determinants, and *emm* types. Diagn Microbiol Infect Dis 64:295–299
Molinari G, Talay SR, Valentin-Weigand P, Rohde M, Chhatwal GS (1997) The fibronectin-binding protein of *Streptococcus pyogenes*, SfbI, is involved in the internalisation of group A streptococci by epithelial cells. Infect Immun 65:1357–1363
Molinari G, Chhatwal GS, Sela S, Jadoun J (1999) Role played by the fibronectin-binding protein SfbI (Protein F1) of *Streptococcus pyogenes* in bacterial internalisation by epithelial cells. J Infect Dis 179:1049–1050
Moyle PM, Olive C, Good MF, Toth I (2006a) Method for the synthesis of highly pure vaccines using the lipid core peptide system. J Pept Sci 12:800–807
Moyle PM, Olive C, Ho MF, Burgess M, Karpati L, Good MF, Toth I (2006b) Method for the synthesis of multi-epitopic *Streptococcus pyogenes* lipopeptide vaccines using native chemical ligation. J Org Chem 71:6846–6850
Moyle PM, Olive C, Ho MF, Good MF, Toth I (2006c) Synthesis of a highly pure lipid core peptide based self-adjuvanting triepitopic group A streptococcal vaccine, and subsequent immunological evaluation. J Med Chem 49:6364–6370
Neely MN, Pfeifer JD, Caparon M (2002) Streptococcus-zebrafish model of bacterial pathogenesis. Infect Immun 70:3904–3914
Nir-Paz R, Korenman Z, Ron M, Michael-Gayego A, Cohen-Poradosu R, Valinsky L, Beall B, Moses AE (2010) *Streptococcus pyogenes emm* and T types within a decade, 1996–2005: implications for epidemiology and future vaccines. Epidemiol Infect 138:53–60
Oehmcke S, Podbielski A, Kreikemeyer B (2004) Function of the fibronectin-binding serum opacity factor of *Streptococcus pyogenes* in adherence to epithelial cells. Infect Immun 72:4302–4308
Okada N, Pentland AP, Falk P, Caparon MG (1994) M protein and protein F act as important determinants of cell-specific tropism of *Streptococcus pyogenes* in skin tissue. J Clin Invest 94:965–977
Okada N, Liszewski MK, Atkinson JP, Caparon M (1995) Membrane cofactor protein (CD46) is a keratinocyte receptor for the M protein of the group A streptococcus. Proc Natl Acad Sci USA 92:2489–2493
Okamoto S, Tamura Y, Terao Y, Hamada S, Kawabata S (2005) Systemic immunization with streptococcal immunoglobulin-binding protein Sib 35 induces protective immunity against group A *Streptococcus* challenge in mice. Vaccine 23:4852–4859
Olive C, Clair T, Yarwood P, Good MF (2002a) Protection of mice from group A streptococcal infection by intranasal immunisation with a peptide vaccine that contains a conserved M protein B cell epitope and lacks a T cell autoepitope. Vaccine 20:2816–2825
Olive C, Batzloff MR, Horváth A, Wong A, Clair T, Yarwood P, Toth I, Good MF (2002b) A lipid core peptide construct containing a conserved region determinant of the group A streptococcal M protein elicits heterologous opsonic antibodies. Infect Immun 70:2734–2738
Olive C, Batzloff M, Horváth A, Clair T, Yarwood P, Toth I, Good MF (2003) Potential of lipid core peptide technology as a novel self-adjuvanting vaccine delivery system for multiple different synthetic peptide immunogens. Infect Immun 71:2373–2383
Olive C, Batzloff M, Horváth A, Clair T, Yarwood P, Toth I, Good MF (2004) Group A streptococcal vaccine delivery by immunization with a self-adjuvanting M protein-based lipid core peptide construct. Indian J Med Res 119(Suppl):88–94
Olive C, Hsien K, Horváth A, Clair T, Taywood P, Toth I, Good MF (2005) Protection against group A streptococcal infection by vaccination with self-adjuvanting lipid core M protein peptides. Vaccine 23:2298–2303
Olive C, Sun HK, Ho MF, Dyer J, Horváth A, Toth I, Good MF (2006) Intranasal administration is an effective mucosal vaccine delivery route for self-adjuvanting lipid core peptides targeting the group A streptococcal M protein. J Infect Dis 194:316–324

Olive C, Schulze K, Sun HK, Ebensen T, Horvath A, Toth I, Guzman CA (2007) Enhanced protection against *Streptococcus pyogenes* infection by intranasal vaccination with a dual antigen component M protein/SfbI lipid core peptide vaccine formulation. Vaccine 25:1789–1797

O'Loughlin RE, Roberson A, Cieslak PR, Lynfield R, Gershman K, Craig A, Albanese BA, Farley MM, Barrett NL, Spina NL, Beall B, Harrison LH, Reingold A, Van Beneden C, Team ABCS (2007) The epidemiology of invasive group A streptococcal infection and potential vaccine implications: United States, 2000–2004. Clin Infect Dis 45:853–862

Olsen RJ, Watkins ME, Cantu CC, Beres SB, Musser JM (2011) Virulence of serotype M3 Group A Streptococcus strains in wax worms (*Galleria mellonella* larvae). Virulence 2:111–119

Pandey M, Sekuloski S, Batzloff MR (2009a) Novel strategies for controlling *Streptococcus pyogenes* infection and associated diseases: from potential peptide vaccines to antibody immunotherapy. Immunol Cell Biol 87:391–399

Pandey M, Batzloff M, Good MF (2009b) Mechanism of protection induced by group A Streptococcus vaccine candidate J8-DT: contribution of B and T-cells towards protection. PLoS ONE 4:e5147

Pandiripally V, Gregory E, Cue D (2002) Acquisition of regulators of complement activation by *Streptococcus pyogenes* serotype M1. Infect Immun 70:6206–6214

Park HS, Cleary PP (2005) Active and passive intranasal immunizations with streptococcal surface protein C5a peptidase prevent infection of murine nasal mucosa-associated lymphoid tissue, a functional homologue of human tonsils. Infect Immun 73:7878–7886

Park H-S, Francis KP, Yu J, Cleary PP (2003) Membranous cells in nasal-associated lymphoid tissue: a portal of entry for the respiratory mucosal pathogen group A *Streptococcus*. J Immunol 171:2532–2537

Perez-Casal J, Okada N, Caparon MG, Scott JR (1995) Role of the conserved C-repeat region of the M protein of *Streptococcus pyogenes*. Mol Microbiol 15:907–916

Pichichero ME, Casey JR (2007) Systematic review of factors contributing to penicillin treatment failure in *Streptococcus pyogenes* pharyngitis. Otolaryngol Head Neck Surg 137:851–857

Piepmeier E Jr, Hammett-Stabler C, Price M, Peters J, Kemper G, Davis M Jr (1995) Myositis and fasciitis associated with group A beta hemolytic streptococcal infections: development of a rabbit model. J Lab Clin Med 126:137–143

Pizza M, Scarlato V, Masignani V, Guiliani MM, Aricó B, Comanducci M, Jennings GT, Baldi L, Bartolini E, Capecchi B, Galeotti CL, Luzzi E, Manetti R, Marchetti E, Mora M, Nuti S, Ratti G, Santini L, Savino S, Scarselli M, Storni E, Zuo P, Broeker M, Hundt E, Knapp B, Blair E, Mason T, Tettelin H, Hood DW, Jefferies AC, Saunders NJ, Granoff DM, Venter JC, Moxon ER, Grandi G, Rappouli R (2000) Identification of vaccine candidates against serogroup B meningococcus by whole-genome sequencing. Science 287:1816–1820

Polly SM, Waldman RH, High P, Wittner MK, Dorfman A (1975) Protective studies with a group A streptococcal M protein vaccine. II. Challenge of Volunteers after local immunization in the upper respiratory tract. J Infect Dis 131:217–224

Potter EV, Stollerman GH, Siegel AC (1962) Recall of type specific antibodies in man by injections of streptococcal cell walls. J Clin Invest 41:301–310

Prakash K, Dutta S (1991) Antibodies to streptococcal opacity factor in a selected Indian population. J Med Microbiol 34:119–124

Pruksakorn S, Galbraith A, Houghten RA, Good MF (1992) Conserved T and B cell epitopes on the M protein of group A streptococci. Induction of bactericidal antibodies. J Immunol 149:2729–2735

Pruksakorn S, Currie BJ, Brandt ER, Martin D, Galbraith A, Phornphutkul C, Hunsakunachal S, Manmontri A, Good MF (1994) Towards a vaccine for rheumatic fever: identification of a conserved target epitope on M protein of group A streptococci. Lancet 344:639–642

Purushothaman SS, Park H, Cleary PP (2004) Promotion of fibronectin independent invasion by C5a peptidase into epithelial cells in group A Streptococcus. Indian J Med Res 119(Suppl):44–47

Rakonjac JV, Robbins JC, Fischetti VA (1995) DNA sequence of the serum opacity factor of group A streptococci: identification of a fibronectin-binding repeat domain. Infect Immun 63:622–631

Rantz LA, Randall E, Rantz HH (1949) Immunization of human beings with group A hemolytic streptococci. Am J Med 6:424–432

Richter SS, Heilmann KP, Beekmann SE, Miller NJ, Miller AL, Rice CL, Doern CD, Reid SD, Doern GV (2005) Macrolide-resistant *Streptococcus pyogenes* in the United States, 2002–2003. Clin Infect Dis 41:599–608

Ringdahl U, Sjöbring U (2000) Analysis of plasminogen-binding M proteins of *Streptococcus pyogenes*. Methods 21:143–150

Rodreiguez-Ortega MJ, Norais N, Bensi G, Liberatori S, Capo S, Mora M, Scarselli M, Doro F, Ferrari G, Garaguso I, Maggi T, Neumann A, Covre A, Telford JL, Grandi G (2006) Characterization and identification of vaccine candidate proteins through analysis of the group A Streptococcus surface proteome. Nat Biotechnol 24:191–197

Sabharwal H, Michon F, Nelson D, Wealing D, Fuchs K, Manjarrez RC, Sarkar A, Uitz C, Viteri-Jackson A, Suarez RSR, Blake MS, Zabriskie JB (2006) Group A streptococcus (GAS) carbohydrate as an immunogen for protection against GAS infection. J Infect Dis 193:129–135

Saetre T, Hoiby EA, Aspelin T, Lermark G, Lyberg T (2000) Acute serogroup A streptococcal shock: a porcine model. J Infect Dis 182:133–141

Sanderson-Smith M, Batzloff MR, Sriprakash KS, Dowton M, Ranson M, Walker MJ (2006) Divergence in the plasminogen-binding group a streptococcal M protein family: functional conservation of binding site and potential role for immune selection of variants. J Biol Chem 281:3217–3226

Sanderson-Smith M, Dinkla K, Cole JN, Cork AJ, Maamary PG, McArthur J, Chhatwal GS, Walker MJ (2008) M protein-mediated plasminogen binding is essential for the virulence of an invasive *Streptococcus pyogenes* isolate. FASEB J 22:2715–2722

Schmidtchen A, Frick IM, Andersson E, Tapper H, Björck L (2002) Proteinases of common pathogenic bacteria degrade and inactivate the antibacterial peptide LL-37. Mol Microbiol 46:157–168

Schulze K, Guzman CA (2003) Identification of the domains of the fibronectin-binding protein I of *Streptococcus pyogenes* responsible for adjuvanticity. FEMS Immunol Med Microbiol 37:173–177

Schulze K, Medina E, Talay SR, Towers RJ, Chhatwal GS, Guzman CA (2001) Characterization of the domain of fibronectin-binding protein I of *Streptococcus pyogenes* responsible for elicitation of a protective immune response. Infect Immun 69:622–625

Schulze K, Medina E, Chhatwal GS, Guzman CA (2003) Stimulation of long-lasting protection against *Streptococcus pyogenes* after intranasal vaccination with non adjuvanted fibronectin-binding domain of the SfbI protein. Vaccine 21:1958–1964

Schulze K, Medina E, Guzman CA (2006) Intranasal immunization with serum opacity factor (SOF) of *Streptococcus pyogenes* fails to protect mice against lethal mucosal challenge with a heterologous strain. Vaccine 24:1446–1450

Severin A, Nickbarg E, Wooters J, Quazi SA, Matsuka YV, Murphy E, Moutsatsos IK, Zagursky RJ, Olmsted SB (2007) Proteomic analysis and identification of *Streptococcus pyogenes* surface-associated proteins. J Bacteriol 189:1514–1522

Shet A, Kaplan EL, Johnson DR, Cleary PP (2003) Immune response to group A streptococcal C5a peptidase in children: implications for vaccine development. J Infect Dis 188:809–817

Shet A, Kaplan E, Johnson D, Cleary PP (2004) Human immunogenicity studies on group A streptococcal C5a peptidase (SCPA) as a potential vaccine against group A streptococcal infections. Indian J Med Res 119(Suppl):95–98

Siegel AC, Johnson EE, Stollerman GH (1961) Controlled studies of streptococcal pharyngitis in a pediatric population 1. Factors related to the attack of rheumatic fever. New Engl J Med 265:559–566

Siegert J, Sastalla I, Chhatwal GS, Medina E (2006) Vaccination equally enables both genetically susceptible and resistant mice to control infection with group A streptococci. Microb Infect 8:347–353

Simerska P, Abdel-Aal AB, Fujita Y, Moyle PM, McGeary RP, Batzloff MR, Olive C, Good MF, Toth I (2008a) Development of a liposaccharide-based delivery system and its application to the design of group A streptococcal vaccines. J Med Chem 51:1447–1452

Simerska P, Abdel-Aal AB, Fujita Y, Batzloff MR, Good MF, Toth I (2008b) Synthesis and in vivo studies of carbohydrate-based vaccines against group A streptococcus. Biopolymers 90:616–911

Sitkiewicz I, Nagiec MJ, Sumby P, Butler SD, Cywes-Bentley C, Musser JM (2006) Emergence of a bacterial clone with enhanced virulence by acquisition of a phage encoding a secreted phospholipase A2. Proc Natl Acad Sci USA 103:16009–16014

Skinner JM, Caro-Aguilar IC, Payne AM, Indrawati L, Fontenot J, Heinrichs JH (2011) Comparison of rhesus and cynomolgus macaques in a *Streptococcus pyogenes* infection model for vaccine evaluation. Microb Pathog 50:39–47

Stafslien DK, Cleary PP (2000) Characterisation of the streptococcal C5a peptidase using a C5a-green fluorescent protein fusion protein substrate. J Bacteriol 182:3254–3258

Stalhammar-Carlemalm M, Areschoug T, Larsson C, Lindahl G (1999) The R28 protein of *Streptococcus pyogenes* is related to several group B streptococcal surface proteins, confers protective immunity and promotes binding to human epithelial cells. Mol Microbiol 33:208–219

Stalhammar-Carlemalm M, Areschoug T, Larsson C, Lindahl G (2000) Cross-protection between group A and group B streptococci due to cross-reacting surface proteins. J Infect Dis 182:142–149

Steer AC, Law I, Matatolu L, Beall BW, Carapetis JR (2009) Global *emm* type distribution of group A streptococci: systematic review and implications for vaccine development. Lancet Infect Dis 9:611–616

Stockbauer KE, Magoun L, Liu M, Burns EHJ, Gubba S, Renish S, Pan X, Bodary SC, Baker E, Coburn J, Leong JM, Musser JM (1999) A natural variant of the cysteine protease virulence factor of group A Streptococcus with an arginine-glycine-aspartic acid (RGD) motif preferentially binds human integrins a_Vb_3 and $a_{IIb}b_3$. Proc Natl Acad Sci USA 96:242–247

Sumby P, Barbian KD, Gardener DJ, Whitney AR, Welty DM, Long RD, Bailey JR, Parnell MJ, Hoe NP, Adams GG, DeLeo FR, Musser JM (2005) Extracellular deoxyribonuclease made by group A Streptococcus assists pathogenesis by enhancing evasion of the innate immune response. Proc Natl Acad Sci USA 102:1679–1684

Sumby P, Tart AH, Musser JM (2008) A non-human primate model of acute group a Streptococcus pharyngitis. Methods Mol Biol 431:255–267

Sun H, Ringdahl U, Homeister JW, Fay WP, Engelberg NC, Yang AY, Rozek LS, Wang X, Sjöbring U, Ginsburg D (2004) Plasminogen is a critical host pathogenicity factor for group A streptococcal infection. Science 305:1283–1286

Talay SR, Valentin-Weigand P, Jerlstrom PG, Timmis KN, Chhatwal GS (1992) Fibronectin-binding protein of *Streptococcus pyogenes*: sequence of the binding domain involved in adherence of streptococci to epithelial cells. Infect Immun 60:3837–3844

Terao Y, Kawabata S, Kunitomo E, Murakami J, Nakagawa I, Hamada S (2001) Fba, a novel fibronectin-binding protein from *Streptococcus pyogenes*, promotes bacterial entry into epithelial cells, and the *fba* gene is positively transcribed under the Mga regulator. Mol Microbiol 42:75–86

Terao Y, Okamoto S, Kataoka K, Hamada S, Kawabata S (2005) Protective immunity against *Streptococcus pyogenes* challenge in mice after immunization with fibronectin-binding protein. J Infect Dis 192:2081–2091

Terao Y, Mori Y, Yamaguchi M, Shimizu Y, Ooe K, Hamada S, Kawabata S (2008) Group A streptococcal cysteine protease degrades C3 (C3b) and contributes to evasion of innate immunity. J Biol Chem 283:6253–6260

Timmer AM, Kristian SA, Datta V, Jeng A, Gillen CM, Walker MJ, Beall B, Nizet V (2006) Serum opacity factor promotes group A streptococcal epithelial cell invasion and virulence. Mol Microbiol 62:15–25

Tsai PJ, Kuo CF, Lin KY, Lin YS, Lei HY, Chen FF, Wang JR, Wu JJ (1998) Effect of group A streptococcal cysteine protease on invasion of epithelial cells. Infect Immun 66:1460–1466

Tsao N, Tsai WH, Lin YS, Chuang WJ, Wang CH, Kuo CF (2006) Streptococcal pyrogenic exotoxin B cleaves properdin and inhibits complement-mediated opsonophagocytosis. Biochem Biophys Res Commun 339:779–784

Tse H, Bao JYJ, Davies MR, Maamary PG, Tsoi H, Tong AHY, Ho TCC, Lin CF, Gillen CM, Barnett TC, Chen HK, Lee M, Yam W, Wong C, Ong CY, Chan Y, Wu C, Ng T, Lim WWL, Tsang THF, Tse CWS, Dougan G, Walker MJ, Lok S, Yuen K (2012) Molecular characterization of the 2011 Hong Kong scarlet fever outbreak. J Infect Dis 206:341–351

Turner CE, Kurupatia P, Wilesa S, Edwards RJ, Sriskandan S (2009) Impact of immunization against SpyCEP during invasive disease with two streptococcal species: *Streptococcus pyogenes* and *Streptococcus equi*. Vaccine 27:4923–4929

Ulrich RG (2008) Vaccine based on a ubiquitous cysteinyl protease and streptococcal pyrogenic exotoxin A protects against *Streptococcus pyogenes* sepsis and toxic shock. J Immune Based Ther Vaccines 6:8

Vajdy M, Singh M (2005) The role of adjuvants in the development of mucosal vaccines. Expert Opin Biol Ther 5:953–965

Virtaneva K, Graham MR, Porcella SF, Hoe NP, Su H, Graviss EA, Gardner TJ, Allison JE, Lemon WJ, Bailey JR, Parnell MJ, Musser JM (2003) Group A Streptococcus gene expression in humans and cynomolgus macaques with acute pharyngitis. Infect Immun 71:2199–2207

Virtaneva K, Porcella SF, Graham MR, Ireland RM, Johnson CA, Ricklefs SM, Babar I, Parkins LD, Romero RA, Corn GJ, Gardner DJ, Bailey JR, Parnell MJ, Musser JM (2005) Longitudinal analysis of the group A Streptococcus transcriptome in experimental pharyngitis in cynomolgus macaques. Proc Natl Acad Sci USA 102:9014–9019

Von Pawel-Rammingen U, Björck L (2003) IdeS and SpeB: immunoglobulin-degrading cysteine proteinases of *Streptococcus pyogenes*. Curr Opin Microbiol 6:50–55

Waldman RH, Lee JD, Polly SM, Dorfman A, Fox EN (1975) Group A streptococcal M protein vaccine: protection following immunization via the respiratory tract. Dev Biol Stand 28: 429–434

Wasson VP, Brown EE (1943) Immunization against rheumatic fever. J Pediatr 23:24–30

Watson RF, Rothbard S, Swift HF (1946) Type-specific protection and immunity following intranasal inoculation of monkeys with group A hemolytic streptococci. J Exp Med 84: 127–142

Wilson MG, Swift HF (1931) Intravenous vaccination with hemolytic streptococci. Am J Dis Child 42:42–51

Wizemann TM, Adamou JE, Langermann S (1999) Adhesins as targets for vaccine development. Emerging Infect Dis 5:395–403

Yokoyama Y, Harabuchi T (2002) Intranasal immunization with lipoteichoic acid and cholera toxin evokes specific pharyngeal IgA and systemic IgG responses and inhibits streptococcal adherence to pharyngeal epithelial cells in mice. Int J Pediatr Otorhinolaryngol 63:235–241

Young DC, Breese BB, Errington AF, Stoeffler W (1946) Failure of type specific *Streptococcus pyogenes* vaccine to prevent respiratory infections. US Nav Med Bull 46:709–718

Young MH, Aronoff DM, Engleberg NC (2005) Necrotizing fasciitis: pathogenesis and treatment. Expert Rev Anti Infect Ther 3:279–294

Zaman M, Abdel-Aal AB, Phillipps KS, Fujita Y, Good MF, Toth I (2010) Structure-activity relationship of lipopeptide Group A streptococcus (GAS) vaccine candidates on toll-like receptor 2. Vaccine 28:2243–2248

Zaman M, Skwarczynski M, Malcolm JM, Urbani CN, Jia Z, Batzloff MR, Good MF, Monteiro MJ, Toth I (2011) Self-adjuvanting polyacrylic nanoparticulate delivery system for group A streptococcus (GAS) vaccine. Nanomedicine 7:168–173

Zaman M, Abdel-Aal AB, Fujita Y, Phillipps KS, Batzloff MR, Good MF, Toth I (2012) Immunological evaluation of lipopeptide group A streptococcus (GAS) vaccine: structure-activity relationship. PLoS ONE 7:e30146

Zhu H, Liu M, Sumby P, Lei B (2009) The secreted esterase of group a streptococcus is important for invasive skin infection and dissemination in mice. Infect Immun 77:5225–5232

Zimmerman RA, Krushak DH, Wilson E, Douglas JD (1970) Human streptococcal disease syndrome compared with observations in chimpanzees. II. Immunologic responses to induced pharyngitis and the effect of treatment. J Infect Dis 122:280–289

Clinical Management of the Most Common Group A β-Hemolytic Streptococcal Infections

Edward L. Kaplan

Abstract Group A streptococcal (*Streptococcus pyogenes*) infections remain important causes of medical and public health morbidity and mortality even during the early twenty-first century. Although most often concentrated in socially/economically disadvantaged populations, the problems remain significant in both industrializing and industrialized countries. The many M/*emm* types of GAS contribute to herd immunity in populations and also affect the control of streptococcal infections in these populations. Although this bacterium remains among the most susceptible to most antibiotics, it is evident that antibiotics alone have not solved the group A streptococcal medical and public health problems, even in those places where access to medical care is readily available. It is likely that the current streptococcal problems will remain difficult to manage and will remain essentially unchanged until the broad implementation of a cost-effective group A streptococcal vaccine, likely some years in the future.

Contents

E. L. Kaplan (✉)
Department of Pediatrics, University of Minnesota Medical School,
MMC 296 420 Delaware St. SE, Minneapolis, MN 55422, USA
e-mail: kapla001@umn.edu

Current Topics in Microbiology and Immunology (2013) 368: 243–252
DOI: 10.1007/82_2012_306

Published Online: 22 January 2013

1 Epidemiology of Group A Streptococcal Infections

It has been recognized for many decades that the group A β-hemolytic streptococcus (*Streptococcus pyogenes*; GAS) is widely spread around the world and throughout diverse populations. What has not been fully appreciated is the fact that the epidemiology of this organism is very dynamic. This is a highly transmissible bacterium, and it has the capacity to spread rapidly and widely among susceptible individuals or in a susceptible population. For these reasons, understanding the epidemiology of these infections is necessary for developing their effective clinical management, as well as for the public health containment of streptococcal infections and their suppurative and nonsuppurative sequelae.

As stated in the first chapter of this volume (Ralph and Carapetis 2012), the group A streptococcus may affect different populations in different ways. This became more evident during the last half of the twentieth century and during the first decade of the twenty-first century. The published descriptions of multiple epidemics of upper respiratory tract infections, scarlet fever, acute poststreptococcal glomerulonephritis, and rheumatic fever completely overshadowed the bacterium's endemicity and the constantly changing epidemiology not only among populations, but also among individuals within any given population. More recently, a more complete recognition of not only endemicity within a single population, but also the seeming intra- and intercontinental spread has changed the way these infections and—more specifically—specific strains of the group A streptococcus are managed. While it can be stated that the infections themselves remain similar in the human host (the only recognized natural reservoir for GAS) no matter where they live, the disease expression in some populations such as those in the developing world appear to be somewhat different. This likely is the result of environment and infrastructure, rather than being strictly of biological origin.

A very significant impact on this medical and public health problem was the introduction at the close of the twentieth century of molecular techniques for identifying and characterizing these organisms [especially typing of the *emm* gene which codes for the M protein virulence factor on the surface of the cell (Beall et al. 1996)]. The ability of these newer techniques to identify outbreaks and clusters has been immensely helpful to both medical and public health professionals, as well as to basic scientists investigating the many remaining mysteries associated with GAS. The fact that one no longer has to be concerned about "non-typable" strains, as was often the case in the past when only T typing and serotyping were available, has proven immensely important. However, it must be recognized that despite these advances, the clinical and the public health management of GAS infections and their sequelae (both suppurative and non-suppurative) still remain problematic in all populations around the world.

An anonymous physician once remarked that group A streptococcal infections are an occupational disease of school children. Why the organism seems to have a predilection for this age group remains incompletely understood. While not totally invulnerable to GAS infections, very young children seem to experience fewer

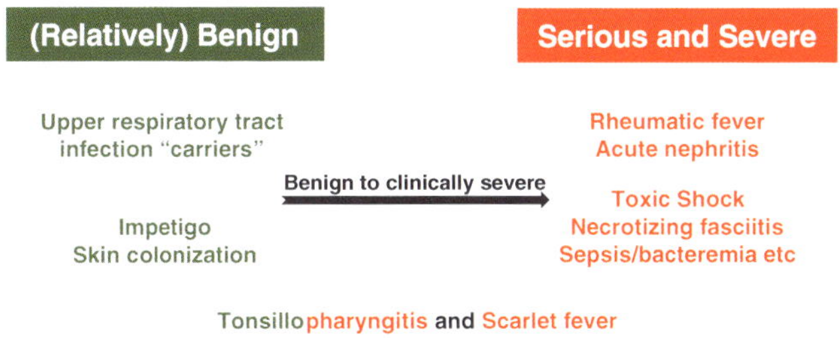

Fig. 1 The spectrum of group A streptococcal infections in humans

GAS infections. However, this does not obviate the fact that it was shown some years ago in prospectively followed families that the average child has had three documentable group A streptococcal upper respiratory tract infections by the age of 13 years (Dingle et al. 1964), While these infections are more infrequent in older teenagers and in healthy adults, these infections and their complications remain a significant problem, for example, among military recruits no matter which country they are studied in. Emphasizing this problem is the observation that there were more than 25,000 cases of acute rheumatic fever in the United States Navy alone during World War II (Streptococcal Disease Laboratory 1951).

Group A streptococci have the potential to verify their threat through a broad spectrum of clinical and epidemiological manifestations. Each of these manifestations presents its own unique medical and public health issues. Whether "relatively benign" or life-threatening (either suppurative or non-suppurative sequelae), required approaches to medical therapy and public health prevention can be quite different.

The predilection, especially for children, has resulted in a spectrum of various manifestations such as that shown in Fig. 1. These represent the major GAS infections and complications to be discussed.

2 Clinical and Public Health Management of Group A Streptococcal Infections

To many, the diagnosis "streptococcal infection" is almost synonymous with group A streptococcal tonsillopharyngitis. This is because it is the most common presenting form of GAS infections. Yet, the management of these infections remains controversial. There are differences in the approach to management not only from country to country, but even among physicians and public health authorities in the same geographic area of a given region. Many countries and medical/public health organizations continue to publish their own individual guidelines for management of streptococcal pharyngitis (Gerber et al. 2009; The Red Book 2012; Shulman et al. 2012; Matthys et al. 2007). Why this is the case remains intriguing to observers of this infection.

There are several approaches to management of GAS infections which have medical and scientific support and should be non-controversial. While the original reason for treating GAS infections of the upper respiratory tract was originally primarily directed to prevention of a specific non-suppurative sequel (i.e., rheumatic fever), it has become clear that treatment of individuals by eradicating the organism from the upper respiratory tract also has the beneficial effect of reducing spread among close contacts, in the home, in the school, or in other susceptible populations. For those reasons, and also since it has been shown that truly infected individuals (see below) usually clinically improve spontaneously whether treated or not, but still are susceptible to rheumatic fever and remain likely to spread the organism, the scientifically documented goal of antibiotic therapy is full eradication of the streptococcus from the upper respiratory tract.

There has never been a clinical isolate of group A streptococci which is even remotely resistant to penicillin (Macris et al. 1998). Even if one goes back almost eight decades and studies isolates from that time, the MIC of *Streptococcus pyogenes* to penicillin has not changed (Macris et al. 1998); the organism is as susceptible today as it has been in the past. The reason why is not understood. This promotes penicillin as an inexpensive and usually effective antibiotic which is available worldwide.

For individuals who are truly allergic to the penicillins, the classically recommended alternative antimicrobial agent has been the macrolide group. However, beginning in the 1960s and 1970s when the majority of GAS were shown to develop resistance to erythromycin in Japan, the problem of macrolide resistance has spread to many parts of the world where these drugs are dispensed—often too frequently and haphazardly. There are multiple examples of this macrolide phenomenon around the world. Even included now is the azalide, azithromycin, which has also been associated with increased resistance by group A streptococci (Kaplan and Cornaglia 2005).

Management of group A streptococcal upper respiratory tract infection with tetracyclines and with sulfa drugs continues to be contraindicated, also because of antimicrobial resistance (Matthys et al. 2007, IDSA).

During the past two or three decades numerous new (and often more expensive) antimicrobial agents have been introduced to treat GAS pharyngitis. Some of these have even been proposed to be more effective than the penicillins. This has further promoted diversity among national and regional recommendations for treatment of this common infection. The use of such antibiotics can be credited, likely in large part, to campaigns by PHARMA.

In a similar vein, there are now available antibiotic regimens using less than the 10 days of therapy which has been the classically verified regimen for decades. Equivalency studies arguably purport to show equivalent and sometimes even superior efficacy in eradication of GAS from the upper respiratory tract. However, there are also published data which demonstrate greater effectiveness in treating for 10 days, when compared to only a "shortened" course (Kaplan et al. 2001). This important question remains incompletely addressed, as is seen in several guidelines.

A study was published in 2002 which further intensified the controversy for using penicillin as a first line of therapy for group A streptococcal tonsillopharyngitis (Kaplan and Johnson 2001). In that study which used generally recommended doses of intramuscular benzathine penicillin G (to avoid the issue of patient adherence), it was found that in 37 % of patients with pharyngitis and culture proven GAS in the throat (not using rapid antigen detection tests)the organism was not eradicated during the month following benzathine penicillin G therapy.

A number of possible explanations emerged is shown in Table 1 below:

Table 1 Possible Explanations for Failure to Eradicate Group A Streptococci with Penicillin

Are there changes in antibiotic susceptibility?
Is there "tolerance" to penicillin?
Is there a problem with the manufacture of penicillin?
Is compliance/adherence documented?
What is meant by the term co-pathogenicity? Does this have any role (beta-lactamase, bacteriocins)?
Is there a role for the group A streptococcal carrier state?

2.1 The group A streptococcal upper respiratory tract carrier state?

Space does not allow detailed discussion of each of the possibilities shown in Table 1. However, many agree that the first five bullet points listed in the figure represent less likely explanations. It is the last option in the list, the group A upper respiratory tract "carrier" state, that merits more discussion.

To clarify what is precisely meant by the "carrier state", a definition of both conditions is required (Kaplan 1980). There is published evidence to the define *true or bona fide infected individual* as a human host from whom the streptococcus can be recovered from the throat and the host experiences an immunologic response to a streptococcal somatic or extracellular antigen(s) indicating host recognition of the presence of the GAS. In contrast, a "carrier" has been defined as the presence of the group A streptococcus in the throat, but for as yet unknown reasons there is no immunologic response to these streptococcal antigens.

This symbiotic relationship between a normally pathogenic bacterium and its human host has been a major controversial issue for management and control of group A streptococcal infections for decades. Either the asymptomatic presence of GAS in the pharynx, or the streptococcal presence in a human who is symptomatic due to another etiologic agent (most commonly a respiratory virus) has presented an enigma to clinicians, public health authorities, as well as to basic scientists.

There are convincing data that this upper respiratory tract carrier state exists (Kaplan et al. 1971; Johnson et al. 2010). Furthermore, there are published data which quite strongly suggest that the carrier state is often quite difficult to eradicate with penicillin therapy, even in recommended doses (Kaplan et al. 1981).

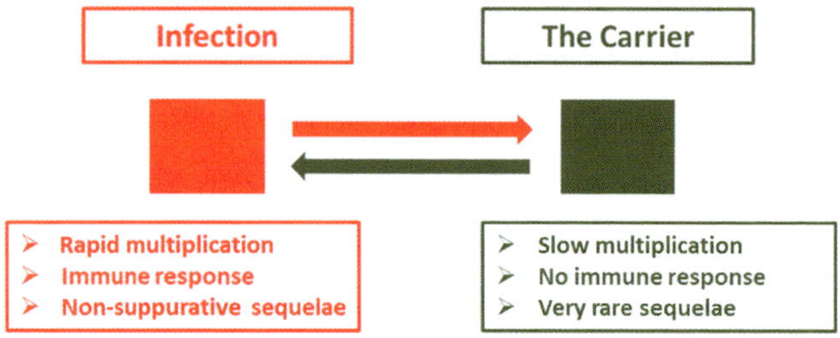

Fig. 2 Two aspects of GAS in the upper respiratory tract

A major reason for this need for differentiation of the two conditions relates to their ability to result in sequelae, either epidemiological or infection-related. (See Fig. 2).

One approach to understanding the failure to eradicate the GAS from the human upper respiratory tract has been to further study the relatively recently described phenomenon of ingestion of viable group A streptococci by respiratory tract epithelial cells (LaPenta et al. 1994). Although relatively little is known about GAS intracellular survival, assuming that this phenomenon is related to the creation of the streptococcal upper respiratory tract carrier state, it is logical to assume that one important reason for persistence of the organisms is the difficulty exhibited by penicillin to penetrate the bacterial cell wall. This has been demonstrated by in-vitro studies (Kaplan et al. 2006).

Much remains to be learned about this condition, especially how to differentiate clinically the carrier state from the truly infected individual at the time of the presenting illness; this differentiation represents considerable difficulty for clinicians (Kaplan et al. 1971). Throat cultures, even quantitative ones, do not reliably allow a differentiation (Kaplan et al. 1971). Streptococcal antibody determinations are only useful in retrospect in this situation. Therefore many times, maybe even in a majority of instances, the physician feels compelled to treat patients with antibiotics, even those who very likely represent the upper respiratory tract carrier state.

The principle of antibiotic treatment of patients with streptococcal pharyngitis or tonsillitis is generally accepted, although there is a small minority who believe that in the presence of reduced rheumatic fever incidence, as is the case in most industrialized countries, antibiotic therapy is not obligatory. The vast majority of public health organizations and professional societies recommend antibiotic therapy. Disagreements are most often about a specific antibiotic, the prescribed dose, and the duration of therapy.

Special comment is warranted about the use of intramuscular benzathine penicillin G (N,N' dibenzlethylenediamine dipenicillin G) for treatment of group A streptococcal upper respiratory tract infections (also known as primary rheumatic fever prevention). There are convincing data in the literature from several decades ago demonstrating its superior efficacy when compared with oral penicillin preparations (Wood et al. 1964). During the past several decades data have been

Table 2 Most often selected antibiotics for treating GAS tonsillopharyngitis

Preferred initial antibiotic therapy	Alternative antibiotics
Oral penicillin V for 10 days (Oral amoxycillin for taste in young children.)	Oral first-generation cephalosporins for 10 days (Some use these as initial therapy. Some use these as an alternative for macrolides/azalides in penicillin allergic person who has not had an immediate reaction to penicillins.)
Oral macrolides/azalides when there is reduced concern for resistance in a community	
Intramuscular benzathine penicillin G	Although approved by some regulatory agencies, data supporting the efficacy of short course oral antibiotic therapy remain controversial

published which cause questions about the earlier reports. Recent data have suggested that serum penicillin levels are not as high nor do they last as long (Kaplan et al. 1989; Broderick et al. 2011). Such studies have raised important questions about this important aspect of the antibiotic treatment in the management of these infections. Furthermore, there are no recent data from studies primarily designed for evaluating children. When data are more complete and updated, the proper role for this valuable form of repository penicillin can be confirmed.

It is inappropriate to generalize across countries and across professional societies, but Table 2 reflects agreement about use of those antibiotics:

During a discussion of management of group A streptococcal respiratory tract infections, it is also appropriate to comment briefly about issues relating to the not infrequent outbreaks of group A streptococcal respiratory tract infections. Whether the outbreaks occur in a well-defined population (e.g., a school, daycare facility, residential facility, or military recruit training base) or in different and most often different populations in the same city or region, it is necessary that a carefully conceived public health approach be implemented for the purposes of defining the extent of the problem. Effective management of this situation requires prompt public health action. In many such outbreaks, broad use of cultures and subsequent antibiotic therapy or even just across population, antibiotics have proven effective in terminating outbreaks. In such situations, the use of intramuscular benzathine penicillin G has been reported to effectively eliminate the problem.

The other often considered "benign" group A streptococcal infections (Fig. 1) are those of the skin and soft tissues. These streptococcal infections, similar to upper respiratory tract infections, affect large numbers of individuals. While quite common in children, especially younger ones, superficial skin infections also affect adults.

The relationship of superficial skin infections (impetigo, pyoderma) to throat infections has been extensively studied. For many years it was believed that impetigo was largely a staphylococcal infection. But studies from around the world during the late 1960s and early 1970s revealed a very significant role for group A streptococci in the etiology of pyoderma (Wannamaker 1970). After prospective studies, it was recognized that certain M types of group A streptococci appeared to have a predilection for skin infections and other M types appeared more often in throat causing infection there. Further studies revealed that rather than streptococci being spread from the throat to the skin, the opposite was true (Ferrieri et al. 1972). A lag period of about 10 days allowed spread from skin to throat. While this finding may appear not to be clinically important, there are still those who believe that rheumatic fever can be caused by skin infection alone (McDonald et al. 2006). Most experts agree that conclusions about this controversial medical and epidemiological issue remain to be more carefully defined.

While it appears that streptococcal impetigo/pyoderma seldom progresses to more serious invasive systemic infections, the risk of this complication following streptococcal pyoderma has not been well-defined.

A major non-suppurative complication of pyoderma is pyoderma-associated acute post streptococcal glomerulonephritis. The latent period with this form of nephritis is slightly longer than the mean interval after streptococcal pharyngitis (3 weeks vs. 2 weeks) (Anthony et al. 1969). Contrary to the situation with prevention of rheumatic fever following antibiotic therapy of streptococcal pharyngitis, it appears that prompt antibiotic treatment of streptococcal impetigo does not prevent post streptococcal glomerulonephritis.

Various antibiotic treatments for impetigo/pyoderma have been studied including oral penicillin, benzathine penicillin, macrolides, and first-generation cephalosporins with varying reports of success. Early studies in an animal (hamster) model of group A streptococcal impetigo indicated that local therapy was not as efficacious as systemic antibiotic therapy (Dajani et al. 1971). However, it has been demonstrated that prophylaxis for prevention of impetigo using intramuscular benzathine penicillin G was successful in preventing impetiginous lesions for several weeks after injection in a susceptible population (Ferrieri et al. 1974).

On the other hand, the antibiotic management of other soft tissue infections like group A streptococcal pyomyositis and necrotizing fasciitis is a more complex problem, usually also requiring surgical debridement (often multiple procedures)in addition to intravenous antibiotics, and other supportive therapy (Stevens 1992). While considerable attention has been paid to specific GAS strains (e.g., *emm* types 1 and 3) as being especially associated with these very serious infections, review of the literature reveals numerous necrotizing fasciitis-associated GAS *emm* types, frequently associated with those *emm* types which are most prevalent in the community. Morbidity and mortality are significant, especially among the very young and the elderly. In children a common preceding event is chicken pox (varicella); in adults common complicating factors include diabetes mellitus.

Antibiotic management of these infections, especially necrotizing fasciitis, usually includes intravenous penicillin and clindamycin, either alone or in combination (Stevens et al. 1992, IDSA). Supportive treatment for accompanying shock, should it occur, is essential.

Group A streptococcal infections remain an important cause of morbidity and mortality in both industrializing and industrialized countries even at the beginning of the twenty-first century (Ralph and Carpetis 2012). It is obvious that these infections remain problematic despite the increased availability of antibiotics and the fact that the organism remains quite susceptible to penicillin. The infections remain a challenge for physicians (including infectious disease specialists), public health practitioners, and epidemiologists, as well as for basic scientists. These conclusions have made it evident that major advances are needed to assist in the management of the various infectious manifestations (Fig. 1). Among these are the heightened translation of advances in the basic science laboratory to the clinic and to the bed side, and in particular the availability and implementation of a cost-effective group A streptococcal vaccine.

References

Anthony BF, Kaplan EL, Wannamaker LW, Briese FW, Chapman SS (1969) Attack rates of acute nephritis following type 49 streptococcal infection of the skin and of the respiratory tract. J Clin Invest 48:1697

Beall B, Facklam R, Thompson T (1996) Sequencing *emm* -specific PCR products for routine and accurate typing of group A streptococci. J Clin Microbiol 34:953–958

Broderick Michael P, Hansen Christian J, Russell Kevin L, Kaplan Edward L, Blumer Jeffrey L, Faix Dennis J (2011) Serum penicillin G levels are lower than expected in adults within two weeks of administration of 1.2 million units. PLoS ONE 10:1–6

Dajani AS, Hill PL, Wannamaker LW (1971) Experimental infection of the skin in the hamster stimulating human impetigo. II. Assessment of various therapeutic regimens. Pediatrics 48:83–90

Dingle JH, Badger GF, Jordan, WS Jr (1964) Illness in the home. The Press of Western Reserve University, Cleveland, Ohio, pp 97–117

Ferrieri P, Dajani AS, Wannamaker LW (1974) A controlled study of penicillin prophylaxis against streptococcal impetigo. J Infect Dis 129:429–438

Ferrieri P, Dajani AS, Wannamaker LW, Chapman SS (1972) Natural history of impetigo. I. Site sequence of acquisition and familial patters of spread of cutaneous streptococci. J Clin Invest 51:2851–2862

Gerber MA, Baltimore RS, Eaton CB, Gewitz M, Rowley AH, Shulman ST, Taubert KA (2009) The American Heart Association. Prevention of rheumatic fever and treatment of acute streptococcal pharyngitis. Circulation 2009(119):1541–1551

Johnson DR, Kurlan R, Leckman J, Kaplan EL (2010) The human immune response to streptococcal extracellular antigens: clinical, diagnostic, and potential pathogenetic implications. Clin Infect Dis 50:481–490

Kaplan EL (1980) The group A streptococcal upper respiratory tract carrier state: an enigma. J Pediatr 97:337

Kaplan EL, Cornaglia G (2005) Persistent macrolide resistance among group A Streptococci: the lack of accomplishment after 4 decades. Clin Infect Dis 41:609–611

Kaplan EL, Johnson DR (2001) Unexplained reduced efficacy of oral penicillin V and intramuscular benzathine penicillin G in the eradication of group A streptococci from children with acute pharyngitis. Pediatrics 108:1180–1186

Kaplan EL, Top F Jr, Dudding BA, Wannamaker LW (1971) Diagnosis of streptococcal pharyngitis: the problem of differentiating active infection from the carrier state in the symptomatic child. J Infect Dis 123:490

Kaplan EL, Gastanaduy AS, Huwe BB (1981) The role of the carrier in treatment failures following antibiotic therapy for group A streptococci in the upper respiratory tract. J Lab Clin Med 98:326

Kaplan EL, Berrios X, Speth J, Siefferman T, Guzman B, Quesny F (1989) Pharmacokinetics of benzathine penicillin G: serum levels during the 28 days after intramuscular injection of 1,200,000 units. J Pediatr 115:146–150

Kaplan EL, Gooch WM III, Notario GF, Craft JC (2001) Macrolide therapy of group A streptococcal pharyngitis: ten days of macrolide therapy (clarithromycin) is more effective in streptococcal eradication than five days (azithromycin) in adults. Clin Infect Dis 32:1798–802
Kaplan EL, Chhatwal GS, Rohde M (2006) Reduced ability of penicillin to eradicate ingested group A streptococci from epithelial cells: clinical and pathogenetic implications. Clin Infect Dis 43:1398–1406
LaPenta D, Rubens C, Chi E, Cleary PP (1994) Group A streptococci efficiently invade human respiratory epithelial cells. Proc Natl Acad U S A 91:12115–12119
Macris M, Hartman N, Murray B, Klein RF, Roberts RB, Kaplan EL, Horn D, Zabriskie JB (1998) Studies of the continuing susceptibility to penicillin of group A streptococcal strains isolated over a period spanning eight decades. Pediatr Infect Dis J 17:377–381
Matthys J, DeMeyere M, van Driel ML, DeSutter A (2007) Differences among international pharyngitis guidelines: not just academic. Ann Fam Med 5:436–443
McDonald MI, Towers RJ, Andrews RM, Benger N, Currie BJ, Carapetis JR (2006) Low rates of streptococcal pharyngitis and high rates of pyoderma in Australian aboriginal communities where acute rheumatic fever is hyperendemic. Clin Infect Dis 43:683–689
Ralph AP, Carapetis JR (2012) Group A streptococcal diseases and their global burden. Curr Top Microbiol Immunol. doi 10.1007/82_2012_280
Shulman ST, Bisno AL, Clegg HW, Gerber MA, Kaplan EL, Lee G, Martin JM, Van Beneden C (2012) Clinical practice guideline for the diagnosis and management of group A streptococcal pharyngitis: 2012 update by the infectious disease society of America. Clin Infect Dis
Stevens DL (1992) Invasive group A streptococcus infections. Clin Inf Dis 14:2–11
Stevens DL, Bisno AL, Chambers HF, Everette ED, Dellinger P, Goldstein EJC, Gorbach SL, Hirschmann JV, Kaplan EL, Montoya JG, Wade JC (2005) Practice guidelines for the diagnosis and management of skin and soft-tissue infections. Clin Infect Dis 41:1373–1406
Streptococcal Disease Laboratory, Francis E. Warren Air Force Base; Commission on Acute Respiratory Diseases; Armed Forces Epidemiological Board and Department of Preventive Medicine. School of Medicine. Western Reserve University. Prevention of Rheumatic Fever. U.S. Armed Forces Med. J. 1951; 2:607–613
The Red Book (2012) Report of the Committee on Infectious Diseases The American Academy of Pediatrics. IL USA, Chicago
Wannamaker LW (1970) Differences between streptococcal infections of the throat and of the skin. New Eng J Med 282:23
Wood HF, Feinstein AR, Taranta A, Epstein JA, Simpson R (1964) Rheumatic Fever in children and adolescents. III, Comparative effectiveness of three prophylaxis regimens in preventing streptococcal infections and rheumatic recurrences. Ann Intern Med 60(Suppl 5):31–40

Index

Current Topics in Microbiology and Immunology (2013) 368: 253–255
DOI: 10.1007/978-3-642-36340-5

Printed by Printforce, the Netherlands